The Casa Cookbook

Part 1: Recipes for XPS Data Processing

Neal Fairley

and

Alan Carrick

The Casa Cookbook - Part 1: Recipes for XPS Data Processing

By Neal Fairley and Alan Carrick

British Library Cataloguing in Publication Data

A catalogue record for this book is available from the British Library
ISBN: 0-9549533-0-4

Published by Acolyte Science, Kinderton Close, High Legh, Knutsford, Cheshire, WA16 6LZ U.K.
http://www.acolytescience.co.uk

Visit the Casa web site: http://www.casaxps.com

Printed and bound in England by Pelican Press, Manchester, on Cyclus Offset paper with ISO 14001and NAPM Certificates

Contents

Contents

Contents

Contents

Contents

Contents

Foreword

CasaXPS is designed to take data from a wide range of instruments and their data systems using native file formats wherever possible, convert these formats to the ISO 14976 ("VAMAS") format and provide a single environment in which to process XPS and AES data. For many analysts problem solving using multiple techniques is common-place and while the origins of CasaXPS are clearly XPS the expansion of the system to cover AES (and, in Part II of this Series, SIMS) is both natural and desirable. In addition, XPS image processing has also become increasingly important, and to fulfil the potential of the modern imaging instruments, new features and new algorithms have been added to CasaXPS.

Many of the capabilities of CasaXPS require an understanding of why as well as how an operation is performed and therefore a large part of this CasaXPS manual is devoted to explaining the context for a sequence of data reduction steps. Procedures for these steps are addressed within sections describing specific examples, but the essential functions of CasaXPS routines and options and the overall structure of the program are presented in separate reference chapters. The practical examples are necessarily wordy and to compensate to some degree the reference section is pictorial in essence.

Seasoned Auger and XPS practitioners may wonder at the apparent omission of separate chapters dealing with topics such as "Depth Profiling". The approach is deliberate as it is felt more appropriate to cover issues of quantitation vs depth in relation to the (more fundamental) processing mechanisms of peak fitting, PCA, SVD, and the like. It is hoped that the index will provide an adequate guide to specific applications orientated questions in this regard.

It is also important to note that Chapter 19 does not intend to supplant existing excellent books which deal in detail with all aspects of XPS as an analytical technique (some of which are listed in the Bibliography), but simply to provide a convenient introduction to the basics for those from disciplines such as biochemistry, earth sciences, and so on who may not otherwise have easy access to this information.

Acknowledgement

CasaXPS wishes to thank all those connected with the production of this manual and the software system it describes.

Section 1 - Quick Summary

CHAPTER 1

Introduction

Computer Aided Surface Analysis for X-ray Photoelectron Spectroscopy (CasaXPS) has been designed and written for analysts and research scientists who use XPS for surface characterisation.

In contrast to the special purpose control systems that form part of commercial XPS instruments, CasaXPS offers a compact, portable, efficient and user-friendly processing system to anyone with an IBM compatible (Pentium) PC running Microsoft Windows 95 (or later) or a suitable emulator[1]. It incorporates much of - and in most cases more than - the processing functionality of the instrument-linked packages without recourse to unfamiliar operating systems or hardware or proprietary file formats. It is designed from the outset on the basis of the ISO 14976 Surface Chemical Analysis Standard Data Transfer Format and so by design has a universal, manufacturer-independent, "cross platform" approach, relying on nothing other than a reasonable adherence to the ISO standard. Spectra collected in the standard format may be selected, viewed and processed in a simple yet powerful way and the results of CasaXPS data reduction, presented in a variety of graphical and tabular formats, are available for incorporation into and use as data by other popular Microsoft pack-

1. Microsoft XP systems are the most likely platform (in 2005) for CasaXPS, but in these days of affordable high performance systems, such as the Macintosh G5, emulators such as Virtual PC (for Mac OS X) should be seriously considered

ages such as "Word" or "Excel". Additional spectrum input filters for other commercial file formats are also provided, supplementing the ISO standard and broadening its usefulness.

Casa XPS has been written entirely in native C++ using the Microsoft Developers Studio programming environment. It employs Microsoft Foundation Class (MFC) libraries to provide a standard graphical user interface as well as much of the data management required by the system.
The software suite as a whole may be regarded as being structured from five interacting components:

- Data display and browser (an "experiment window")
- Element library (a "dialog window" as are the others)
- Graph annotation
- Data processing
- Quantification.

This publication provides a detailed description of the system, its operation, command and function set, and its use in practice in "real life" analytical applications.

The CasaXPS Cookbook's objectives are to provide both a guide to the software and its use and some background to the techniques of surface analysis as well as the principle behind the mathematics used in the evaluation of the analytical data.

CasaXPS is more than "just another data system": it is hoped that, similarly, this "Cookbook" provides more than just "recipes", provoking serious thought about procedures and possibilities in one of the most challenging of experimental disciplines.

Getting Started

Installing CasaXPS

You must install the application from the CasaXPS CD onto your hard disk. "Drag and Drop" installation is all that is required for Windows XP or '98. Instructions are available, if required, in the "Read Me" file on the installation CD and some further details in "Appendix 6: CasaXPS Installation Notes" on page 358. Make sure that you have your serial number to hand when you install the application, or else the package will revert to "Demonstration Mode" and will not allow you to save the results of your work. The serial number is available from your registration documents (which may be e-mailed to you or included with the CD). You can "validate" a registration at any time after installation, turning a "demo" version into a fully licensed product, by using the "About" button in the Toolbar.

CasaXPS - Organisation, work flow, and information

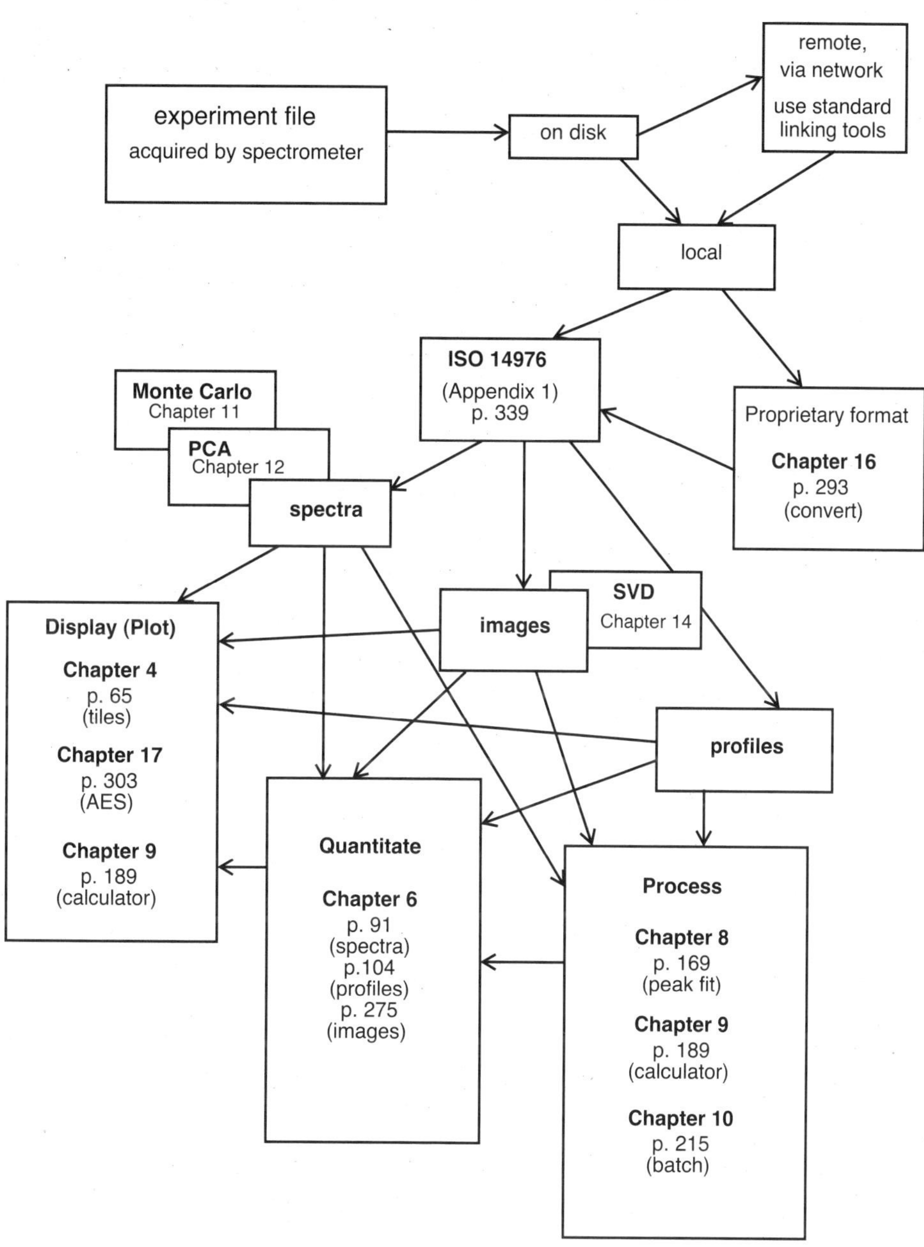

If you are using an emulator package on a processor other than an Intel Pentium, ensure that you obey the emulator's instructions for installation of applications packages, that you understand the limitations (if any - e.g. mouse button availability) of the system, and that you have sufficient space and processing power for the convenient use of CasaXPS.
There is further, more detailed, information on program installation (and removal) in "Appendix 6: CasaXPS Installation Notes" on page 358

Starting CasaXPS

Simply click on the desktop short-cut (if, as recommended, you installed one) or double click on the programme icon in "Windows Explorer". Ensure that you have stored your ISO 14976 data files in an easily accessible directory (or have the demo data on the installation CD available) Ensure also that you have an appropriate element library available (one needs to be installed before the first use of "Library" system - see "Element Library Files" on page 320 - a simple "generic" version is available on the installation disk, but check with your instrument operator which is the most appropriate file for the data acquired by that particular instrument). At some point you may wish to install and use the on-line "help" files (see note on your CD inlay) and you should have a ".def" directory prepared if you wish to store configuration files for future use (see e.g. p. 79, p. 143).

Terminology

CasaXPS components and windows appear on screen as shown in Figure 1•1, where the annotation defines their names as used throughout this book.
It is assumed always that the user is familiar with the "normal" IBM PC and Microsoft Windows operation and terminology.
The following section should be used as a "hands on" introduction to CasaXPS to provide an overview of key features. Subsequent sections enlarge briefly on the role of the system modules, and the remaining chapters deal in depth with specific topics and applications.

The Program Frame

1. The program frame contains experiment frame windows.
2. Each experiment frame is divided into to two splitter windows.
3. The left-hand-side displays the spectra while the right-hand-side offers a data array (data browser) showing the logical structure of the experiment.

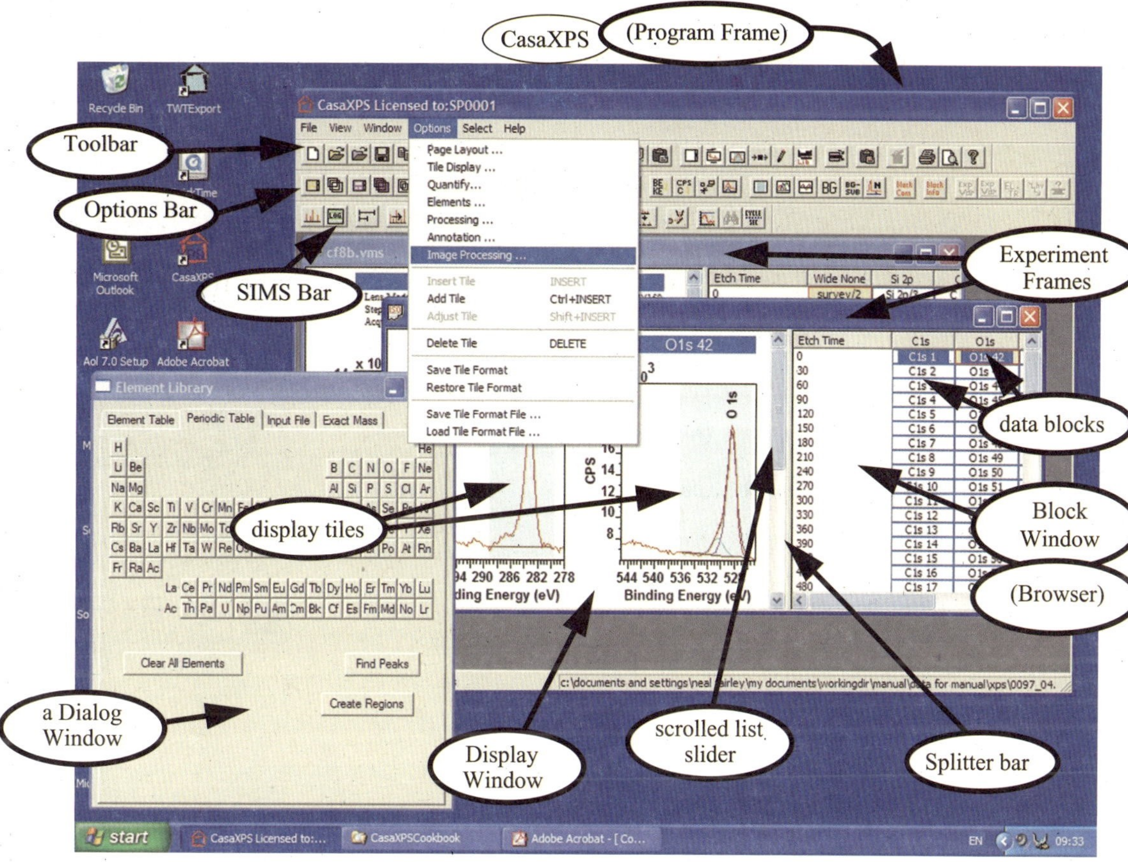

FIGURE 1•1 A Windows XP™ screen showing CasaXPS in operation

How do I....?

(These commands are cross referenced to fuller descriptions in Chapter 3 starting on page 49: U = Upper (main) Toolbar, L = Lower (Options Bar) , 2•3 indicates 3rd button in 2nd group, and so on)

Display spectra

L 1•1-5 (page 58)

Select spectrum block(s) in right-hand-side splitter window (browser - Figure 1•1)

- Press a toolbar button

Zoom in and out

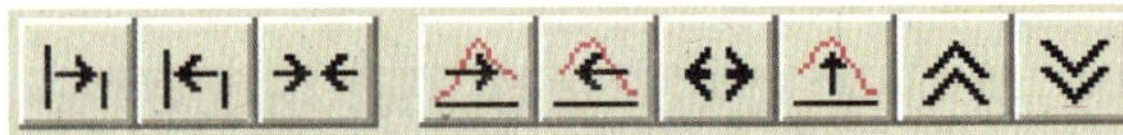

L 2•1-3 (page 58),
L 3•1-6 (page 59)

- Mark zoom box on spectrum by dragging cursor over spectrum.
- Left-click inside box to zoom in; clear the zoom box by clicking outside the box.
- Crtl right-click to zoom out.
- Adjust zoom position using toolbar: step-left, step-right, expand abscissa, rescale Y and increase Y scale.

Access the options dialog windows

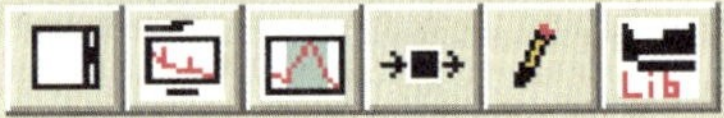

U 2•1-6 (page 55)

Save my work

U 1•4-8 (page 52)

Create or open a new file

U 1•1-3 (page 51)

Toggle the display options

L 3•1-4 (page 59)
L 4•1-6 (page 60)

Alter ISO file parameters

L 5•1-5 (page 60)

Quantify spectra

- Zoom into peak on spectrum
- Press Create button (**U 2•3**)
- Adjust region limits using mouse and dragging the vertical bars
- Select fields to edit (press return to accept changes)
- Background types are L = Linear, S = Shirley, T = Tougaard, N = none, D = default and more.
- Annotate spectra with region table using Annotation Dialog

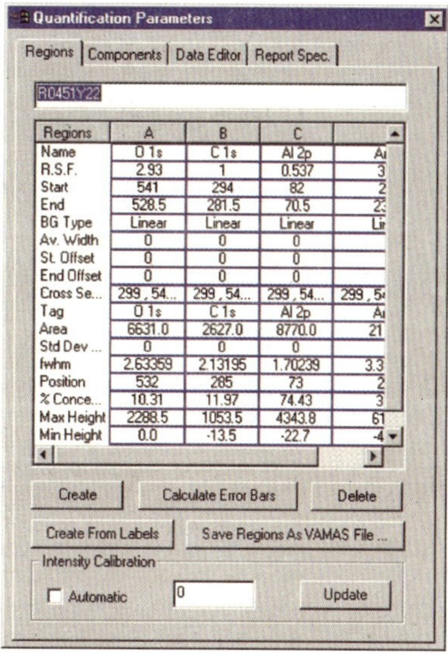

Create quantification regions from labels

U 2•3

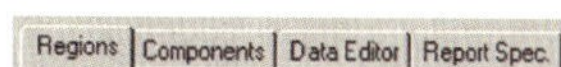

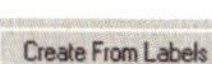

- Identify the peaks using the element library
- Annotate the peaks using the Peak Labels property page of the annotation dialog window. Only annotate those peaks for which quantification regions are required.
- On the Regions property page of the Quantification Parameters dialog press the Create from Labels button.

Create a peak model

- Press Create button **(U 2•3)**
- Adjust component position/height using mouse on peak top
- Adjust FWHM using mouse on peak side
- Select fields to edit (press return to accept changes)
- Create more peaks if required
- Press "Fit Components" to auto-fit the peak parameters
- Annotate spectra with colour coded table using Annotation Dialog (page 13)

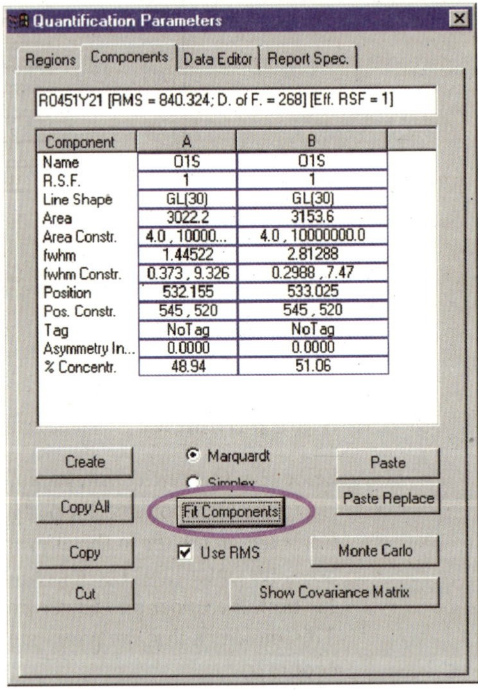

Constrain Peak Parameters

- Select the Quantification Parameters dialog window **(U 2•3)**
- Go to the Components Property Page
- Create a set of peaks
- Define the area relationships using Area Constraint field and the column headings and a multiplicative factor
- Define the FWHM relationships using the FWHM Constraint field and the

column headings using a multiplicative factor

- Define the position relationships using the Position Constraint field and the column heading using an additive offset

Component	A	B	C
Name	Peak 1a	Peak 1b	Peak 1c
R.S.F.	1	1	1
Line Shape	SGL(10)T(3)	SGL(10)T(3)	SGL(20)T(2...
Area	4514.4	2248.2	2212.0
Area Constr.	1.0 , 10000...	A * 0.498	A * 0.49
fwhm	1.11884	1.1	1.1
fwhm Constr.	0.9436 , 1.2	0.9 , 1.1	0.9846 , 1.1
Position	285.011	285.435	289.249
Pos. Constr.	288.499 , 2...	A + 0.424	A + 4.2379
Tag	NoTag	NoTag	NoTag
Asymmetry In...	0.1058	0.1058	0.1608
% Concentr.	36.76	18.31	18.01

Fix a fitting parameter

- Select the Quantification Parameters dialog window (**U 2•3**)
- Go to the Components Property Page
- Enter a # character in the appropriate constraint field.
- Press the enter key. The current value for the chosen parameter is entered for both the upper and lower limits for that parameters constraint interval. This indicates that the parameter will not be used in any optimization procedure.

Release a fixed fitting parameter

- Select the Quantification Parameters dialog window (**U 2•3**)
- Go to the Components Property Page
- Enter a # character in the appropriate constraint field, but append a numerical value after the # character.
- Press enter. The current value for the parameter plus and minus the numerical value entered after the # character will become the upper and lower limits for the constraint interval for that parameter.

Perform a Monte Carlo error analysis

- Select the Quantification Parameters dialog window (**U 2•3**)
- Go to the Components Property Page
- Create a peak model
- When the peak fit results in an acceptable residual, press the Monte Carlo button
- Wait whilst the Monte Carlo simulation procedure completes the task
- Save the error matrix and/or the error distribution or copy and paste the results through the clipboard into Excel.

Edit spikes

- Click on spectrum to see specific energy range
- Right-click on abscissa to invoke edit dialog window (below) ***or***
- Hold down the Control and Shift key together to point and click a new value from the spectrum. ***or***
- Replace data using the defined backgrounds from the current quantification regions.

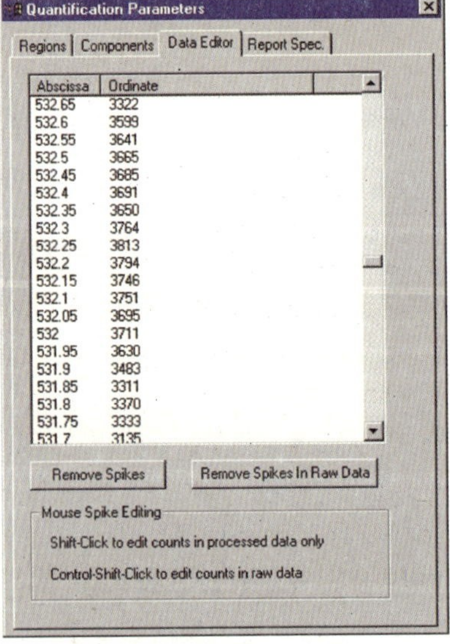

Generate quantification reports

- Select spectrum block(s) in the browser (Figure 1•1)
- Press desired quantification report type (on Report Spec Property Page, see page 95)
- Copy report to clipboard or save to file.

Annotate spectra with text

- Enter text
- Choose attributes (Position Type, Font, Vertical Text

(U 2•5)

- Press Apply
- Adjust position using Annotation History

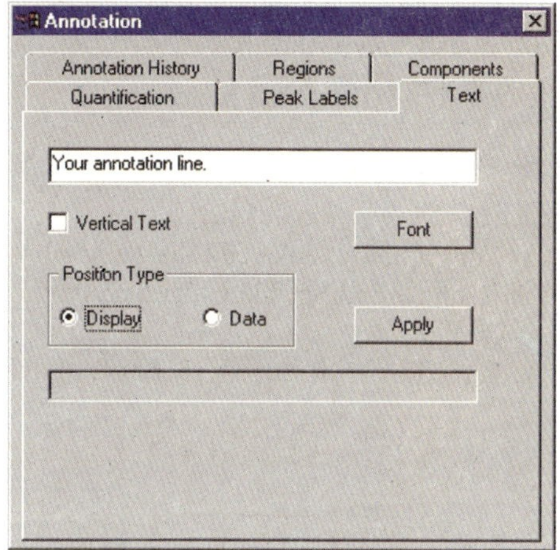

Annotate spectra with regions table

as above, from "Regions" tab

- Choose position and typeface attributes
- Press Apply
- Adjust table position using Annotation History and the drag boxes on spectra.

Annotate spectra with synthetic components table

as above, from "Components" tab

(U 2•5)

- Choose position and typeface attributes
- Press Apply
- Adjust table position using Annotation History and the drag boxes on spectra.

Annotate spectra with quantification table

(U 2•5)

as above, from "Quantification" tab

- Choose position and typeface attributes
- Press Apply
- Select spectrum blocks in right-hand-side splitter frame (Figure 1)
- Adjust table position using Annotation History and drag boxes on spectra.
- Note that the table is generated from the current selection of spectrum blocks.

Annotate spectra with peak labels

(U 2•5)

- Select labels in scrolled list (Crtl Left-Click to add to selection)
- Choose position and typeface attributes
- Press Apply
- Adjust label positions using Annotation History and drag boxes on spectra.

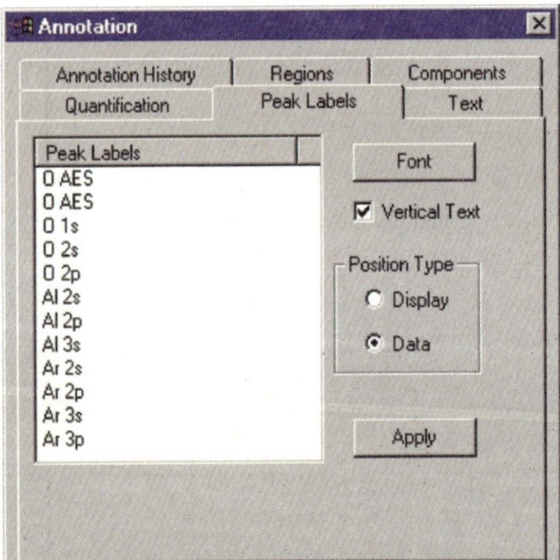

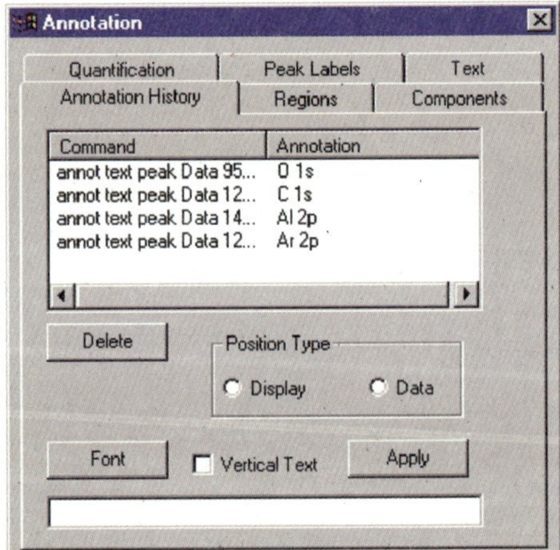

Move and alter annotation attributes

(U 2•5)

- Select row in table of annotation entries
- Adjust attributes
- Press Apply

To move

- Select row
- Left-click on the spectrum *or*
- Drag annotation position box on spectrum.

Adjust tile display parameters (U 2•1-2)

- Adjust settings
- Press OK or Apply buttons

Create an inset tile (L 1•1-5)

- Display a spectrum in the left-hand-side of the experiment-frame.
- Draw a zoom box using the left-hand mouse button.
- Press the Insert (Ins) key on the keyboard. A new Inset Tile will be created within the zoom box.
- Adjust the display parameters for the Inset Tile using the toolbar buttons and normal display dialog windows.

Note: The Inset Tile is identical in functionality to the parent tile and can be used to define regions and components in just the same way as the parent tile.

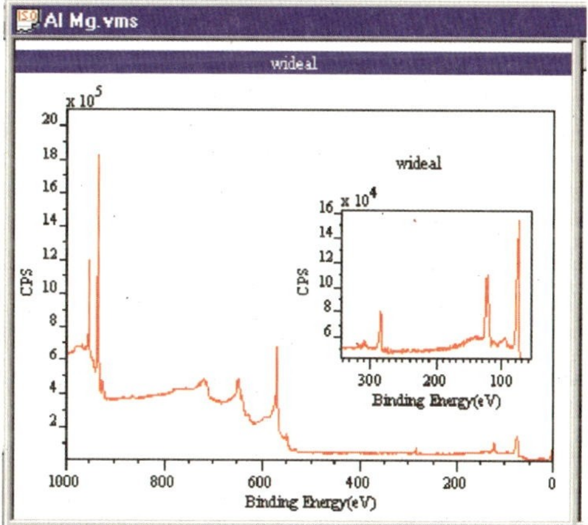

Delete a tile

To remove a tile from the display list:

- Select the tile to be deleted using the left-hand mouse button.
- Press the Delete keyboard key.

Process spectra (U 2•4)

- select the appropriate function

Energy calibrate spectra (U 2•4)

- Use mouse to mark a box from the measured peak position to the true peak position
- Adjust values if required
- Select tick boxes Regions and/or Components if required
- Press Apply to adjust calibration for active tile
- Press "Apply to Selection" to calibrate the selected set of spectrum data blocks

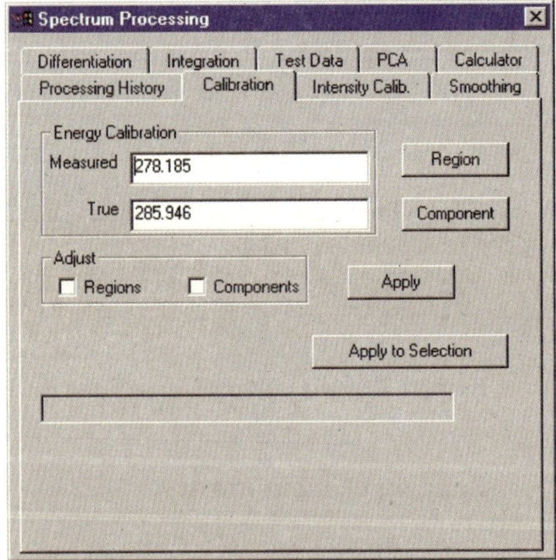

Identify peaks (U 2•6)

- Point at a peak on the active tile and left-click
- Scroll to the most likely peak assignment in the scrolled list
- Select the name field in the scrolled list
- If required, select the name field a second time to remove peak markers from spectrum
- Repeat until all peaks have been identified
- When finished, remove all the markers from the display using the "Clear all Elements" button (see below)

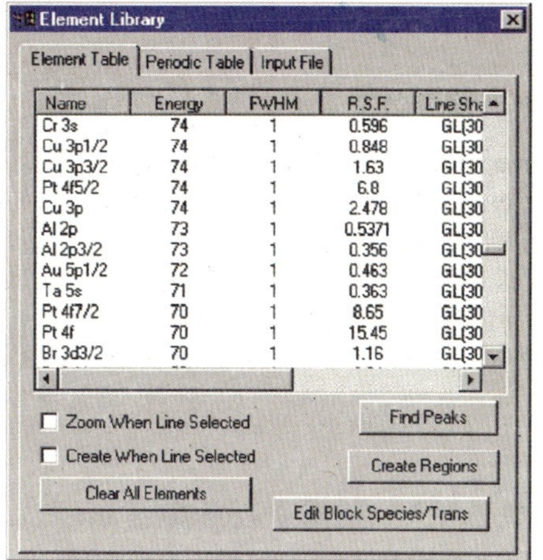

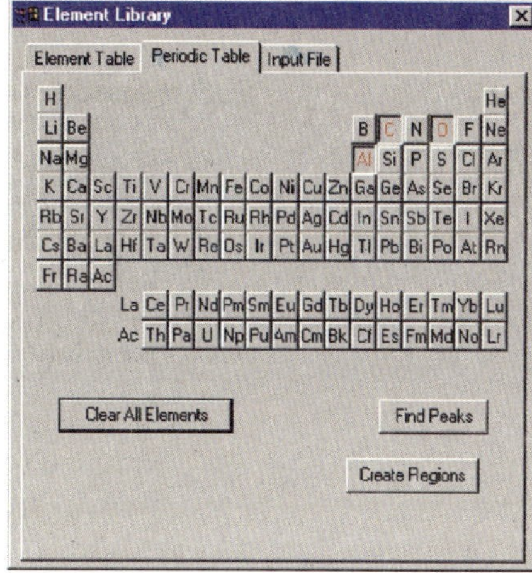

Create quantification from element markers (U 2•6)

- Press the Find Peaks button on the Element Table or Periodic Table Property Pages
- Remove or add elements as required using periodic table buttons
- Press the Create Regions push button

Exiting Regions will be over-written by the new set of quantification regions.

Load an element library (U 2•6)

- Browse your disk for the appropriate element library
- Overwrite the current library by pressing Load button

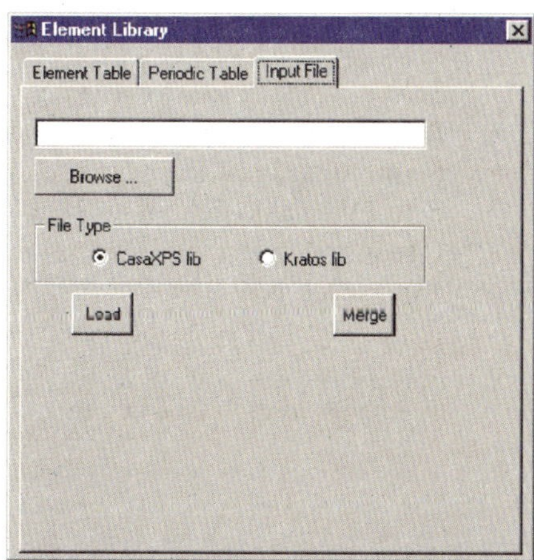

Save a bitmap to the clipboard (U 1•5)

- Prepare the current display tile (Figure 1•1)
- Press toolbar button (or Crtl C)
- In Word, Excel or other application, select Paste (or Crtl V)

Save an enhanced meta file of the active tile to the clipboard

(U 1•6)

- Prepare the active display tile (Figure 1•1)
- Press toolbar button
- In Word or Excel select Paste (or Crtl V)

NB Vertical text in Enhanced Meta File (EMF) format does not display correctly in some versions of Word and/or Excel.

Save an EMF of ALL the tiles to the clipboard (U 1•7)

- Prepare the active display tile (Figure 1•1)
- Press toolbar button
- In Word or Excel select Paste (or Crtl V)

NB Vertical text in EMF format does not display correctly in some versions of Word and/or Excel.

Save a TAB spaced ASCII file containing peak fitted data

(U 1•9)

- Prepare peak-fits
- Select spectrum blocks in right hand browser window (Figure 1•1)
- Press toolbar button
- Select file name to save data
- Read text file into Excel or other program

Transfer my work to other spectra (U 3•1)

- Prepare processing, annotation, regions and/or components for a spectrum
- Select the set of spectra in the right hand browser window (Figure 1•1) for which the current spectrum's operations require propagating

- Right-click the mouse with the cursor over the displayed spectrum
- Select (with checkboxes) the appropriate actions on the Browser Operations "Propagate" dialog box
- Check that the list of spectrum blocks is your intended set of spectra (look at both name and file)
- Press the OK button
- Display your selected spectra and visually check the results.

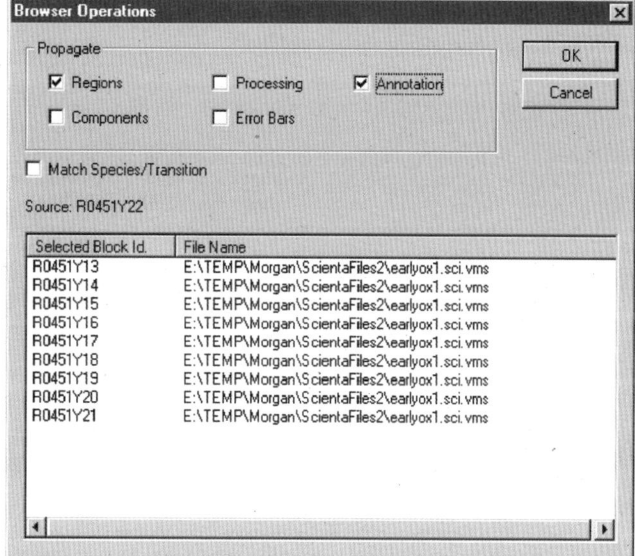

Copy spectra to a new file (U 1•1)

- Create an empty experiment frame by pressing the New toolbar button
- Select spectrum block names in the right hand splitter frames for each file involved. Use the Crtl key at the same time as the mouse to add to the selection.
- Right-click the mouse with the cursor over the right hand browser window of the empty experiment frame window.
- Check the list of files.
- Press the OK button.

(see also Adding and Deleting Blocks, page 56)

Display normalized overlaid spectra (L 5•6)

- Copy the spectra into a new file (if data is spread over several files).
- Select the spectra in the right hand browser window (Figure 1•1) and press the overlay toolbar button.
- Normalize the display by pressing the toolbar button.
- Move the normalization point by holding down the Shift key and left clicking on the spectrum display tile.
- Terminate the normalized display by pressing the normalize button a second time.

(L 1•2) 

(L 5•6) 

Satellite subtraction (U 2•3)

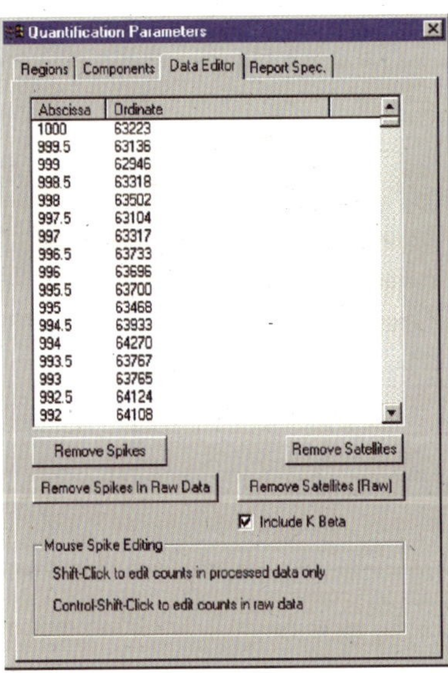

- Define regions for each photoelectric peak
- Press Remove Satellites push button to temporarily remove satellites from the spectrum or press Remove Satellites (Raw) to *permanently* remove the satellites from the data.

Temporarily removed satellites may be reinstated using the Reset option on the Processing History property page of the Spectrum Processing dialog window.

Toggle edit mode for experimental variable label (L 6•8)

Switch between Experimental Variable and CasaXPS row labels. In Edit Mode, double-click a row to edit label. When in Edit Mode, quantification reports are generated using row labels rather than experimental variable values.

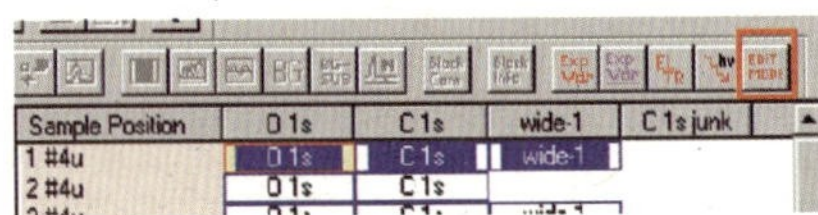

Section 2 - What CasaXPS Can Do

CHAPTER 2

Basic Concepts & Case Studies

The Main Window and Experiment Frames

The CasaXPS program window (Figure 2•1) manages sets of data files in the same way Excel™ manages books of spreadsheets. Each ISO 14976 file is opened into a separate window (an Experiment Frame) within the main program window, where these Experiment Frame windows can be maximised, minimised, tiled or cascaded to aid access to multiple open data files. The data files themselves may contain complex sets of spectra related by some experimental variable, such as etch time or temperature change, or may contain merely a single spectrum. It is the role of the Experiment Frame to manage the logical structure of these data files, where the right-hand-side of the Experiment Frame organises the data into an array of items labelled by the names given to the ISO data blocks, which are aligned with respect to the experimental variable and the element/transition fields associated with each spectrum.

This array of items appearing in the right-hand-side browser offers a means of selecting individual spectra for display (in the left-hand-side pane) or batch processing. Figure 2•1 shows the main program window, where two opened data files are displayed. In one a depth profile is constructed (b-tp.exp.vms): in the other a simple file containing just a single Si 2p spectrum is shown. The third window (CasaXP1) is empty.

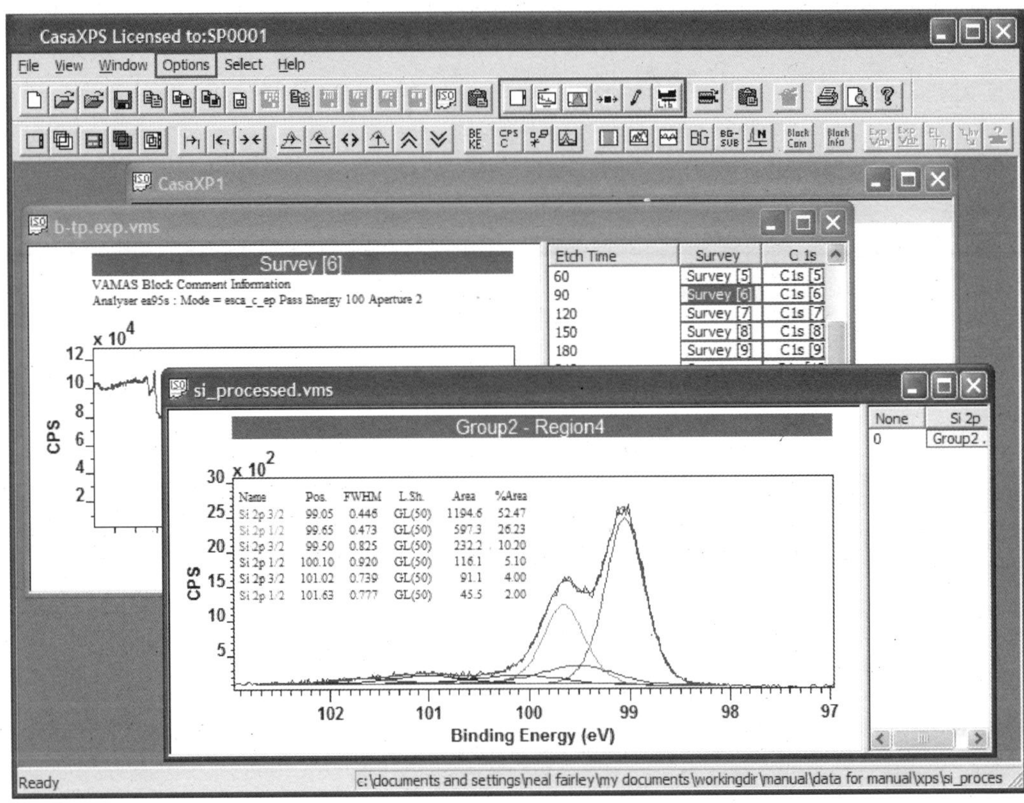

FIGURE 2•1 The main program window of CasaXPS showing an empty Experiment Frame (CasaXP1) and two ISO14976 files (b-tp.exp.vms and si_processed.vms)

Toolbars and Menus

The Experiment Frame is divided into two sections. The left-hand-side displays the spectra in familiar graphical form while the right hand side offers a means of browsing through the spectra (data blocks) in a file. Actions and options associated with each of these sections within an Experiment Frame are available from the menus and toolbar buttons, initially positioned along the top of the CasaXPS main program frame. Clicking in either of the sections within the Experiment Frame will change the availability of both menu options and toolbar buttons. As focus moves between a graphical display and data browser, the colour of the buttons changes to reflect the options that apply to the particular item selected by the mouse.

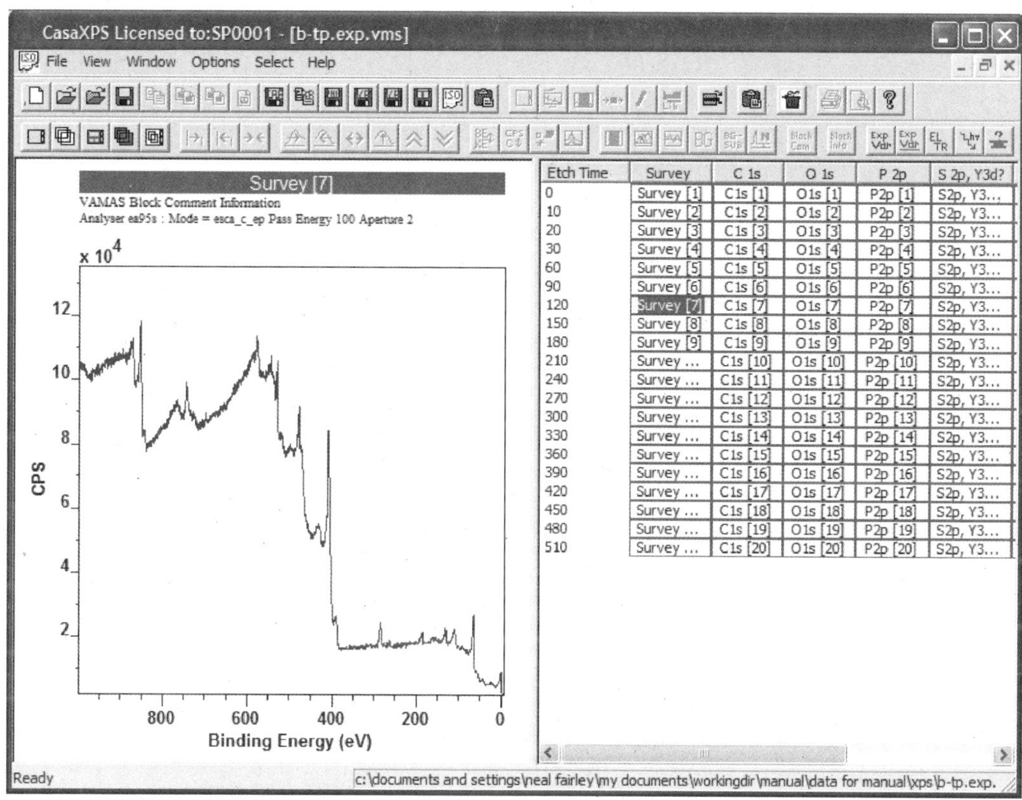

Etch Time	Survey	C 1s	O 1s	P 2p	S 2p, Y3d?
0	Survey [1]	C1s [1]	O1s [1]	P2p [1]	S2p, Y3...
10	Survey [2]	C1s [2]	O1s [2]	P2p [2]	S2p, Y3...
20	Survey [3]	C1s [3]	O1s [3]	P2p [3]	S2p, Y3...
30	Survey [4]	C1s [4]	O1s [4]	P2p [4]	S2p, Y3...
60	Survey [5]	C1s [5]	O1s [5]	P2p [5]	S2p, Y3...
90	Survey [6]	C1s [6]	O1s [6]	P2p [6]	S2p, Y3...
120	Survey [7]	C1s [7]	O1s [7]	P2p [7]	S2p, Y3...
150	Survey [8]	C1s [8]	O1s [8]	P2p [8]	S2p, Y3...
180	Survey [9]	C1s [9]	O1s [9]	P2p [9]	S2p, Y3...
210	Survey ...	C1s [10]	O1s [10]	P2p [10]	S2p, Y3...
240	Survey ...	C1s [11]	O1s [11]	P2p [11]	S2p, Y3...
270	Survey ...	C1s [12]	O1s [12]	P2p [12]	S2p, Y3...
300	Survey ...	C1s [13]	O1s [13]	P2p [13]	S2p, Y3...
330	Survey ...	C1s [14]	O1s [14]	P2p [14]	S2p, Y3...
360	Survey ...	C1s [15]	O1s [15]	P2p [15]	S2p, Y3...
390	Survey ...	C1s [16]	O1s [16]	P2p [16]	S2p, Y3...
420	Survey ...	C1s [17]	O1s [17]	P2p [17]	S2p, Y3...
450	Survey ...	C1s [18]	O1s [18]	P2p [18]	S2p, Y3...
480	Survey ...	C1s [19]	O1s [19]	P2p [19]	S2p, Y3...
510	Survey ...	C1s [20]	O1s [20]	P2p [20]	S2p, Y3...

FIGURE 2•2 Depth Profile.

Many of the menu options are also offered via the toolbar buttons. Figure 2•1 indicates the relationship between the Options menu and the set of toolbar buttons used to access the dialog windows that control the display, annotation, processing, quantification and element identification. Pressing one of the toolbar buttons enclosed by the bounding box in Figure 2•1 either opens or brings to the front the dialog window corresponding to the menu options under the Options menu. In addition to menu accelerators, toolbar buttons also change the display state and represent accelerators for options available on the Tile Display dialog window, while others provide the means for displaying sets of spectra in the left-hand-side. Adjusting the view of the data via zoom and interval stepping are also offered on the toolbars.

The mechanisms for analysing spectra will now be explained using the data shown in Figure 2•2 as a set of case studies.

Case Study 1: Quantification of a survey spectrum

The data shown in Figure 2•2 is a depth profile where etch cycles are interleaved with acquisition of a spectra over a range of energy intervals. Survey spectra are acquired at each depth, and one of these spectra will be used to illustrate the quantification procedure.

Converting the Data

These data were originally in SpecsLab I file format. The original SpecsLab I data are in an ".exp" ASCII file and must be converted to the ISO 14976 file format through CasaXPS. Simply select the Convert option on the File menu (also available on the topmost toolbar, Figure 2•3) and open the .exp file via the "Convert to VAMAS" file dialog window. A new file with a .vms file extension is written into the same directory as the original .exp file. Note, write permission for that directory and sufficient disk space is required for a successful format conversion.

FIGURE 2•3 File handling toolbar buttons: the Convert button is outlined.

Browsing the Spectra

Any depth profile experiment typically appears in the browser section of an Experiment Frame in a similar format to that seen in Figure 2•2. The keyboard arrow keys can be used to step around the array of data blocks tabulated in the browser, and the block selected by the arrow keys is also shown in the (left hand) display section. Additionally, the mouse can be used to select and display a specific block using "double click" and "marquee select" actions over the block or blocks of interest. Using one of these mechanisms, the survey spectrum displayed in Figure 2•2 has been selected for quantification.

Element Identification

Quantification using relative sensitivity factors requires that each element within a spectrum be identified and included in the quantification table. The process of identifying the elemental composition relies on the Element Library dialog window (selected from the menu option indicated in Figure 2•4), which displays the set of transitions prepared and stored in an ASCII file called CasaXPS.lib (located in the same directory as the CasaXPS.exe executable file). Once loaded, the transitions in the CasaXPS.lib file are displayed via the Element Table property page on the Element Library dialog window (Figure 2•4). Element markers are added and removed from the display by clicking on the name field in the table on the Element Table property page and by matching the spectral features against the lines within the element library, the composition for the surface is established. If the mouse is used to point at a feature on the spectrum and the left mouse button clicked, the Element Table scrolls to the energy indicated by the mouse. Alternatively, element markers can be displayed using the Periodic Table property page on the Element Library dialog window. The Periodic Table interface offers an array of buttons, one per element, and these buttons allow markers to be toggled on or off the display.

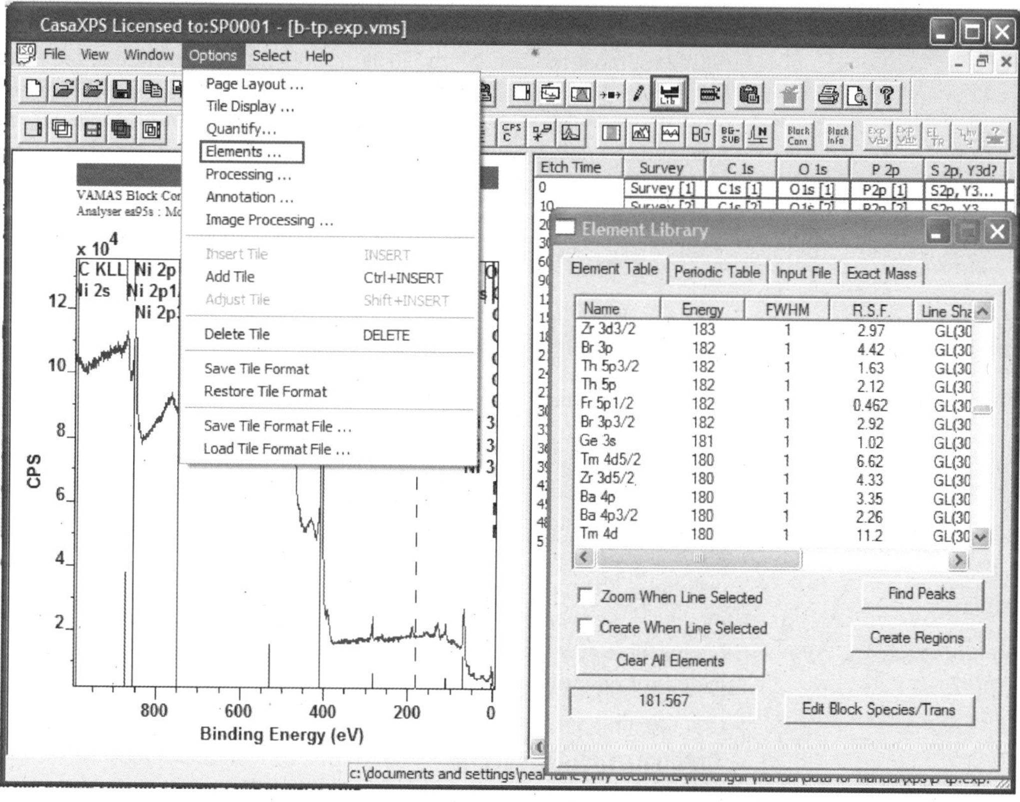

FIGURE 2•4 Element Library dialog window showing Element Table property page.

A further mechanism, sometimes useful, is automatic peak identification. The Find

Peaks button on both the Element Table and the Periodic Table property pages offers a means of placing markers on a spectrum. For each element found in the library with matching peak patterns to that observed in the spectrum, an element marker is placed on the spectrum. In the case of complex spectra, the Find Peaks button provides a starting point from which the true elemental composition can be hand crafted using the Periodic Table property page.

Quantification

Once all the spectral features have been identified and corresponding markers displayed on the spectrum, a quantification table is created and displayed over the spectrum. This is achieved by pressing the Create Regions button on either the Element Table (Figure 2•4) or the Periodic Table (Figure 2•5) property pages. Quantification regions are created for each element using the transition with the largest relative sensitivity factor in the element library, subject to not being overridden for an element by transitions specified in the CasaXPS_quant.lib library file (also located in the same directory as the CasaXPS.exe executable file). Regions will only be created for an element provided a peak within the data is identified. As a result, manual adjustments to the number of elements used in the quantification is sometimes necessary.

Manual Adjustment of Quantification Regions

Although quick, the automatic method for creating quantification regions may not account for all elements required to quantify a sample and it is also wise to check the background limits. Such adjustments to the quantification regions are made using the Quantification Parameters dialog window (available from the Options menu at the head of the main window - see Figure 2•8 on page 30). If the Regions property page on the Quantification Parameters dialog window is the active topmost property page, the start and end limits of the background for each region can be modified under mouse control. Regions are displayed with vertical grey bars whenever the region limits can be altered using the mouse. When making such adjustments, it is often necessary to zoom into the energy interval surrounding a quantification region. If the Reset toolbar button is pressed then the entire set of quantification regions is placed on the zoom list and so each time the Zoom-out toolbar button is pressed the display cycles through each of the quantification regions in turn. Region backgrounds can be assessed and adjusted one after another until the full set of regions is visually inspected.

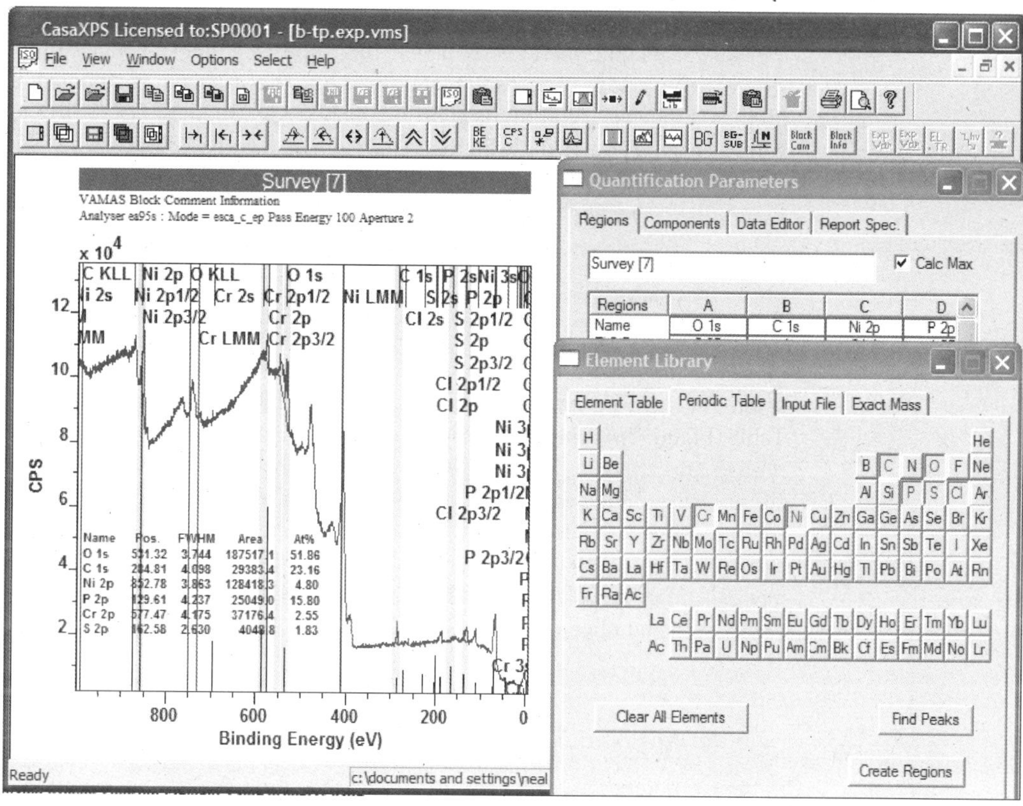

FIGURE 2•5 Quantification generated using the Find Peaks and Create Region buttons.

Summary

To quantify a survey spectrum:

* Display the spectrum in the left hand window of an Experiment Frame.
* Identify spectral features using the element markers available from the Element Library periodic table.
* With the element markers displayed over the spectrum, press the Create Regions button on either of the Element Table or Periodic Table property pages.
* Check the background definitions using the Quantification Dialog window and the Reset/Zoom-out viewing mechanism.

Case Study 2: Display of Spectra and Printed Logo

After selecting a set of spectra in the right-hand-side of an Experiment Frame, the spectra are displayed in the left-hand-side using the set of buttons ringed in Figure 2•6. The first button from the left causes each selected spectrum in the right-hand-side to appear in a separate display tile in the left-hand-side. Any format of tiles currently displayed in the left-hand-side is replaced by the new set of tiles laid out using the information found in the current browser selection, one spectrum per tile.

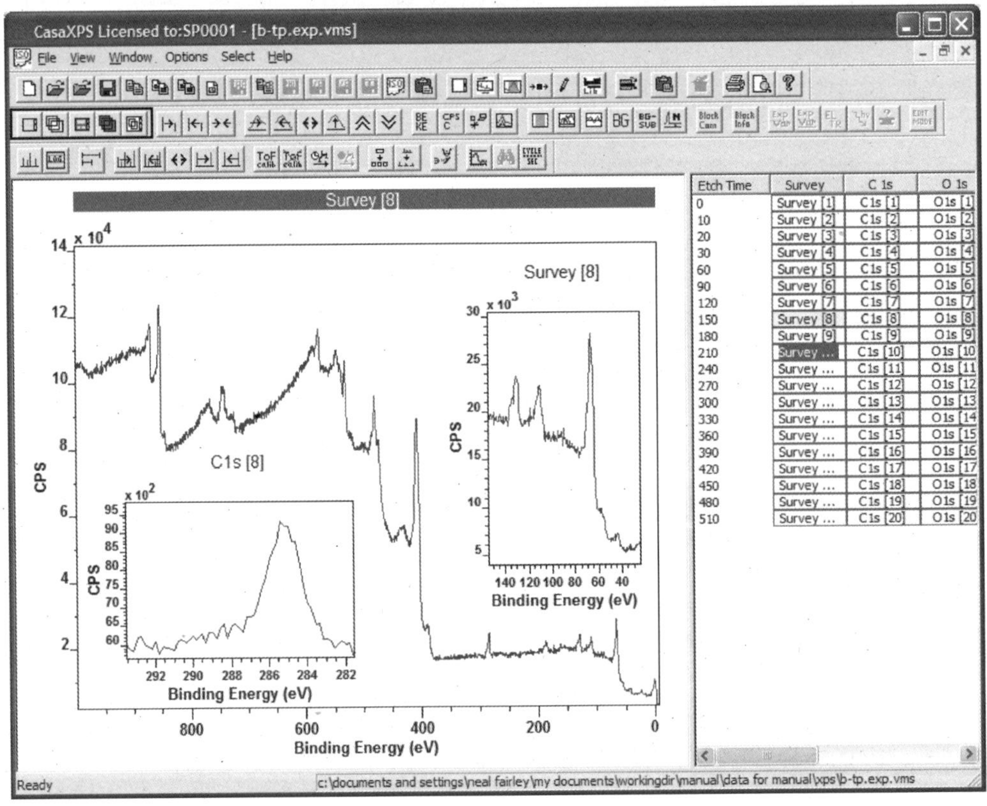

FIGURE 2•6 A Custom Tile Format.

If, however, a particular layout for a set of spectra has been carefully prepared (as in Figure 2•6) and it is desired to retain this display format, then two options are open to the analyst.

One option is to use the New Window menu option under the Window menu

(Figure 2•7). Once selected, a new Experiment Frame opens offering an independent view of the file from the initially opened Experiment Frame. A counter is added to the file name displayed in the heading for each Experiment Frame, indicating and distinguishing between the different Experiment Frames all referencing the same underlying file. Spectra can be browsed and processed independently using multiple windows, thus allowing the original display layout to be preserved whilst still permitting the data to be examined.

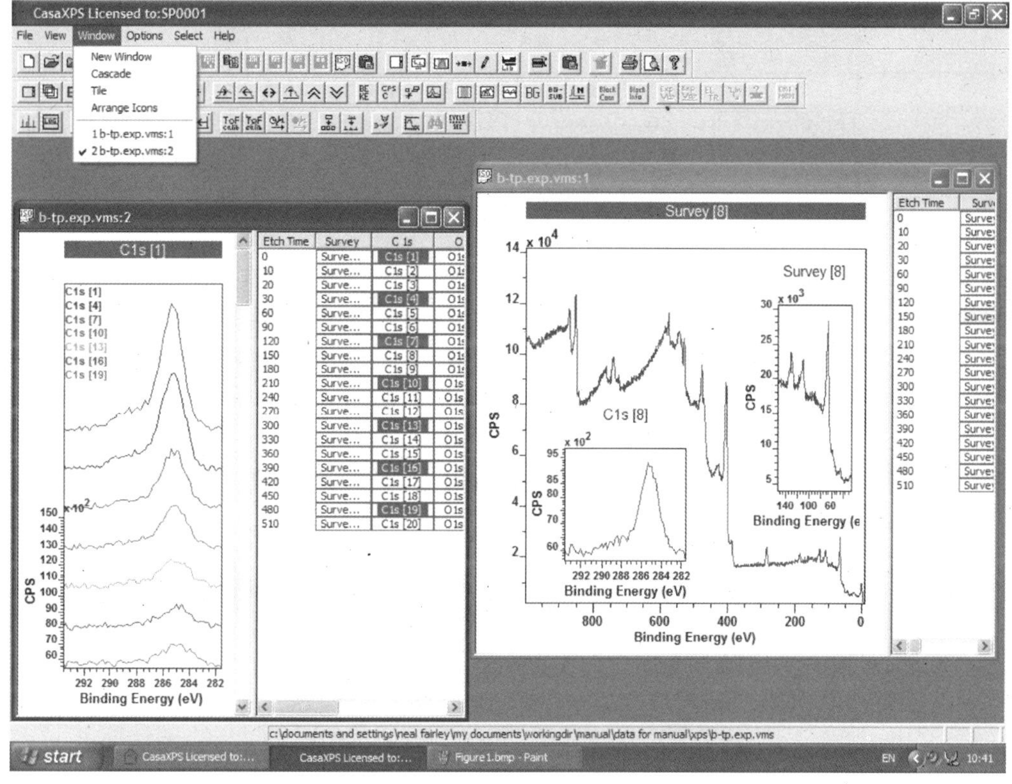

FIGURE 2•7

As an alternative, within each Experiment Frame, a given tile format can be stored and retrieved using the Options menu (Figure 2•8). Once a format is established, the Save Tile Format menu option takes a copy of all the tiles currently displayed in the left-hand-side scrolled list. New spectra can then be displayed in the left-hand-side, however the original tile format may be restored to the scrolled list by selecting the Restore Tile Format menu option at any time. Each Experiment Frame maintains a slot for storing a tile layout, so if more than one format requires temporarily saving then a combination of creating new windows via the Window menu and saving the current tile layout for each new window is possible.

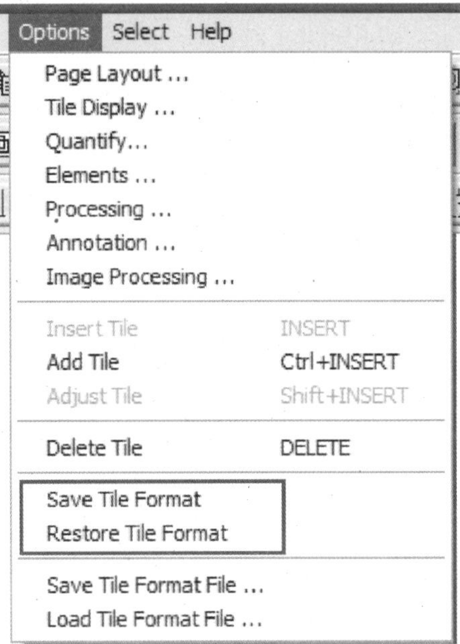

FIGURE 2•8 Save and Restore tile format within a session.

Adding a Logo to Printouts

A bitmap logo can be added to the printout of a tile format. If a bitmap is placed in a file called logo.bmp located in the CasaXPS.DEF directory then that bitmap is printed (Figure 2•9) in front of the footnote string (read from the PrintFootNote.txt file). The size of the bitmap is dependent on the resolution of the printer used to create the hard copy. Appropriately sized bitmaps can be prepared in the Paint program usually found on the Programs/Accessories menu under the Start button of the Desktop. Note that limiting the number of colours reduces the size of the bitmap. CasaXPS reads the bitmap logo file at start-up and so 24-bit bitmaps is wasteful of resources and also usually unnecessary.

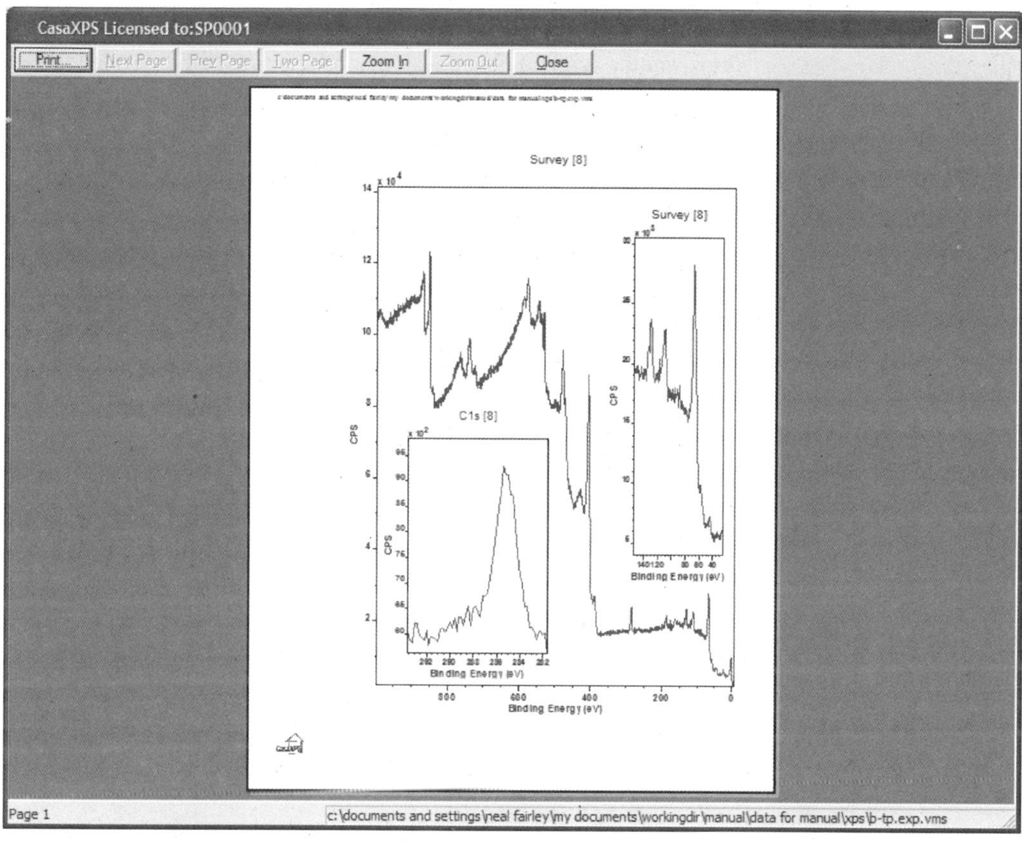

FIGURE 2•9 Print preview window showing logo position on the printed page.

Case Study 3: Satellite Subtraction

Spectra acquired using non-monochromatic X-ray anodes include features in the photoelectric spectrum corresponding to minor resonance lines in the X-ray spectrum from the X-ray gun anode. Figure 2•10 shows a portion of a survey spectrum in which the smaller satellite feature to lower binding energy of the C 1s line is a result of a minor Mg X-ray resonance. The energy separation and relative intensity of satellite lines with respect to the primary X-ray line are well known (page 315) and so, to a first approximation, these features can be removed from the electron spectrum. The method used to remove these satellite lines is to iteratively generate the satellite subtracted spectra using the convolution of a response function with the current iteration to partition the data into the primary spectrum and the satellite spectrum.

The satellite subtraction procedure requires the identification of each photoelectric line using quantification regions (the contribution to Auger lines are not separated in energy from the primary Auger spectrum and therefore should be excluded from the satellite procedure). These regions define which lines are to be included in the satellite subtraction algorithm and also define the background to the peaks. The background choice determines, in part, the shape of the primary line and a poorly defined background will impair the outcome of the satellite subtraction.

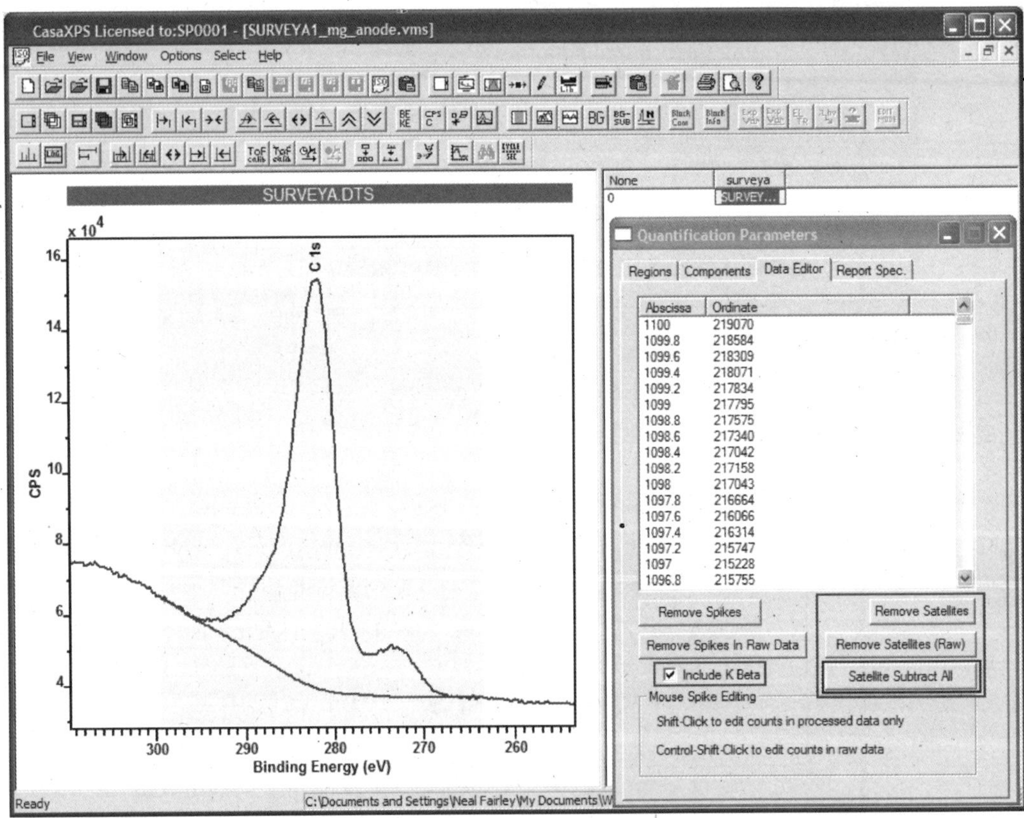

FIGURE 2•10 Data Editor property page on the Quantification Parameters dialog window.

Satellite subtraction does not fit well with the processing regime in CasaXPS. The reason being that the procedure uses the current definition of quantification regions in order to perform the subtraction. The set of regions used in satellite subtraction is typically different from the set of regions used to quantify a spectrum and so it is difficult to track the procedure, in a processing-history sense. As a result, satellite subtraction is performed on the Quantification Parameters dialog window using buttons found on the Data Editor property page. Here, spectra can be processed individually or collectively, where those spectra overlaid in the active tile are the set of spectra processed by the Satellite Subtract All button indicated in Figure 2•10. Spectra modified via the Data Editor property page may either be permanent (i.e. the raw data is edited) or temporary in the sense that resetting the processed data using the Processing History property page will undo the satellite subtraction. If the raw data is edited then the process cannot be reversed and an entry is written into the block comment for that spectrum indicating the data is satellite subtracted. If the raw data is modified it is recommended that the data is saved to a new file name rather than overwriting the original data file.

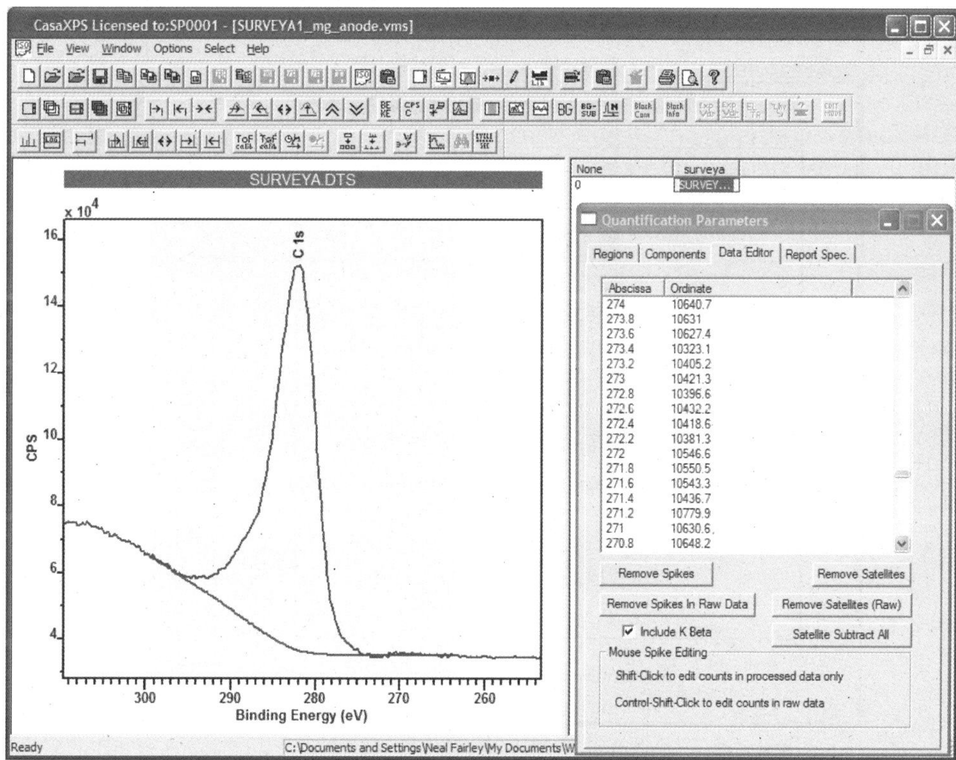

FIGURE 2•11 Same spectrum shown in Figure 2•10 after satellite subtraction.

Figure 2•11 shows the data in Figure 2•10 following satellite subtraction. Here the Satellite Subtract All button was used, where the resulting dialog box (Figure 2•12) indicates how many spectra will be processed and asks whether or not the raw data should be edited. Select the No button unless it is desired to permanently modify the data.

Summary

- Create quantification regions on each photoelectric line in a spectrum
- Display all spectra for which satellite subtraction is required into the active tile.
- On the Data Editor Property Page select Satellite Subtract All button.
- Answer the dialog question in Figure 2•12.
- If raw data was modified, save the spectra to a new file name.

FIGURE 2•12

Case Study 4: Quantification Tables and Areas

It is sometimes necessary (and usually desirable) to understand the source for the quantification results reported in the quantification tables generated by CasaXPS. Corrections to the data for instrument transmission and sample mean free path have a direct result on the area reported for synthetic components and intensities determined from integration regions. The following discussion attempts to uncover the way these corrections influence the values reported at the various stages when analysing spectra.

The area intensity for the Si 2p spectrum shown in Figure 2•13 can be computed from the sum over the area parameters for the optimised set of synthetic components or alternatively the area measured using the integrated signal between the background and the data (determined using the Regions property page). The situation for this spectrum is very simple, namely, symmetrical Gaussian-Lorentzian line-shapes arranged as doublet pairs sit on a Shirley background, where no corrections for mean-free-path nor instrument transmission are involved. As a result, the reported intensity for the area value is identical for both methods of reporting. That is, the sum of the synthetic component area parameter is entirely equivalent to the value reported on the Regions property page (Figure 2•13).

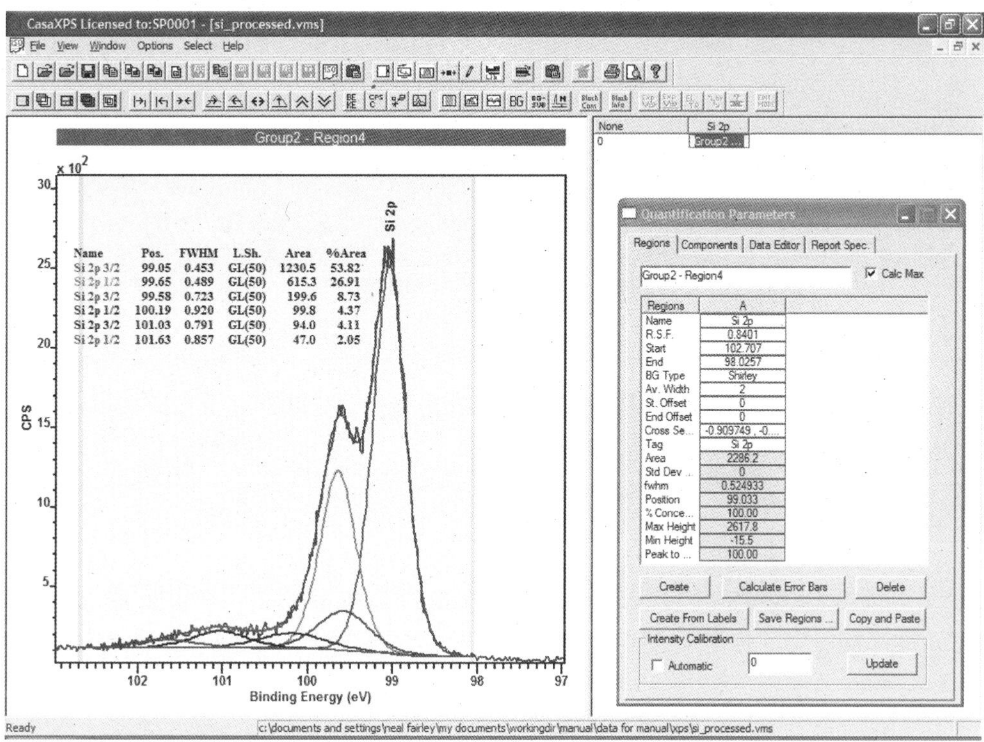

FIGURE 2•13 Si 2p spectrum intensity analysed using synthetic components. No mean-free-path correction applied.

With no corrections made to the data, the area reported on the Regions property page (Figure 2•13) is 2286.2 CPSeV. If however a mean-free-path correction is applied to the quantification process, as specified in the Intensity Calibration field at the bottom of the Regions property page, then the situation changes. The area field determined by the quantification region is normalised with respect to the energy dependent adjustment, namely kinetic energy raised to the power of the text-field entry, which in this case (Figure 2•14) is equal to –0.5.

Note that the components table annotated on the spectrum still reports uncorrected areas. The reason for this is the components table is reporting synthetic parameter values not quantification values. The normalised area calculated from the quantification region is now reported to have a value of 61.4. Note that the area value for the Regions property page is computed from the data rather than, in the case of the components, an input parameter. If the table displayed on the spectrum is required to correspond to the results of the analysis, then the table over the spectrum must come from the Quantification property page on the Annotation dialog window (Figure 2•15). The quantification annotation table then reports the results of the component analysis and so normalised areas are reported.

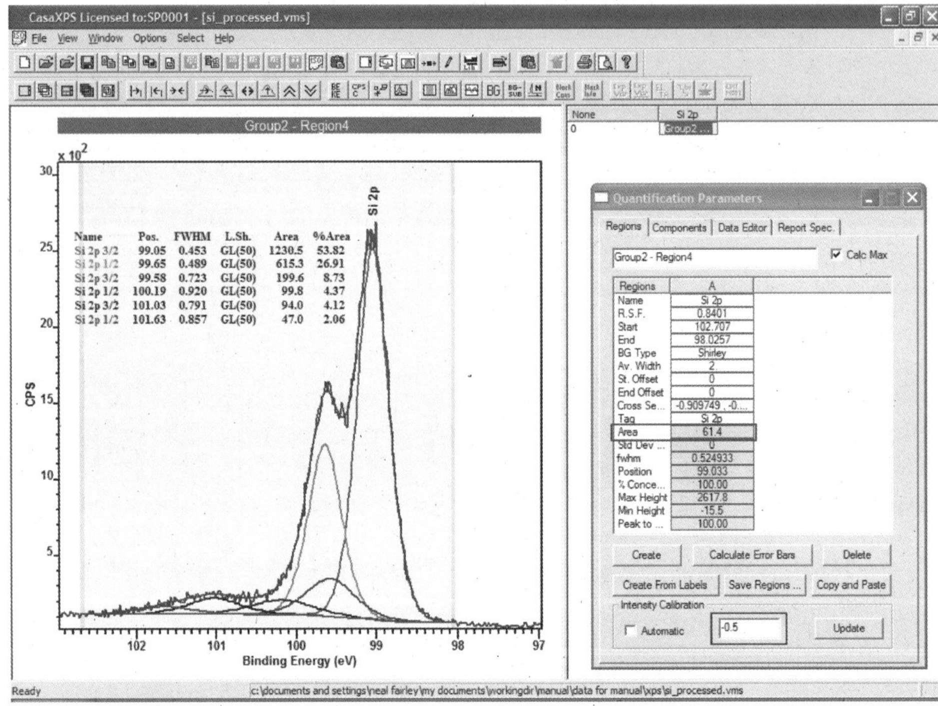

FIGURE 2•14 Mean-free-path correction applied to the results of quantification.

A further observation is that the atomic concentration is an output for all tables and therefore any corrections to areas used in the computation have been made prior to reporting these values. Note also, the output area value in a standard quantification report is normalised with respect to mean-free-path and transmission correction (if present), however only the percentage concentration values are computed using the relative sensitivity factors.

Facts relating to quantification:

- Results of calculations are always corrected for any mean-free-path and/or transmission factors. That is, energy dependent adjustments are included in area values *calculated* during quantification.

- Input parameters are not adjusted for energy dependent parameters. That is component areas entered on the Components property page are *uncorrected* area values.
- Reported areas in **Standard Reports** *do not* include relative sensitivity

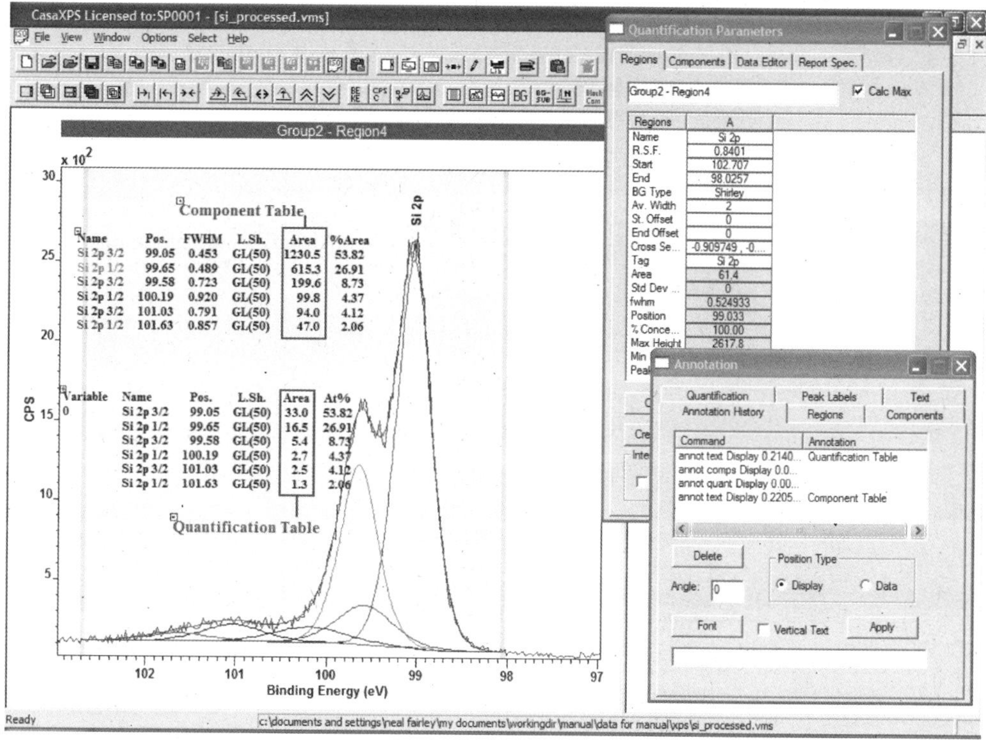

FIGURE 2•15 The components table (above) represents the set of parameters used to specify the synthetic components, while the quantification table (below) offers the normalised areas used to calculate the atomic concentrations.

adjustments, however percentage concentrations *do include* relative sensitivity adjustments.

- Area values in **Custom Reports** *are* corrected for mean-free-path, transmission and relative-sensitivity factors. Including the relative-sensitivity factor in the Custom Report table allows the intensities to be plotted in a meaningful way equivalent to a similar plot for the table of atomic concentrations.
- Area values are always determined using background-subtracted data. If values are required relative to a baseline then the background type **zero** should be used.

Case Study 5: Displaying Spectra & Saving Display Formats

Spectra are displayed in a scrolled list on the left hand side of the Experiment Frame. The selected spectrum blocks from the Experiment Frame Browser are displayed as a scrolled list of display tiles or inserted into the active display tile using the toolbar buttons indicated by the outlined box in Figure 2•6 on page 28. The procedures for displaying spectra are as follows:

Mouse and Keyboard Operations:

1. Arrow keys on the Keyboard. The scrolled-list is re-initialized with a single tile displaying the adjacent spectrum block to the block previously displayed in the Active Tile.
2. Double-click the spectrum block name. A spectrum block selected using the left-hand mouse button double-click action results in the scroll-list re-initializing to a single display tile contain the data from the selected spectrum block.
3. Double-click the spectrum block experimental variable value. Left-clicking on the experimental variable value selects all the spectrum blocks in the corresponding row. Selecting these same spectrum blocks using a double-click action results in the scrolled-list filling with tiles, one per selected spectrum block each displaying the data from the spectrum block.
4. Combining a range selection with a double-click display action. A range of spectrum blocks may be selected using the Shift and Control Key as normal, however if the last selection within a range selection is performed using a double-click action of the left mouse button, the selected spectrum blocks will be used to re-initialize the scroll-list of display tiles, one spectrum per tile.

Toolbar Display Buttons

Toolbar button or F1 Function Key displays the data blocks currently selected in the right-hand-side of the Experiment Frame.

Toolbar button or F2 Function Key overlays the data blocks currently selected in the right-hand-side of the Experiment Frame into the Active Tile.

Toolbar button displays a new selection of data blocks using the existing display state, one spectrum per display tile. If an equivalent set of blocks is selected and displayed using this toolbar button, the spectra are displayed using the current set of axes ranges in each tile.

Toolbar button overlays the current data block selection in the active tile without any adjustments to the axes display ranges within the active tile.

Toolbar button displays the current data block selection such that each row of blocks in the right-hand-side is overlaid in a display tile. If multiple rows are selected, then the left-hand-side displays a sufficient number of tiles to accommodate all the rows selected in the right-hand-side.

Saving the Formatting of the Display Tiles

Once spectra are displayed in the left-hand-side, the fonts, colours, and other parameters for a given tile are adjusted using the Tile Display Parameters dialog window (Figure 2•17). Many of the display settings on the Display property page shown in Figure 2•17 are also available from the toolbars, and display settings can be toggled between various states from there. Toolbar buttons offer Toolbar Hints ("tool tips") indicating the nature of the action associated with the button when the cursor hovers over them.

Once a set of tiles has been formatted, the entire display state for the scrolled list on the left hand side of the Experiment Frame can be saved to file for use at a later time, if desired. The options for saving and loading the tile format file are available on the File menu and also on the Options menu (see Figure 2•18 on page 41). The Option menu also offers two additional menu options for temporarily saving the tile format within memory rather than writing the format to disk. Formats saved using the options in the upper box in Figure 2•18 are only available within a given CasaXPS session.

The procedure for restoring a tile format file is as follows:

1. Open the spectrum file.
2. Load the corresponding Tile Format File. If the tile format file corresponds to the spectrum file used to prepare the Tile Format File, then the display will return to the state at the point the file was saved. Note however, the tile format file uses the spectrum block index to determine the appropriate spectrum to display in a tile, therefore if the spectrum blocks have been reordered between saving the Tile Format File and reloading the display state again will result in undefined behaviour.

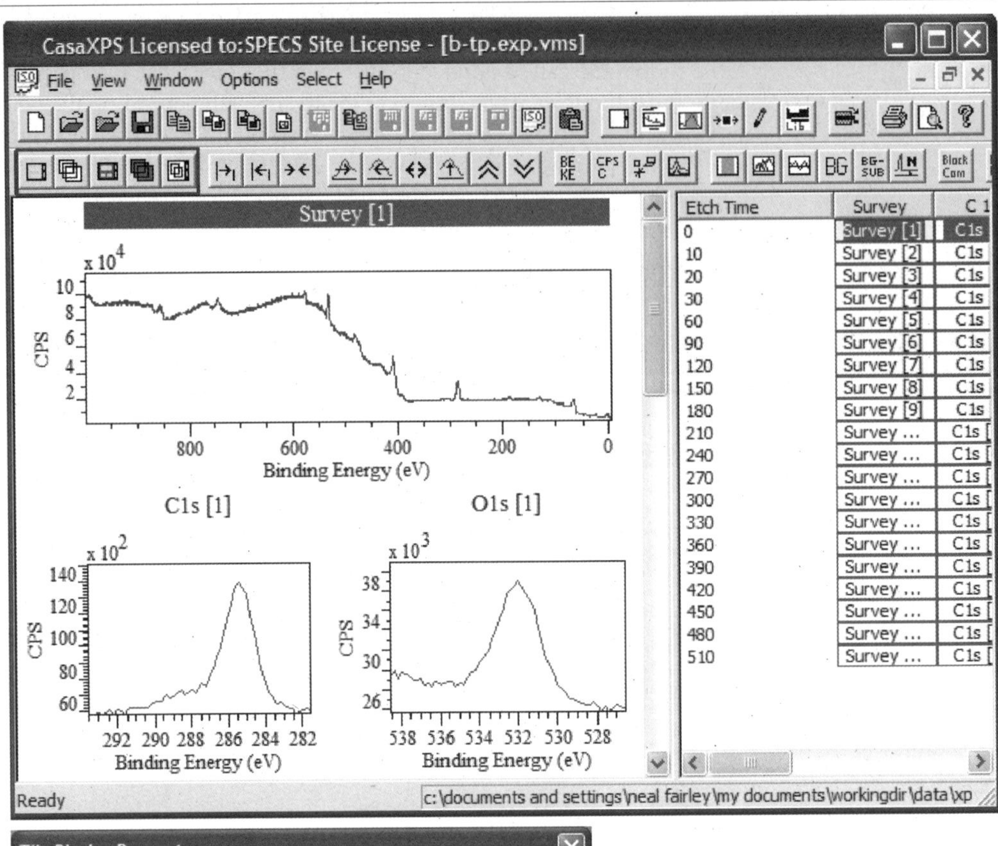

Tile Display Parameters

X Axis | Y Axis | Geometry | Display | Colours | Global

Display Items

- [x] Title
- [ ] Vamas Block Comment
- [ ] Acquisition Parameters
- [x] Annotation
- [x] Quant. Region Labels
- [ ] Component Labels
- [x] Quant. Regions
- [x] Synthetic Components
- [ ] Residual
- [x] Spectrum Background
- [x] Normalise Residual
- [ ] Draw Box on Header
- [x] Vert. Quant. Region Labels
- [ ] Half Tile Display
- [x] Draw Graph Box
- [x] Draw 3D Planes
- [x] Draw Lines
- [ ] Draw Points
- [ ] Fill Below
- [ ] Draw Key
- [ ] Key on Left
- [ ] Description Key
- [ ] Include Exp Variable
- [ ] Display Comps from First Spectrum
- [ ] Dashed Comps
- [ ] Display Spectrum Index
- [ ] Dashed Lines

Header Font

Axes Font

Line Width: 1

Title: Survey [1] | Font

OK | Cancel | Apply

FIGURE 2•16 (above)

FIGURE 2•17 (alongside)

A tile format file can be applied to spectrum files other than the one used to prepare the original tile format file, however the results will depend on the contents of the data file. If the data file contains equivalent spectrum blocks to the original file, that is, the acquisition regions are identical to the original spectrum file and are in the same order as the original spectrum file, then the tile format file will result in an equivalent display of the new data to the display previously observed for the original data. A dialog window will ask if the intensity range should be rescaled appropriately for the new data or remain as prepared for the original file.

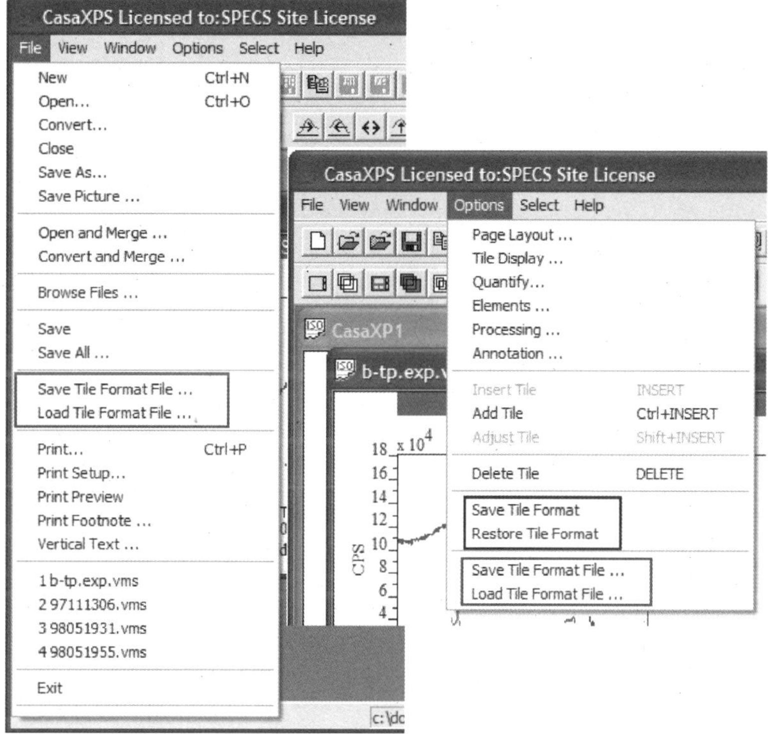

FIGURE 2•18 Saving and restoring tile formats

Case Study 6: Multiple Experimental Variables and ISO Files

During the course of an experiment multiple acquisition sequences may be performed leading to several sets of data related by an experimental variable. That is to say, the experimental variable is a characteristic parameter of the experiment recording, for example, position, temperature or, perhaps more commonly, ion etch times. The ISO file format includes these experimental parameters in each block and, what is more, there may be more than one such experimental variable for a given experiment, therefore the data blocks allow for sets of these experimental variables to be assigned to each spectrum. CasaXPS displays the active experimental variable in the first column of the Experiment Frame data browser (Figure 2•19). When multiple experimental variables are present within the ISO blocks, the active experimental variable can be cycled by using the Next Exp Var menu option on the Select menu shown in Figure 2•19. Cycling through the available experimental variables is also achieved by holding the Control Key down and pressing the F1 function key.

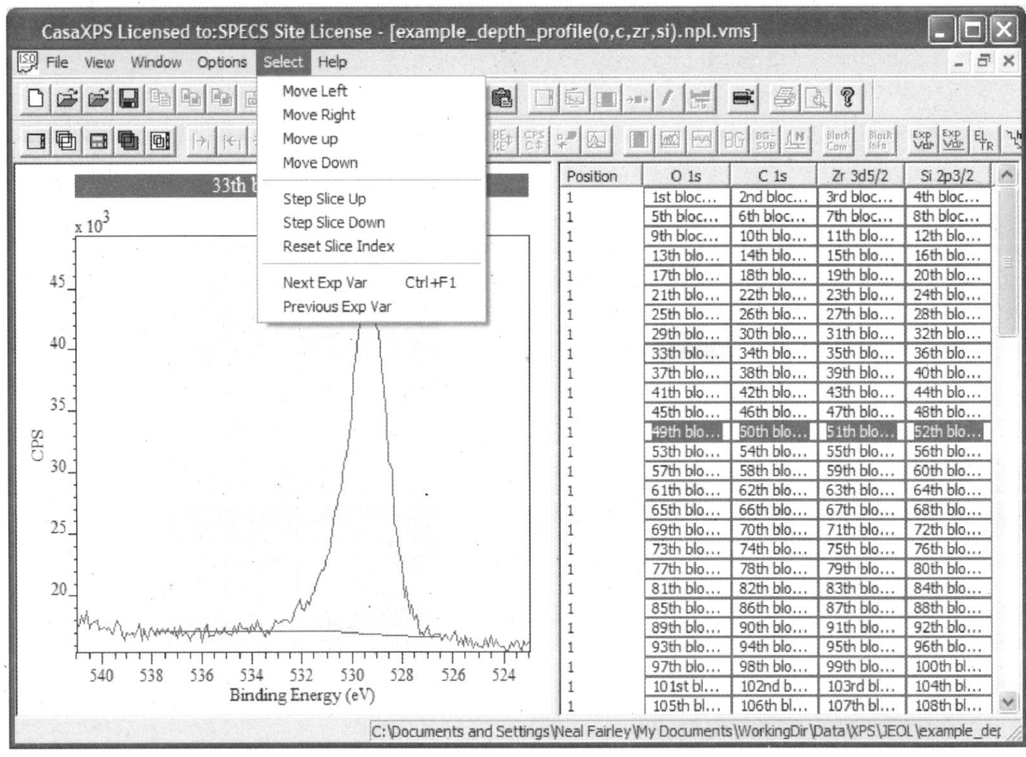

FIGURE 2•19 A depth profile acquired on a JEOL JPS-9200. The experimental variable in the Figure is displaying the position index rather than the etch time. Both etch time and position index are included in the JEOL files exported in spectrum format.

The same data file seen in Figure 2•19 is also displayed in Figure 2•20 where the active experimental variable has been switched to etch time.

Saving and Restoring the Display State from File

CasaXPS maintains a processing history within each data block of an ISO file for saving data reduction steps however display information providing the framework for visualizing spectra from a file as a set of tiles must be saved and restored from a CasaXPS specific file type. The tile file format ASCII files (.tff) includes fields for preserving a given display between sessions and also offers a means of defining a display template suitable for files of similar data for which identical presentation is required.

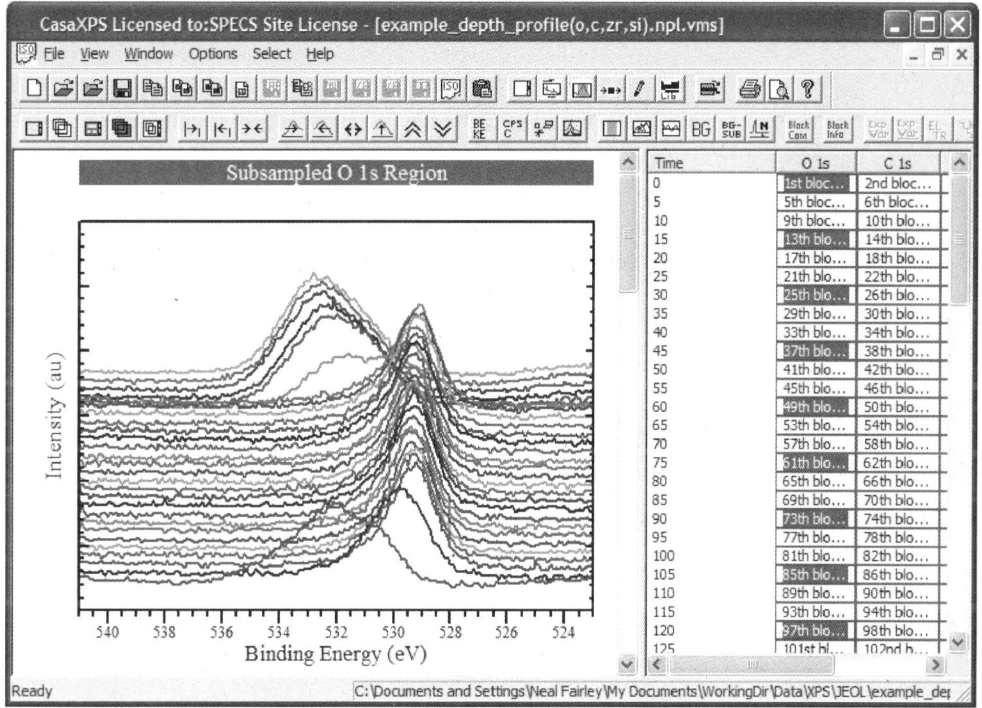

FIGURE 2•20 Depth profile from JPS-9200: experiment variable is etch time

The display section of the Experiment Frame (Figure 2•20) shows a sub-sampled set of O 1s regions at regular time intervals. Displays of this nature are precisely the type of display formats which can now be saved to file. The scrolled list of tiles as well as the current number of tiles per page may be saved to file using the options on the File menu (Figure 2•21), namely Save Tile Format File and Load Tile Format File. Note, these same options are also repeated on the Options menu, where similar options are available for temporarily saving and restoring a tile format within a CasaXPS session.

To save a tile format:

1. Prepare the scrolled list in terms of number of tiles per page and the required display for each tile in the list.
2. Select the Save Tile Format File … menu option on the File menu.
3. Using the File name field on the resulting File Dialog window enter the name of the desired tile format file. The default tile format file name will be the original ISO file where the .vms extension is replaced by a .tff file extension.

To restore a tile format:

1. First load an experiment file.
2. Select the Load Tile Format File … menu option on the File menu.
3. Select the appropriate tile format file for the given file. If the tile format file derives from a different data file, a warning dialog window appears offering the name of the file for which the tile display was prepared.

A tile format file can be applied to other files for which the spectra appear in the same block structure as the original file used to save the tile format. The spectra must appear in the same order with respect to data block index and the data must be equivalent to the original file.

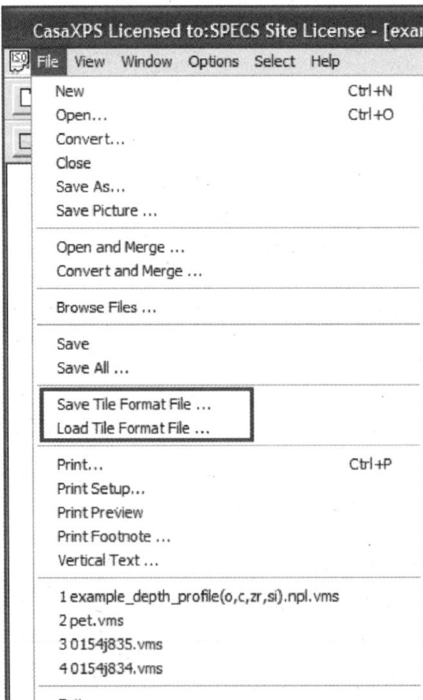

FIGURE 2•21 Tile format controls

Case Study 7: Cu Auger Line Curve Fitting

Problem: Given an "unknown" spectrum of oxidised copper, determine the relative contributions from Cu_2O, CuO and Metallic Cu to the unknown.

Four spectra are displayed in Figure 2•22, three of which are believed to contribute to the fourth spectrum. The three known - "reference" - spectra of metal and oxides can be used to construct line-shapes forming the basis for a nonlinear optimization, the results of which provide an estimate for the relative proportions of these materials in the unknown.

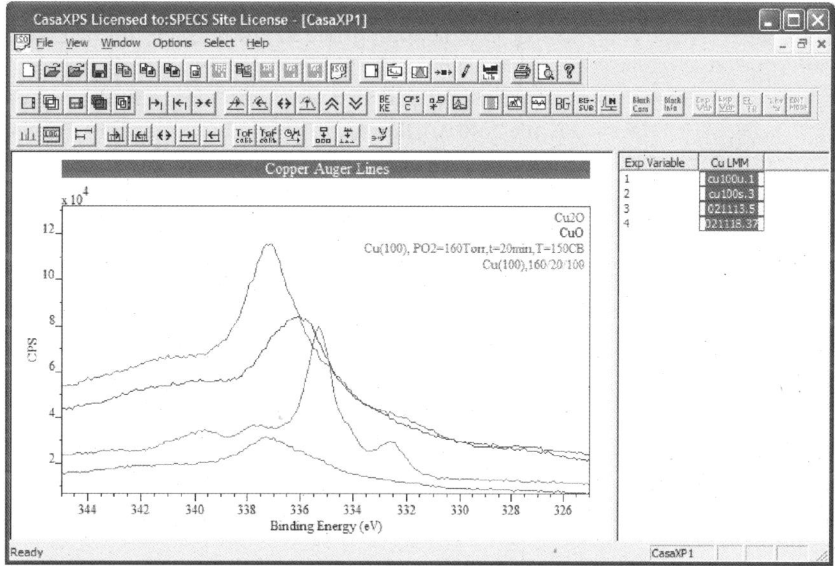

FIGURE 2•22 Cu Auger lines. Data supplied by: Institute of Physics, Tampere University of Technology, Finland.

The following steps are required to perform the analysis:

Construct background subtracted line-shapes from the data.

For each of the spectra required by the peak-fit, add an appropriate background and create a new spectrum block in the data file consisting of the background subtracted data.

- Add a background using the Regions property page on the Quantification

Parameters dialog window (Options menu “Quantify …”).

- Subtract the background from the data using the Test Data property page on the Spectrum Processing dialog window (Options menu “Processing …”).
- Copy the processed data to a new spectrum block by right-clicking over the right-hand-side of the Experiment Frame and selecting the Data Only tick-box (Figure 2•24).
- Adjust the names of the new data using the Edit Info toolbar button. The Block Ids for the new line-shape spectrum block must start with a % character (Figure 2•25).

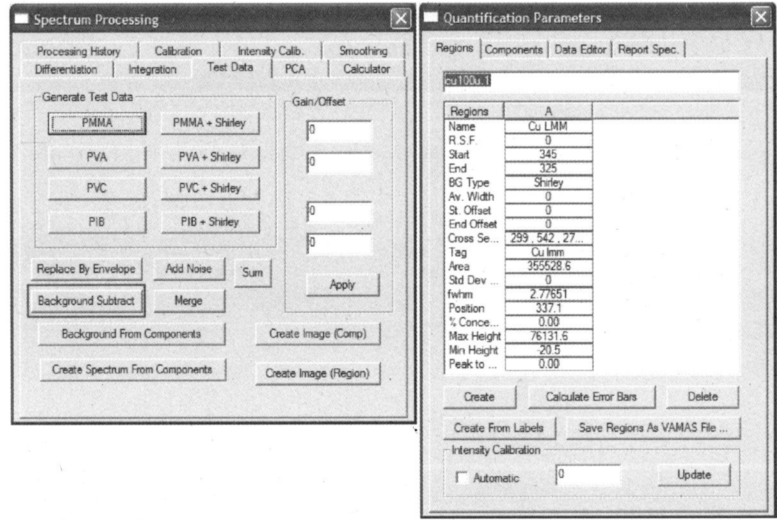

FIGURE 2•23 Preprocess the basis data using the Background Subtract option.(above)

FIGURE 2•24 Copy the processed line-shape to a new spectrum block.(below)

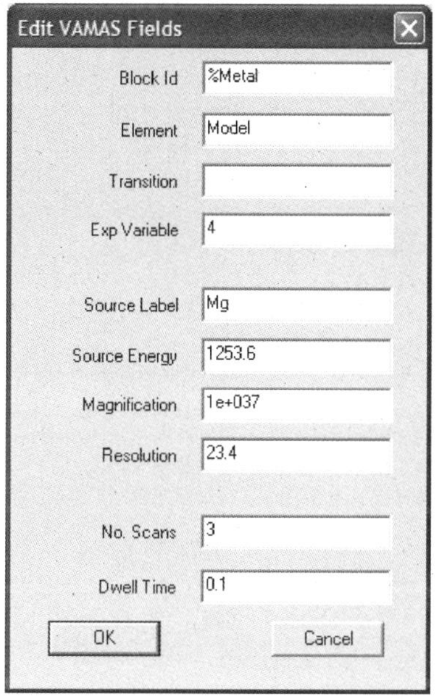

FIGURE 2•25 (left) Edit the line-shape spectrum block and ensure the Block Id starts with a% character.

Exp Variable	Cu LMM	Model
1	cu100u.1	%Cu2O
2	cu100s.3	%CuO
3	021113.5	
4	021118.37	%Metal

FIGURE 2•26 (above) Right-hand-side of the Experiment Frame following creation and renaming of the Block Id, Element and Experimental Variable fields in Figure 2•23

Define a peak model on the unknown spectrum in terms of the new line-shapes.

- Using each of the original basis spectra fit the new line-shape derived from that data to the basis spectra. This provides initial starting points for the peak parameters ultimately used to create the model for the unknown spectrum. Create a component on the basis spectrum and enter the Block Id for the appropriate line-shape into the Line-Shape field on the Components property page (see Figure 2•26 for an example).
- Fix the fwhm parameters derived from fitting to the original basis data and also limit the position parameters to a small interval around the value determined for the basis data. Note the fwhm parameter should be close to unity owing to the requirement that the line-shape from data should not be deformed away from the shape of the basis data.
- Copy and Paste the components from the basis spectra onto the unknown spectrum. Manual adjust the components to guide the auto-fitting and press the Fit Components push button on the Components property page.

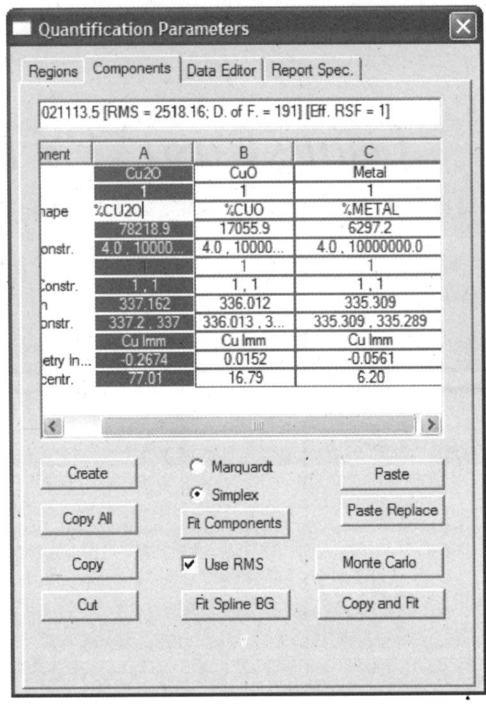

FIGURE 2•27 Model defined from data. Note the Block Id of the line-shapes is entered into the Line-shape field.

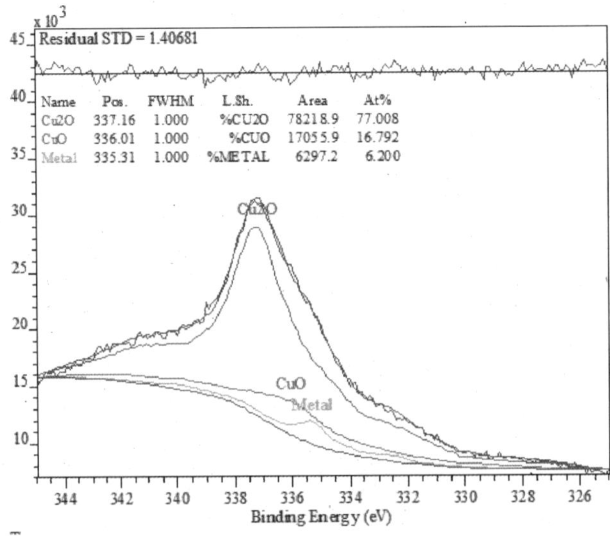

FIGURE 2•28 A possible fit to the unknown spectrum using the line-shapes taken from known spectral forms.

CHAPTER 3

Using CasaXPS Toolbars

Toolbar buttons are an important part of the CasaXPS user interface and provide short-cuts to many of the actions performed using dialog windows. Toolbars are initially displayed at the top of the Main Program window and are activated or deactivated by left-clicking the mouse with the cursor over either the left-hand-side or the right-hand-side of an Experiment Frame (Figure 3•1 on page 50). There are three toolbars - Toolbar, Options Bar, and SIMSBar. Any one of these toolbars can be moved and re-positioned using the mouse or hidden from view using the toggle menu items in the View menu of the Experiment Frame (Figure 3•2). Note that the SIMS toolbar is always present within the system package, but its functionality may not be. Casa SIMS capability and the actions afforded by the SIMS Bar are described in Part II of this Series.

Dedicated XPS / AES users may remove the SIMS Bar from the display (to maximise available display area) by "un-checking" the item in the View list. CasaXPS remembers the View List state between sessions and will not display the SIMS Bar again (or any other of the Toolbars, if desired) until instructed to do so.

The Status Bar item corresponds to the usual "Internet Explorer" status display at the bottom edge of the Main Program Frame. As well a indicating the full (operating system) path to the active Experiment File (on the right), the left hand side of the status line provides a longer version of the "Tool Tip" (function hint) produced when the cursor hovers over a button. Again this Status line may be removed, if display area is critical.

Maximising an Experiment Frame within the Main Program window (Figure 3•1) removes the upper border of the Frame and adds the familiar set of Microsoft window control buttons (minimise, tile, quit) to the Program Frame menu line (above the toolbars). With more than one Experiment file open, "changing focus" of a Program Frame (shifting between (invisible) Experiment Frames) may be achieved either through the "Window" menu or these standard control buttons.

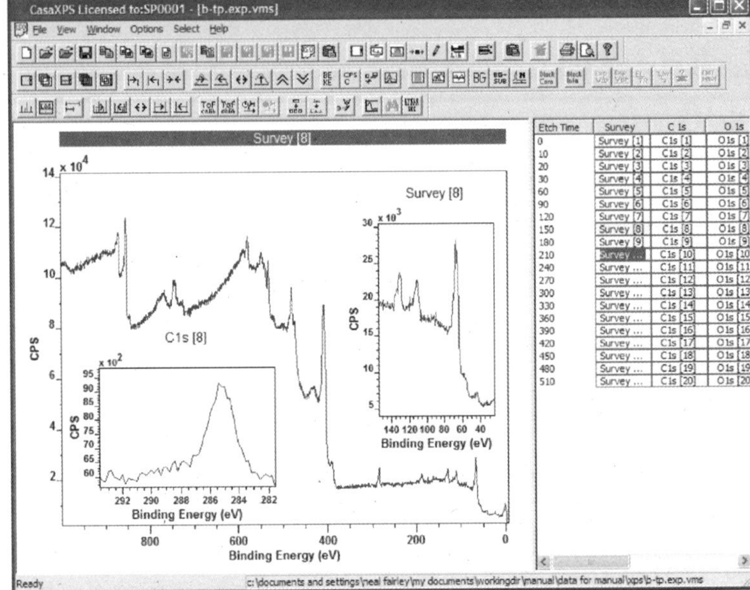

FIGURE 3•1 CasaXPS Program Frame containing a maximised Experiment Frame

Each toolbar's buttons are grouped into subsets based on the type of actions performed by the buttons. The buttons and their associated dialog windows are now described in the context of these groupings.

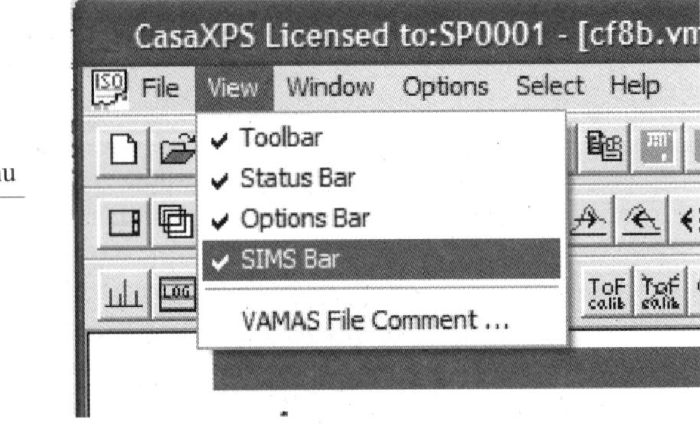

FIGURE 3•2 "View" Menu

Toolbar (upper toolbar, main toolbar)

First Group

(Mechanisms for transferring data into and out of CasaXPS)

Create a new (empty) Experiment Frame window.
An Experiment Frame window is used to manage an ISO 14976 file and is created by the pressing this button or when a .vms file is opened.
Data from other open files can be moved into an empty Experiment Frame, which can serve as a temporary work area for analyzing data or form the basis for collecting useful results for saving separately from the original data files. Copying data blocks between Experiment Frames is facilitated by the "Add" toolbar button (page 56).

Open one or more existing ISO 14976 files into individual Experiment Frames.
A file dialog window appears and this dialog allows .vms file to be selected from any acessible directory. On pressing the Open button, the data in the selected files is displayed in separate Experiment Frames, one per selected file. (See the Open and Merge option on the File menu for a means of collecting multiple files into a single Experiment Frame.)

Convert a proprietary format to ISO 14976 format.
Pressing the Convert toolbar button causes a file dialog window to appear. This dialog offers a set of display filters for viewing files with a particular extension. See "Converting File Formats - using different systems" on page 293 for details associated with converting specific file formats. The file format converter is a fully active feature of the demonstration version of the software.
There are two strategies for converting data files, where either the individual files are converted on a per file basis or alternatively a directory of data files is converted by a single action. The method used for any given format depends on the philosophy behind a given data format and how closely this philosophy matches the design of ISO 14976. The general rule for choosing the strategy is that single Experiment files are created whenever an individual source file contains the same experimental context for multiple acquisitions from a sample (e.g. a survey plus one or more high resolution spectra). Converting a directory of files approach applies to data such as VG Eclipse data, where each spectrum (from the same experiment) appears as a separate file within a directory.
The general approach to converting file formats is to supply a file extension using a

File dialog window. For example, a directory of files containing X/Y pairs in ASCII format can be converted *en masse* into a single Experiment Frame by providing a new file name with a ".dat" file extension via the File Dialog window. The base name for the file name is not important, however the ".dat" extension to the specified file name causes the conversion to proceed on the basis that all the file in the current directory are ASCII X/Y pairs. A new ISO 14976 file is written into the current directory, so it is important that the user has write permission for the directory in question.

Save the current file.
A file dialog is activated that allows the currently active sub-frame to save any changes to the same file or to another file name.

Copy to Clipboard.
When an Experiment Frame is the sub-window with focus, the Copy button places a bitmap of the Active Tile on the clipboard. Once on the clipboard, the bitmap can be pasted into any program such as Microsoft Paint, Word, Excel or any other program accepting bitmaps from the clipboard. On pressing the Copy button, a dialog window appears in which the dimensions in pixels for the bitmap is specified. The plot in the Active Tile is scaled to fit the size of the bitmap defined by the dialog window.
If a Report Frame is the sub-window with focus, then pressing the Copy button results in a text dialog window, from which the tabulated information in the report can be placed onto the clipboard in a Tab spaced format suitable for spreadsheet programs.

Copy EMF to Clipboard from Active Tile
Using the drawing instructions for the Active Tile, copy an EMF to the clipboard. Enhanced Meta Files are a mechanism used by Microsoft to transfer graphical information between programs. An EMF can be displayed in any program capable of interpreting these drawing instructions and so, in principle, an EMF is a compact method for transferring pictures between Microsoft programs. The EMF files generated by CasaXPS are created using the Microsoft Visual C++ compiler, however the underlying standard has changed over the years resulting in anomalous behaviour when a CasaXPS EMF is displayed in different versions of Word (for example). Switches for modifying the behaviour of the EMF file are included in CasaXPS, however it is not possible to create an EMF that works in Word 97 and also in Word 2000. The two main inconsistencies between different versions of Word are: 1) the positioning of vertical text and 2) mirror text display. The former can be adjusted using the **Vertical Text** menu item on the File menu to force the vertical text to appear

correct for one version of Word; unfortunately, if a Word document is subsequently sent to a person with an incompatible version of Word, the graphic, if left as an EMF insert, will appear incorrect to the observer. For this reason, the use of EMF files is not recommended for general use.
The problem of positioning vertical text is not confined to EMF visualization. An analogous issue exists for some printer drivers and therefore the Vertical Text menu item dialog window invoked from the File menu also provides a solution to the incorrect placement of vertical text on printed output.

Copy Scrolled List to Clipboard as an EMF
Using the drawing instructions for the entire set of tiles in the left-hand-side of the Experiment Frame, copy an EMF to the clipboard. See the above discussion relating to EMF from the Active Tile for caveats.

Save TAB spaced ASCII file
Data in a format suitable for import to Excel, Sigma Plot or Origin is output to an ASCII file using the Save TAB spaced ASCII file button. The resulting text file is created from those data blocks currently selected in the right-hand-side of the Experiment Frame, where separate tables are formatted from the data in each selected block. The tables include the binding/kinetic energy, spectral intensities, background, each synthetic component in any peak model defined on the data and the total synthetic envelope. The table for each block output includes columns using both counts-per-second and counts-per-channel.

Copy TAB spaced ASCII to Clipboard
Tabulated data in a similar format to that generated by Save TAB spaced ASCII file is placed on the clipboard with the view to pasting into a spreadsheet or plotting program. The source for the data, however, is the spectrum in the Active Tile rather than the selected blocks in the right-hand-side. Note that while the data displayed in a Clipboard Selection dialog window may appear misaligned, the columns of data are separated by tab spaces and will paste into most spreadsheets property aligned. The Clipboard Selection dialog window also allows the information to be saved to a text file.

Write a PHI MultiPak file.
ISO blocks in CasaXPS may be exported (written to disk) in PHI MultiPak format.

Save data as individual ISO files

Both buttons save the selected blocks as a set of ISO files, where each block becomes a separate file. Some programs are not capable of using ISO files unless the most basic use of the VAMAS format is employed to save the data. These two buttons provide alternative means of saving basic formatted data in an attempt to aid the exchange of data between programs. These options may be useful, for example, if spectra are required as input for the NPL transmission correction software.

Save in QUASES format

QUASES format relates to the program written by Sven Tougaard. A dialog window is invoked which offers two options for exporting the data:

1. The Use Browser Selection tick-box is ticked. For each selected block a QUASES formatted ASCII file is generated, where the file name of the new file(s) indicated the block index and corresponding variable index for the source data.
2. The Use Browser Selection tick-box is not ticked. The ISO file is assumed to be ordered consistent with spectra generated from an image data-set. A range of data blocks and corresponding variables are defined using two pairs of indices entered on the dialog window, such that when the OK button is pressed, the QUASES files are created based on the specified indices rather than the currently selected blocks.

Paste ISO formatted data from clipboard

Some acquisition systems, such as SPECSLAB II place data on the clipboard in ISO (VAMAS) format. If the clipboard is loaded with formatted data then pressing the Paste button on the toolbar will cause the data to appear in a new Experiment Frame in CasaXPS. A temporary file is used to keep a copy of the data on disk, however the data must be saved to an appropriate file following the paste operation or the file will be lost.

Paste data from clipboard

If ISO formatted data is placed on the clipboard, then pressing the Paste data toolbar button performs the same task as the Paste ISO formatted data paste button, except that a File dialog window requires you to chose a disk file for the Experiment Frame. The Paste data button also allows data from RBD's Auger Scan Software to be pasted from the clipboard into CasaXPS. The data appears in a new Experiment Frame and must be saved to a disk file at a later stage.

Second Group

(Toolbar selections invoking dialog windows)

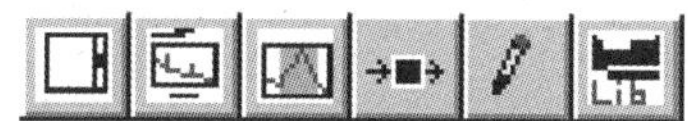

The primary dialog windows for manipulating data are invoked using these toolbar buttons. The Options menu (shown in the margin) also provides access to these same dialog windows.

Page Layout

The scrolled-list of tiles in the left-hand-side of an Experiment Frame represents a sequence of pages and on pressing the Print toolbar button, spectra are printed using the tile layout currently used to view the data in the scrolled-list. A page corresponds to the set of tiles visible at any one time in the left-hand-side of the Experiment Frame. The Page Tile Format dialog window (see Figure 4•2 on page 66) offers a set of default layouts for one or more display tiles per page. Each property page on the dialog window numbered 1 to 16 defines an array of non-overlapping display tiles, where initially the number of tiles in the array corresponds to the value used to label the property page. The format defined by a property page can be adjusted using the three sets of radio buttons. A format is defined by specifying the number of rows of tiles, then for each row the number of tiles to appear in each row. The number of rows can be selected from the Number of Rows/Column radio buttons. For each row from a possible maximum of four rows, the number of tiles within a row is assigned using the Tiles per Row/Column radio buttons. As an alternative, the definition for the tiles per page can be made using columns rather than rows. Switching the page layout between rows and columns is performed using the Tile Arranged in radio buttons.

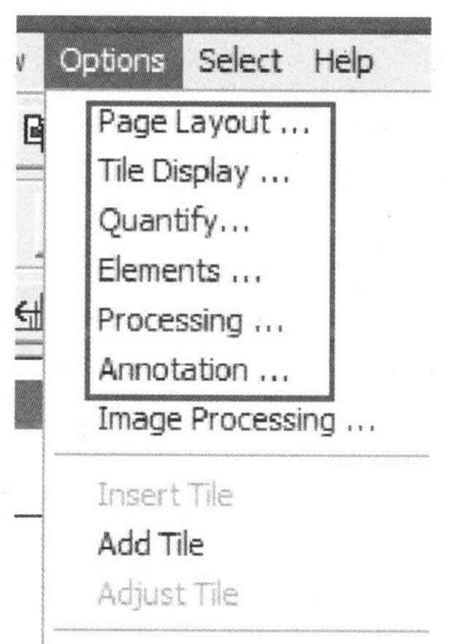

Tile Display Parameters

The Tile Display Parameter dialog window offers a set of property pages for altering the appearance of the Active Tile.

A detailed description of the functions contained within the six "Property Pages" of this dialog window may be found in Chapter 4, "Tile Display Parameters" on page 69

Quantification Parameters

Launches the dialog window for managing the quantification of spectra. The Quantification Parameters window contains four Property Pages which enable quantification (background) region specification, synthetic component creation, data-editing (spike removal) and definition of report specification. See Chapter 6, "Quantification" on page 91.

Spectrum Processing Dialog Window
Displays the dialog window for managing the processing of spectra. Smoothing, differentiation, integration, energy calibration, intensity calibration, PCA, test data, the spectrum calculator, and processing history are accessed as Property Pages from the Processing Dialog window. Most of these operations are important enough to warrant separate sections: for PCA, see Chapters 12 & 13, for the calculator, Chapter 9, and for smoothing, differentiation, and integration, Chapter 15.

Annotation Dialog Window
Accesses the dialog window for managing tile annotation. Features such as quantification tables taken form both regions and synthetic components can be added to the spectrum display via the Annotation Dialog window. These tables are updated as the components and their parameters are adjusted and allow an overall view of the modelling process.

Element Library Dialog Window
Opens the dialog window for managing the element library. The element library provides the means of identifying features within a spectrum and is an integral part of the mechanism for quantifying the data. Regions and synthetic components can be created directly from the energy-ordered scrolled-list of transitions offered on the element library dialog. See Chapters 5 & 17 for further details.

Propagate commands to other spectra
Processing, quantification regions, synthetic components and annotation prepared on a spectrum can be propagated to other spectra requiring the same actions. Once the analysis of spectrum is complete, spectra within the same file or from other files can be selected (using the right-hand-side (browser) of open Experiment Frames) and by clicking on the Active Tile displaying the prepared spectrum then pressing the propagate toolbar button, processing actions associated with the Active Tile are transferred to the selected data blocks. The Browser Operations dialog window invoked by this button displays the block identifier of the source block and also a list of the target blocks. The types of action to propagate are chosen using tick-boxes in the Propagate section of the dialog window.

Add and Delete Blocks from an Experiment Frame
Data blocks can be moved between Experiment Frames by first selecting the desired blocks using the Experiment Frame browser, clicking on the Experiment Frame to

which the data are to be transferred, then pressing this Add Blocks toolbar button. A dialog window appears listing all the data blocks currently selected. On pressing the OK button on the dialog window the data are copied into the previously selected Experiment Frame. The tick-box labeled Processed Data Only switches between copying the raw data plus the processing command from each of the source blocks and copying the processed data only. The consequence of ticking the Processed Data Only tick box is that the new block has no record of the processing steps used to prepare the data and only contains the processed data. For example, if the desire is to copy satellite subtracted data to a new file, the satellites can be removed using the temporary data manipulation option and once all operations have been performed successfully, the satellite subtracted data copied to a new Experiment Frame using the Processed Data Only tick box, thus providing a permanent record of the processed spectra.

Similarly, blocks can be deleted from the current Experiment Frame. Simply, selecting the blocks in the frame browser and pressing the Delete toolbar button ultimately removes the selected blocks from the Experiment Frame. A confirmation dialog window will offer a second chance to reconsider the action, but should the delete action be confirmed, the selected blocks are permanently removed (no "undo") from the Experiment Frame.

An alternative strategy for collecting data blocks from more than one file is to use the Open and Merge menu option on the File menu. The use of this option requires all the desired Experiment files to be gathered together in the same directory (folder). A selection of files all located in the same directory can be made using the Merge VAMAS Files dialog window invoked by the Open and Merge menu option. Use of the Shift key and the Control Key allows multiple files to be selected prior to pressing the Open button. Each file selected will appear in the same new Experiment Frame, where each set of data blocks from the original individual files are assigned a unique index value as the experimental variable.

Print and Print Preview

All the display tiles currently loaded into the left-hand-side of the Experiment Frame are printed when the Print toolbar button is pressed.

It is worth noting that, if Adobe Acrobat is installed on the PC running CasaXPS, the Adobe Acrobat Distiller option appears as a possible printer on the Print dialog window. The Adobe Acrobat Distiller allows the display tiles in the scrolled list in the left-hand-side of the Experiment Frame to be converted to PDF format and saved to disk.

About CasaXPS

Shows the current version number and also enables licensing information to be entered

Options Bar (lower toolbar)

First Group

(Toolbar buttons for displaying data blocks in the scrolled list of display tiles)

The buttons in the first group of the lower toolbar are used to paste spectral regions currently selected in the browser into display tile(s) located in the left-hand-side (scrolled view) of an Experiment Frame.

For each data block selected in the right-hand-side of the Experiment Frame a new display tile is created. A new scrolled list is created in the left-hand-side of the Experiment Frame in which a tile is created for each of the selected blocks and one spectrum/image appears in each tile. The display settings last in use are used to initialize the newly created tiles.

One or more spectra/images are overlaid in the Active Tile. Select the data blocks in the block browser then select the Active Tile from the possible tiles already displayed in the left-hand-side. On pressing this toolbar button, the data from the selected blocks are displayed overlaid in the Active Tile. To alter the data in a given display tile, simply left-click the tile, select the block and press the toolbar button. When more than one block is selected, the order in which the blocks were selected determines the order of the data in the Active Tile.

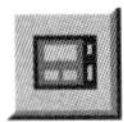 

Identical in operation to the previous two buttons, except that the current energy and intensity ranges, if possible, are used to display the data.

Using the selected data blocks, this toolbar button **initializes** the scrolled list of tiles in the left-hand-side of the Experiment Frame, creating a new tile for each row of blocks within the browser. These new tiles are in turn initialized by overlaying the data from each row of data blocks selected at the time the button is pressed.

Second Group

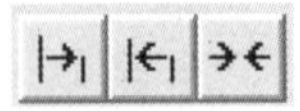

(Toolbar buttons for manipulating the zoom list)

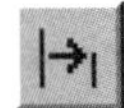

Drag a zoom box over a spectral feature of interest (using the mouse while holding down the left mouse button). Pressing this (Zoom In) button causes the tile display axes to be rescaled to the ranges for energy and intensity defined by the zoom box.

Each successive zoom action causes an additional zoom state to be added to the current zoom list.

After zooming into a region of a display tile, pressing this (Zoom Out) toolbar button steps back through the list of zoom states. Repeatedly pressing the button causes each zoom state to appear in a cyclical order.

Reset or re-initialize the zoom-list. If a spectrum already has quantification regions defined, then pressing the Reset toolbar button initializes the zoom list using the energy interval extracted from the start and end parameters defined for the quantification regions.

Third Group

(Toolbar buttons for moving the current display zoom-state in energy and intensity.)

Once a tile abscissa/ordinate ranges have been zoomed the Step Right button causes the display to step half the abscissa range to the right and auto-scale the ordinate range to fit the data.

Once a tile abscissa/ordinate ranges have been zoomed the Step Left button causes the display to step half the abscissa range to the left and auto-scale the ordinate range to fit the data.

Refit the ordinate range to the visible data. Manually adjusting the visible portion of a spectrum may leave the display in a state such that the data does not fill the display area. This button re-scales the ordinate range to take full advantage of the available space.

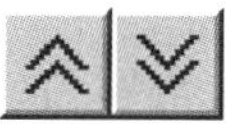

Adjust the Y scale in incremental steps.

Fourth Group

(Toolbar buttons for toggling through display options.)

Toggle the abscissa scale between Binding Energy and Kinetic Energy.

Toggle the ordinate scale between Counts Per Second and Counts.

Cycle the display-state through normal 2-D (flat) spectrum mode, 3-D plot of spectra and projection plots onto Factor Space generated by a Principal Component Analysis (PCA). Factor Space mode is only appropriate when the spectra have been replaced by abstract factors calculated from a PCA.

Cycle through offset plot modes. Offset plot parameters are available on the Display Parameters dialog window on the Y-Axis property page. These options allow a set of overlaid spectra to be stacked with a constant offset between the spectra, displayed in reverse order and reset to the normal overlay mode. Pressing this toolbar button causes the display to cycle through the offset plot settings.

Fifth Group

(More buttons for toggling through display options.)

Cycle through the display state for the quantification regions (Region background colour on / Region label on -> Region background colour off / Region label on -> Region background colour on / Region label off -> Region background colour off / Region label off.)

Toggle On/Off the display of the synthetic components.

Toggle On/Off the display of the residual between the synthetic component envelope and the data.

Toggle On/Off the display of the calculated backgrounds to the spectra.

Toggle On/Off the display of background subtracted spectra.

Toggle On/Off the Normalise display for overlaid spectra. To adjust the position where the overlaid spectra tie together, use the left mouse button together with holding down the Shift-key.

Sixth Group

(Buttons which deal with file parameters.)

Edit/View the block comments associated with the spectrum displayed in the active display tile. Be careful to distinguish between this"extra" information, and the "official" block info fields described below. Comment lines in general do not define an experiment and are not used to define processing. They can be added to display tiles in the form of a header and the first line of the comment can also be used to create a legend for overlaid spectra. (see also ISO 14976 annotation, page 342)

Edit/View the block information fields associated with the spectrum displayed in the active display tile.

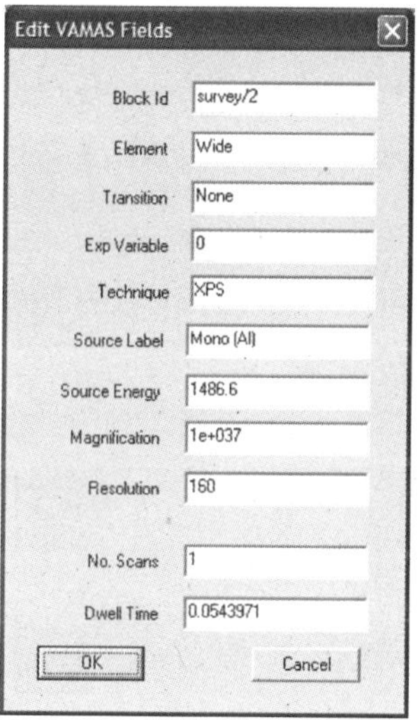

Experimental information is stored in each data block. These "official" block fields take many forms and while in theory should be fully defined by the original data file at acquisition time, the details may be incomplete. For this reason, some of the items are made available by CasaXPS for viewing and editing in a dialog window. Several of these fields are also offered for editing within other more specific dialog windows; however the "Edit VAMAS Fields" dialog represents the most general means of modifying the experimental information.

To view or modify the block fields, display the target block in the Active Tile then press the toolbar button. The most common fields requiring adjustment are the Block Id, the Element and Transition fields plus the experimental variable value. Care should be exercised when editing all fields as changes may affect quantification and energy calibration.

Change the Experimental variable for a spectrum file.

This is a means of globally altering the value for the current experimental variable in all blocks in a file. The radio buttons specify the nature and type for the experimental variable. As an example, consider a depth profile in which narrow scan spectra are repeatedly acquired interleaved by etch cycles of a given duration. Figure 6•18 on page 122 presents the data from such a profile experiment, however ultimately the idea is to plot the trends measured from the spectra as a function of depth rather than etch-time. The simplest way to convert the etch-time to depth is to enter the initial and final depths for the crated etched during the course of the experiment, select the Depth radio button on the dialog window, and then press the OK button. The Depth

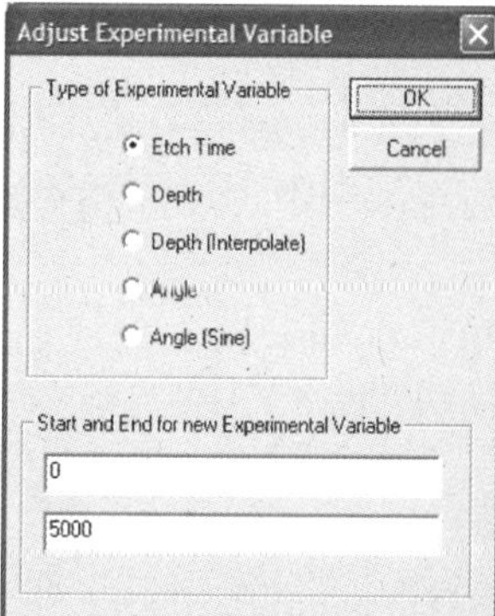

option assumes that the rows are equally spaced in depth: this is not always true. The etch-time between acquisition cycles may not be uniform and therefore a second option Depth (Interpolate) uses the etch time to compute the depth increments between acquisition cycles. The former method is used when the rows are assigned an experimental variable value which does not correspond to etch-time, for example if a set of data were collected from a set of X/Y pair ASCII files without the required etch-time information; in which case, the initial experimental variable may be data set index.

Two other options for setting the experimental variable are Angle and Angle (sine). The former assigns a range of linearly spaced values to the existing rows, while the latter generates row values which are the sine of a linear sequence of steps determined by the start and end values in the range text fields.

The experimental variable fields in a data block help to define the context for an acquisition. Within CasaXPS, data blocks are aligned in the block browser using the value from the current experimental variable from each block and since data can only be quantified with respect to each other when the corresponding blocks all appear in the same row on the right-hand-side, it is important to be able to make adjustments to these values. Experimental variables values, such as those in Figure 6•18 on page 122, appear in a natural sequence of increasing etch-times. Not all such sequences necessarily increase with row position, and generally CasaXPS does not assume any ordering of the data blocks with respect to these values. The only assumption is that blocks with the same experimental variable should appear, whenever possible, on the same row in the block browser. Provided all blocks with the same experimental variable have distinct species/transition fields, the blocks when viewed in the browser will appear in the same row. There is a subtle but important difference between the Edit Experimental Variable by Row dialog window and all other means of changing an experimental variable value in a block, namely, Edit by Row forces blocks within the underlying file structure to be re-ordered with respect to the newly assigned experimental value, whereas all other adjustments to these same values leave the underlying order of the blocks unchanged. Forcing the order for the blocks to change is useful in situations where the structure of the file does not properly reflect the experiment and sadly, not all file formats contain the full context for an experiment, thus necessitating such adjustments.

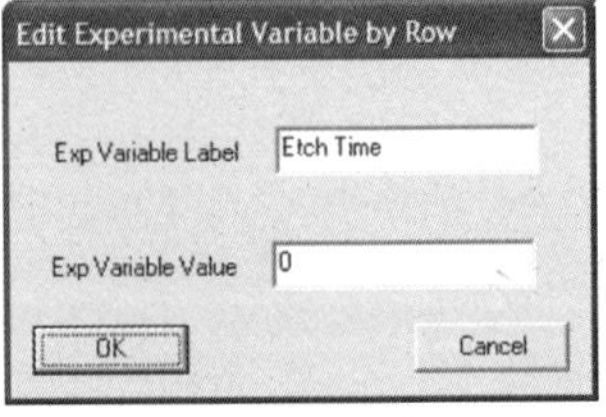

To change one or more experimental variables in a set of blocks, select the appropriate set of blocks in the block browser and press the toolbar button. A new label for the Experimental Variable heading may be entered and a new value assigned via the dialog window. On pressing the OK button, this new value is assigned to the selected blocks and the full set of blocks is reordered to reflect the numerical ordering in the context of the new values.

Edit Species and Transition for the selected blocks in a spectrum file.
Element/transition or species/transition data block fields for the current selection in the block browser are adjusted using the strings entered in the Element and Transition text-fields in the dialog window. The alignment of the blocks displayed in the Experiment Frame into columns is determined from these Species and Transition block strings. A new column appears for each block with distinct pairs of these strings. For example, two blocks identified as Si 2p and Si 2p/1 would appear in separate columns in the browser, and so the use of these strings becomes important for rearranging data in preparation for quantification.

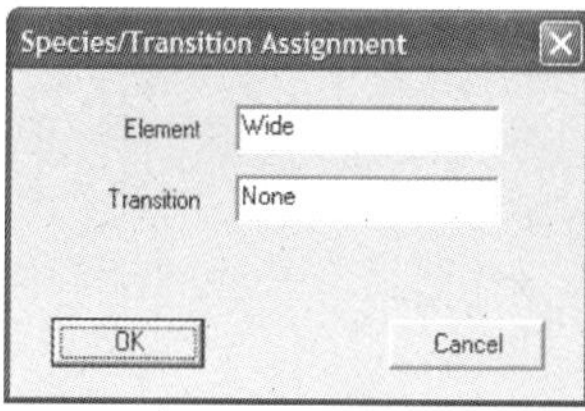

Edit the Excitation Source fields for the selected blocks in a spectrum file.
Other information often missing for data converted via X/Y pair ASCII files relate to the X-ray anode type (Source Label), photon energy (Source Energy), electron optics magnification and pass energy or retard ratio (Resolution). Again, these values can be assigned to the current selection of data blocks in the Experiment Frame, however when assigning a new x-ray source, be sure all the spectra selected have the same Magnification and Resolution settings.

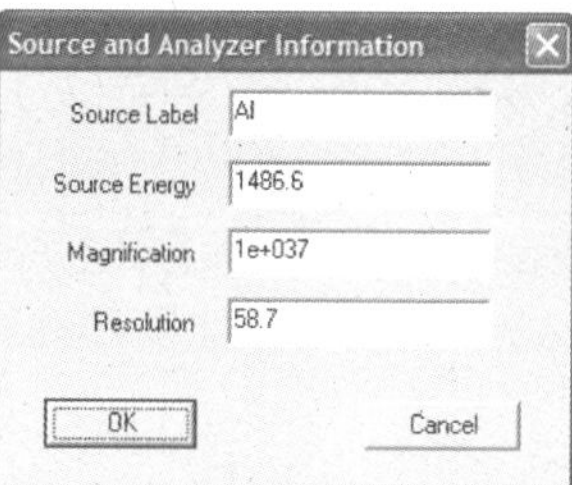

Edit Block/Sample Identifiers
The Block Id is the string that appears in the array of rectangles representing the data blocks in the block browser in the right-hand-side of an Experiment Frame. In addition to the Block Id, the Sample Id is a further sting entered into each block offering a means of characterizing the data using a text string. The Sample Id is used to construct legends for overlaid spectra and also aligning the data blocks in the Experiment Frame when Edit Mode is enabled.
The dialog window offers a pair of strings and two tick-boxes so that a choice can be made with respect to which of the strings is used when the OK button is pressed. The data blocks updated with the new values can be selected in as many Experiment Frames as desired. That is to say, the assignment is not restricted to the blocks selected in the Experiment Frame currently in focus.

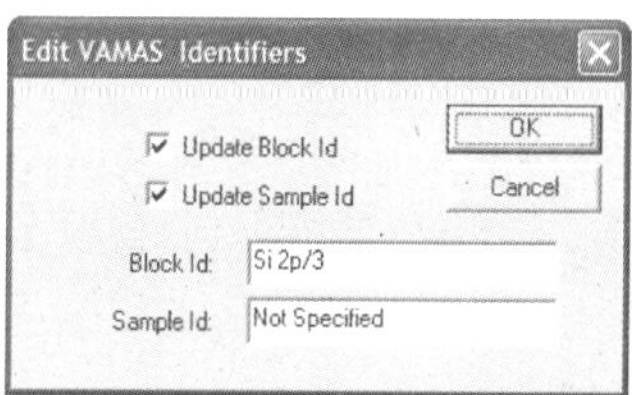

The Edit Mode toolbar button switches the state of the first column in the right-hand-side of the Experiment Frame between using the experimental variable to order the rows and the strings held in the Sample Id. When pressed once, the experimental variable values are replaced by a set of row labels and these row labels may be edited simply by double-clicking the mouse with the cursor over a row label.

Alignment of Blocks in the Browser

The layout of the data blocks in the browser is determined by the order in which the blocks of data are sequenced in the ISO file. When viewed in CasaXPS, the data are organised based on matching two strings: a row string, constructed from either the experimental variable or the string assigned to the Sample Id field, and a column string constructed from concatenating the species (element) and transition fields. On opening a file, blocks with identical row labels but distinct column labels will appear in a single row. Blocks with identical row labels and also identical column labels result in new rows appearing in the browser. Each new row label encountered will result in the start of a new row appearing in the browser. It is worth noting that the rows are not ordered with respect to the numerical value of the experimental variable, although typically the order of the blocks within the file produces such an ordering, this is not necessarily the case. Figure 3•3 and Figure 3•4 illustrate two possible alignments for the blocks from the same file based upon the use of an identical experimental variable or, the more natural, row alignment using the Sample Id from each block.

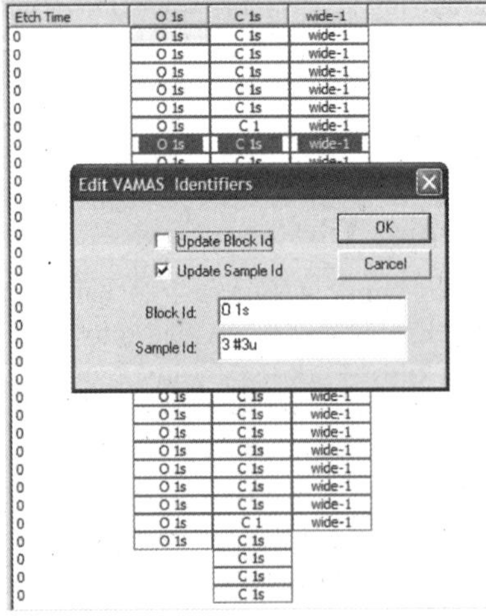

FIGURE 3•3 Browser before realignment

Sample Identifier	O 1s	C 1s	wide-1
1 #4u	O 1s	C 1s	wide-1
2 #4u	O 1s	C 1s	wide-1
2 #4u	O 1s	C 1s	
3 #4u	O 1s	C 1s	wide-1
1 #3u	O 1s	C 1s	wide-1
1 #3u		C 1	
2 #3u	O 1s	C 1s	wide-1
3 #3u	O 1s	C 1s	wide-1
1 #2u	O 1s	C 1s	wide-1
2 #2u	O 1s	C 1s	wide-1
3 #2u	O 1s	C 1s	wide-1
3 #2u		C 1	
1 #4non	O 1s	C 1s	wide-1
2 #4non	O 1s	C 1s	wide-1
1 #3non	O 1s	C 1s	wide-1
2 #3non	O 1s	C 1s	wide-1
1 #2non	O 1s	C 1s	wide-1
2 #2non	O 1s	C 1s	wide-1
3 #4non	O 1s	C 1s	wide-1
4 #3ex	O 1s	C 1s	wide-1
4 4ex	O 1s	C 1s	wide-1
5 4ex	O 1s	C 1s	wide-1
6 4ex	O 1s	C 1s	wide-1
4 2ex	O 1s	C 1s	wide-1
5 2ex	O 1s	C 1s	wide-1
3 2non	O 1s	C 1s	wide-1
3 3non	O 1s	C 1s	wide-1
3 3non		C 1	
4 3non	O 1s	C 1s	wide-1
4 4non	O 1s	C 1s	wide-1
5 4non	O 1s	C 1s	wide-1
5 3non	O 1s	C 1s	wide-1

FIGURE 3•4 .. and after.

CHAPTER 4

Managing the Display using Inset Tiles

An Experiment Frame is a window for managing both the experimental data and the visual presentation of the results. The data browser on the right hand side of an Experiment Frame shows the logical structure of the data with respect to an experimental variable such as etch-time, while the left-hand-side provides the means of presenting the spectra using so called Display Tiles. These Display Tiles sub-divide the graphical section of the Experiment Frame so that one or more spectra can be viewed at the same time (Figure 4•1).

Regular Sets of Display Tiles

The Page Layout dialog window (Figure 4•2) manages the format for regular arrays of Display Tiles. The display area is divided into rows of tiles and each row can be allocated a number of tiles. The display in Figure 4•1 corresponds to the settings shown in Figure 4•2. Here, the number of rows has been chosen to be two, as seen by the radio buttons entitled Number of Rows/Columns. The first row has one tile-per-row while the second has four tiles-per-row. These settings are indicated by the array of radio buttons entitled Tiles per Row/Column. Note that only the top two rows of radio buttons are used to define the format for the two rows in use.

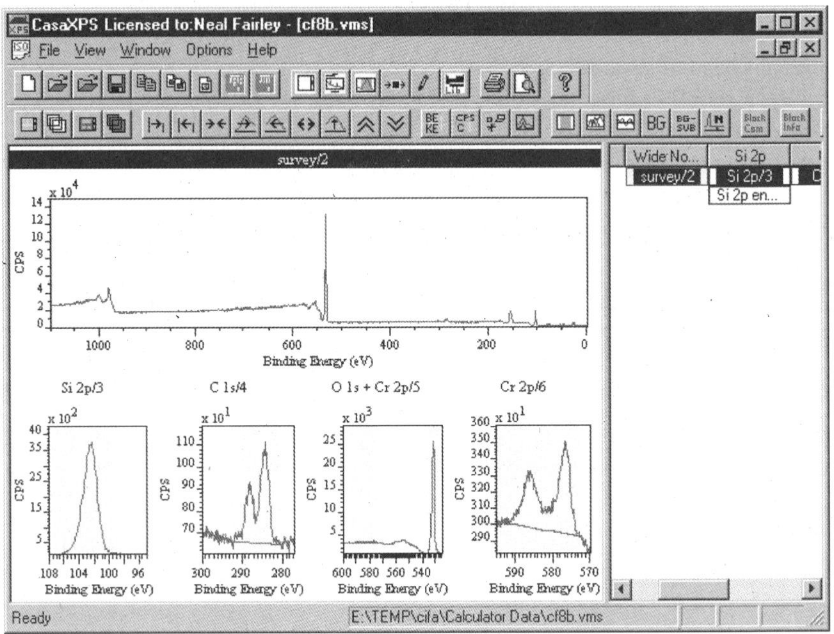

FIGURE 4•1 A maximized Experiment Frame, where spectra are displayed using a regular set of five Display Tiles [1].

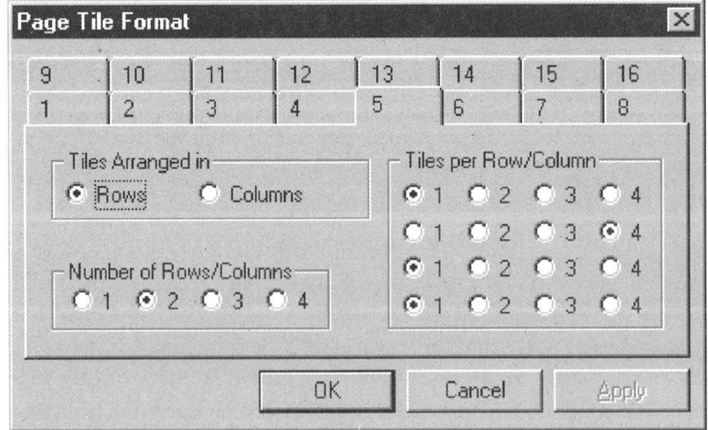

FIGURE 4•2 Page Layout dialog window

Inset Tiles

Often a survey spectrum includes a combination of intense and weak peaks for which a single intensity scale fails to visually convey the information found in the data. Although regular arrays of Display Tiles offers a means of plotting multiple views of the same spectrum, it is sometimes more appropriate to show features inset into the global view of the data (Figure 4•3).

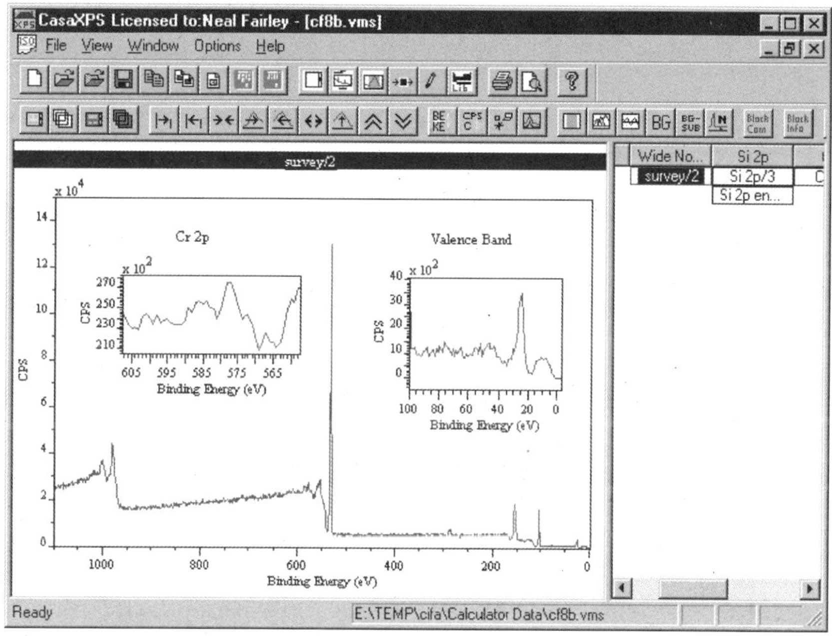

FIGURE 4•3 Survey spectrum displayed using two inset views of the data, but zoomed to show regions of interest.

To Create an Inset Tile

An inset tile is created as follows:

1. Display the spectrum using one of the regular page layout options.
2. Mark a zoom box where the inset should appear.
3. Press the Insert Key on the keyboard or select the Insert Tile option under the Options Menu on the CasaXPS program frame.
4. Use the normal zoom and display adjustment mechanisms to show the data of interest.

Note: Inset tiles are identical to regular tiles. The spectra can be selected and displayed in an Inset Tile in an identical way to the regularly formatted Display Tiles.

Deleting Display and Inset Tiles

To remove a tile, select the tile (using the left mouse button) so that it becomes the active tile and press the Delete Key on the keyboard. The Option menu includes a Delete Tile option and this will also delete the active Display Tile in an identical fashion to the keyboard Delete Key.

Peak Fitting Using Inset Tiles

Figure 4•4 shows a situation where significant structure to the higher binding energy side of a C 1s spectrum is visually dominated by the major peaks in the data envelope. An Inset Tile allows both the minor and major structure in the spectrum to be visualized at the same time and because both Display Tiles are equivalent, the fitting steps may be performed from either of the tiles shown in Figure 4•4. Simply select the Inset Tile when adjustments to the minor peaks require mouse interactions and revert back to the Display Tile when working with the larger peaks.

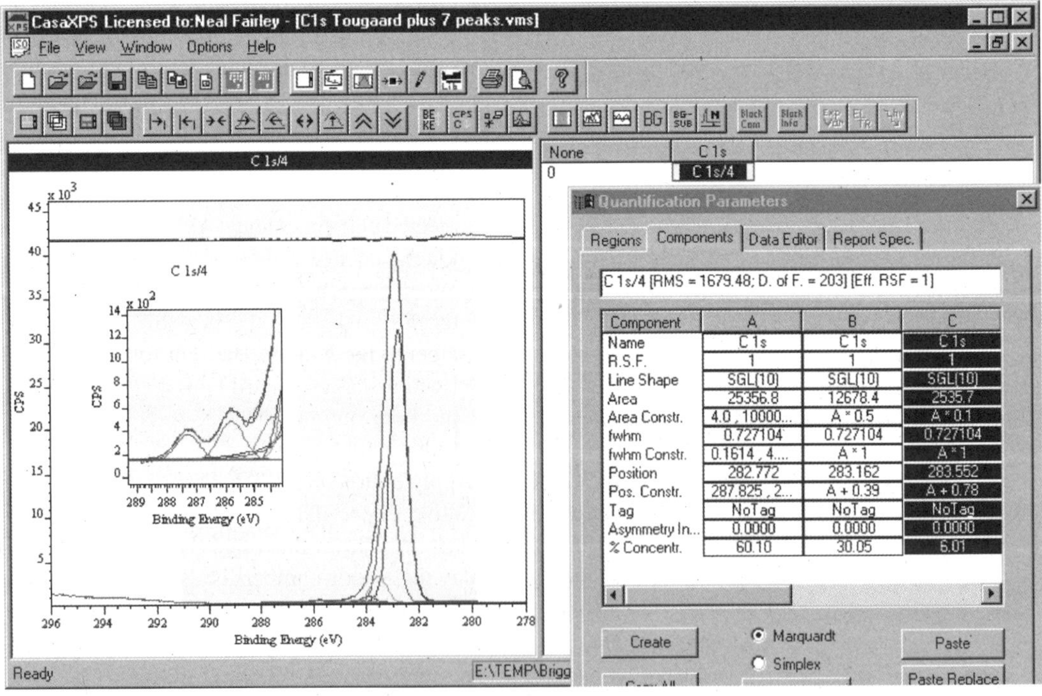

FIGURE 4•4 C 1s spectrum where small and large peaks require fitting to synthetic models [2]

Tile Display Parameters

The Tile Display Parameter dialog window provides a set of six property pages for altering the appearance of the Active Tile: X Axis, Y Axis, Geometry, Display, Colours, and Global

X Axis Property Page

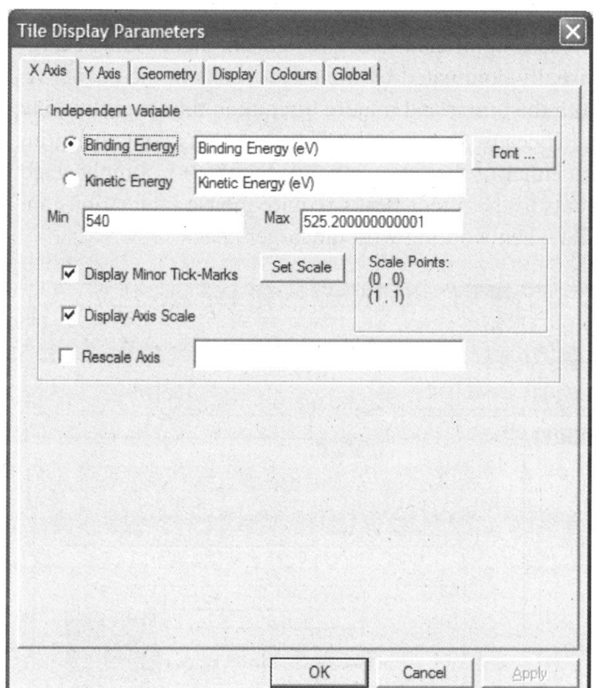

For display purposes, two strings are offered on the X Axis property page and are associated with the choice of binding or kinetic energy display for the energy axis. The convention in XPS is to display the binding energy axis as a decreasing energy scale; however if the core electrons are viewed as lower in energy than the more excited valence band electrons, then it is clear the convention for the binding energy scale is simply missing a minus sign. Thus both kinetic energy and binding energy scale should be seen as increasing energy scales. Binding energy is more commonly used for XPS because the positions of photoelectric lines with respect to the binding energy scale are independent of the X-ray photon energy. The opposite is true for Auger lines, that is, Auger lines always appear at the same positions on the energy scale with respect to the kinetic energy of the ejected electron. Hence AES data is displayed using the kinetic energy scale.

When a spectrum is plotted, the chosen energy type determines which of the two strings are displayed as the label to the X-axis. The font for the X-axis label is defined using a standard Font dialog window invoked by the Font button on the X-Axis property page. Once a font is chosen from the Font dialog window, it is necessary to press the OK or the Apply button on the Tile Display Parameters dialog window before the font in the Active Tile is changed. The energy range over which the spectrum is displayed is determined from the values entered in the Min and Max fields. The font used to display the X-axis numbers is also used to display the Y-axis numbers, therefore the button to make adjustments to the font used on the axes is located on the Display property page rather than on either the X axis or Y axis prop erty pages.

It is also possible to reduce the complexity of the display by either removing the X-axis scale entirely or limiting the tick marks on the X-axis to only the major values. These options are available as tick-boxes on the property page.

Y Axis Property Page

The Y axis label and intensity range are analogous to the X axis property page; however there is an additional label associated with displaying the data using arbitrary units. The tick-box labeled **Arbitrary Units** removes the numerical scale from the intensity display and adds a third string as the Y axis label in preference to both the strings associated with the counts per second and counts per bin radio buttons. **Arbitrary units** is useful when the data has been normalized in some way.

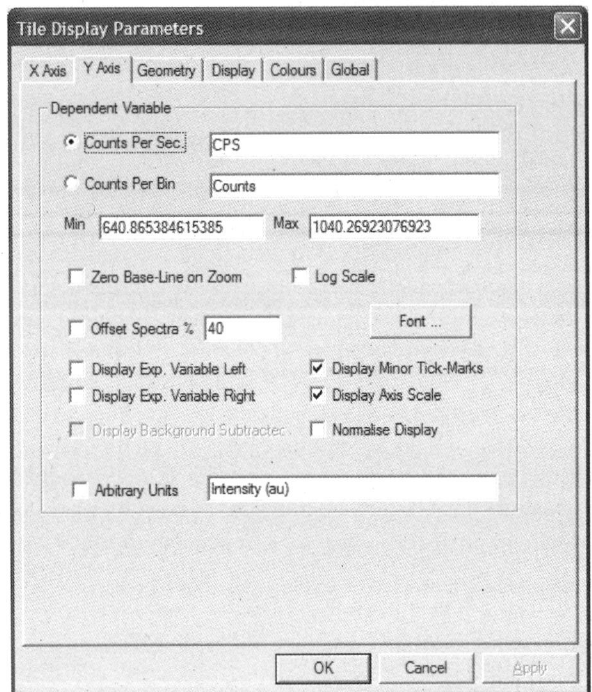

Two frequently used options on the Y axis property page are the **Normalise Display** and **Offset Spectra** tick-boxes (also available from the toolbar buttons).

These options are often used when visually comparing overlaid spectra, where the **Offset Spectra** tick-box creates a set of vertically offset plotting areas within the Display Tile and the **Normalise Display** tick-box adjusts the intensities of each spectrum displayed in the tile relative to the intensity of the first spectrum selected via the right-hand-side of the Experiment Frame. The relative size of the plotting areas and consequently the offset between these plotting areas is specified via the % text-field adjacent to the Offset Spectra tick-box. The value entered in the text-field specifies the percentage size of the plotting areas relative to the size of the axes box in the Display Tile. If the **Normalise Display** option is also used, a vertical line marks the point at which the data is scaled, where the scaling is performed so that the traces all pass through the same point which intersects the vertical line. The position of the vertical line can be adjusted by holding the Shift-key down and then left-clicking the mouse with the cursor indicating the desired tie point. When printed, the vertical line is not included on the output.

Zero Base-Line on Zoom and **Log Scale** are two tick boxes chiefly useful for SIMS data but with possible XPS applications: the former modifies zoom behaviour when using the mouse to enlarge a peak. Default zoom behaviour uses the dimensions of both the width as well as the height of the zoom rectangle, which is drawn following dragging the cursor over a display tile. If the cursor is located inside the defined zoom box when the left mouse button is clicked, then provided the Zero Base-Line on Zoom tick box is *not* ticked, the intensity range defined by the zoom box is used to redisplay the data. Alternatively, if the Zero Base-Line on Zoom tick box *is* ticked, then only the upper-most boundary of the zoom box defines the maximum intensity when the data is redisplayed; the lower limit to the intensity scale is always zero.

Geometry Property Page

The Geometry property page provides the means to switch between the normal 2D graphical display used to display spectra overlaid on a single pair of axes and a 3D plot, where the axes are offset in both X and Y directions to provide a depth to the display determined by the values assigned to the experimental variable for each spectrum. When a set of spectra are plotted using the 3D option, the size and location of the axes used to draw the data are defined in terms of the Front Plane and the Back Plane. When overlaid spectra are drawn, the first spectrum in the list of selected blocks is drawn on a pair of axes positioned and scaled using the Scale X, Scale Y and Shift X parameters in the Front Plane Parameters section. Similarly the last spectrum in the list of selected blocks is drawn using a pair of axes positioned and scaled according to the same parameters as the Front Plane, but with the additional parameter of Shift Y. All other spectra within the selected blocks appear at axes positioned and scaled using a linear interpolation between these two defining planes. The parameters are percentage based, for example, Scale X 75 represents a set of axes where the size of the X direction is 3⁄4 the size of the box which otherwise would have been used to plot the data in 2D mode. Once the Front and Back Planes are scaled to some (reduced) percentage of the surrounding box, the position of the origin of the reduced axes is defined using the convention that a value of -100 shifts the scaled plane all the way to the left of the surrounding box, while +100 shifts the plane entirely to the right. Figure 4•5 shows a display where the Front plane is scaled in both X and Y using 75 and shifted in X by -80.

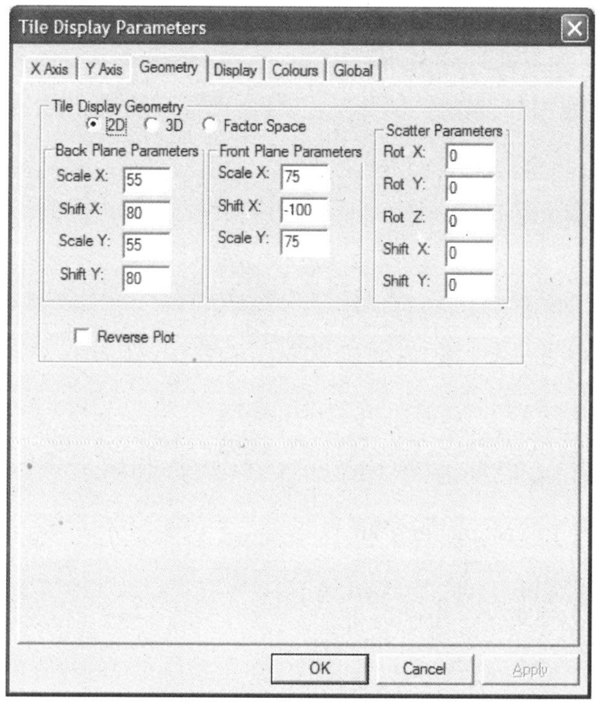

The toolbar button  toggles between the three display modes - 2D, 3D and Factor Space.

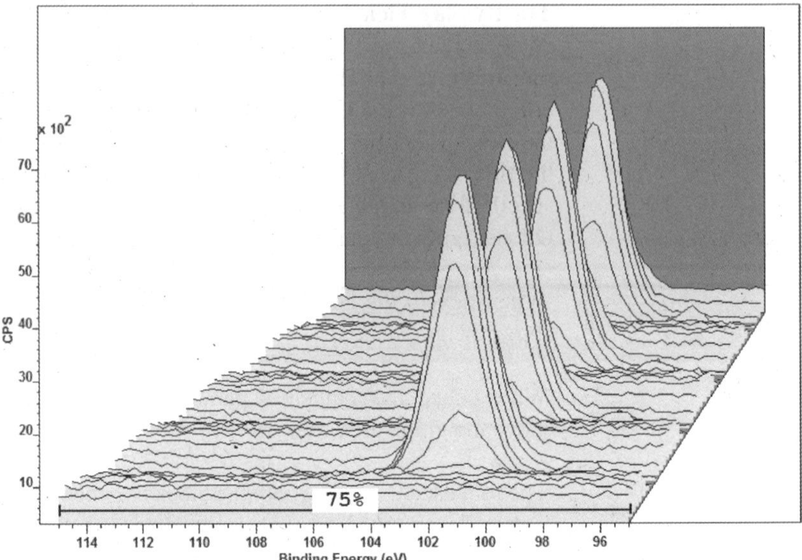

FIGURE 4•5 3D plot of overlaid spectra, where the Front Plane is of size 75% of the enclosing box in both X and Y directions.

Display Property Page

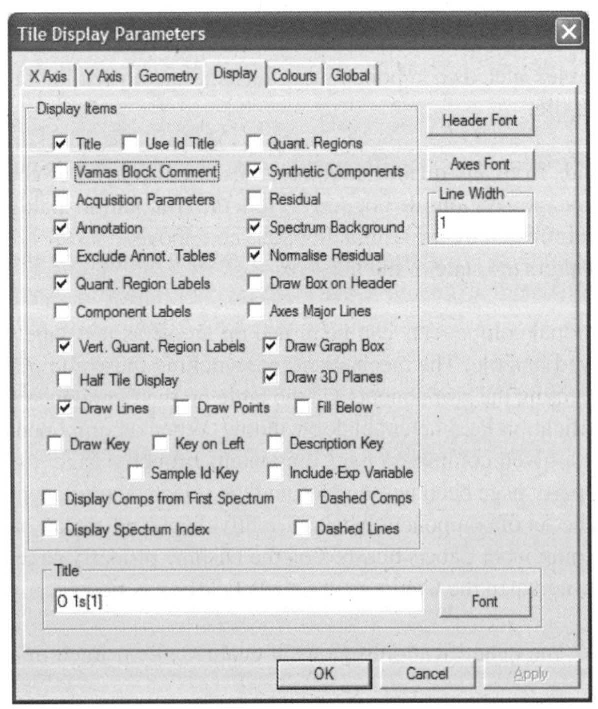

The display characteristics of the Active Tile are adjusted using the Display property page. The extensive array of tick-boxes permits the inclusion or exclusion of information on the drawing area. Some display properties are also toggled using toolbar buttons, where pressing these toolbar buttons cycles through various combinations of the tick-box setting. For example, pressing the button initially removes the background colour used to define the range over which a quantification region computes a background, a second press returns the background colour previously removed but removes the region name label and a third press removes both region colour and also the region name string from the display. A fourth button press returns both the region colour and also the region name label. These toolbar button actions could be explicitly defined using the two tick-boxes Quant. Regions and Quant Region Labels on the Display property page.

The Display Tick-Boxes

Title   Use Id Title

When ticked, the Title tick-box causes the title string to be displayed above the plotting area in a Tile. The source for the title is either the text-field on the Display property page or if the **Use Id Title** tick-box is ticked, then the title is constructed from the Block identifier and Sample identifier strings in the source block. The font for the title string is also defined using a standard Font dialog window. Note, the font specified through the Font dialog window only takes effect following pressing of the OK or Apply buttons on the Tile Display Parameters dialog window.

Vamas Block Comment

When ticked, the comment text lines maintained in the data block are displayed above the graph area.

Acquisition Parameters

When ticked, the acquisition parameters for the data displayed in the tile are printed above the graph area.

Header Font

The font used to display both the block comment text and the acquisition parameters is altered using the Font dialog window invoked by the Header Font button.

Annotation

When ticked, the annotation defined by the Annotation dialog window is displayed on the data in the tile. Note, not all text on the display derives from the Annotation dialog window, e.g. component and region name text.

Exclude Annot. Tables

When ticked, annotation tables such as components, regions and quantification tables are not displayed in the tile.

Quant. Region Labels

When ticked, the name fields from any quantification regions are displayed on the spectra. Labels from quantification regions do not derive from the Annotation dialog window and are therefore unaffected by the Annotation tick-box above.
The toolbar button alters the state of the tick-box

Component Labels

The name fields for individual components can be turned on an off as text labels for each component displayed in a tile. The mechanism for switching these component labels on and off is selecting the components via the table on the Components property page of the Quantification Parameters dialog window. When a component is selected, by left-clicking a given column on the Components property page, the tick-box on the Display property page becomes enabled and the selected component label is toggled on or off. The set of component labels currently displayed can all be turn off by un-ticking the Component Labels tick-box on the Display property page. The labels only appear in a tile when the Component Labels tick-box is ticked.

Vert. Quant. Region Labels

When ticked, the labels for the quantification regions, if enabled, are printed in a vertical direction.

Quant. Regions

When ticked, the energy limits to a quantification region are displayed on a spectrum using a different background colour to the default tile background colour. These colours are defined using the Colour property page on the Tile Display Parameters dialog window. The toolbar button alters the state of the tick-box.

Synthetic Components

When ticked, components defined on a spectrum are displayed.

The toolbar button also toggles the display of the component by altering the state of the Display tick box.

Residual

When ticked, the residual plot representing the difference between the data and the total synthetic envelope is plotted along the top of the tile.

Normalise Residual

When ticked, the residual plot is normalized with respect to the expected noise in the data. Poisson counting statistics imply particular noise characteristics and so the un-normalized residual is largest where the counts are highest. Normalizing the residual produces an even appearance to the residual plot regardless of the counts per channel. If the counts per channel values are unknown, then the residual should be plotted with the Normalise Residual button un-ticked.

Spectrum Background

When ticked, the background to the spectrum computed from the quantification region parameters is displayed on the graph.

Draw Box on Header

When ticked and header information from the VAMAS comment and/or the acquisition parameters are enabled, a box or boxes are drawn around the displayed header information between the title and the axes box.

Axes Major Lines

When ticked, for each major tick mark on the axes, horizontal and vertical dashed lines are drawn (see below).

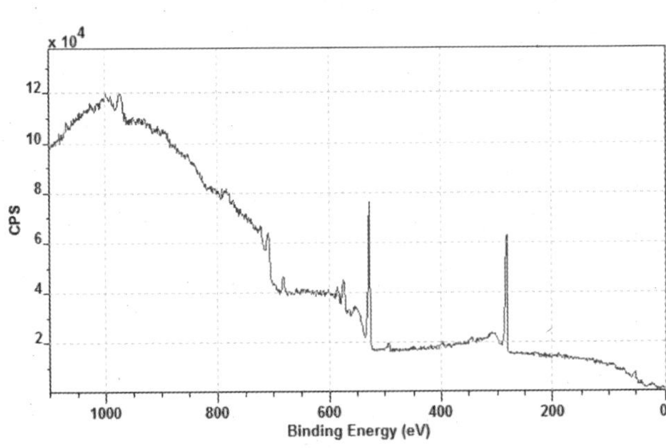

☑ Draw Graph Box

When ticked, the rectangle defining the drawing area (bounding box) in a tile is drawn.

☐ Half Tile Display

When ticked, the graphical area used to display the data is reduced to half the size of the tile area. While the graph itself appears in the top half of the tile, annotation defined in the block can be moved below the graphical area as shown in Figure 4•6, where both a table of components and a text have been placed outside the graphical area used to plot the spectrum.

x 10³

CPS

Binding Energy (eV)

This is annotation text for the spectrum displayed above

Name	Pos.	FWHM	L.Sh.	Area	%Area
C 1s	282.29	1.373	GL(30)	17804.0	71.77
C 1s	283.89	1.373	GL(30)	3378.1	13.61
C 1s	285.29	1.373	GL(30)	1355.6	5.46
C 1s	286.29	1.373	GL(30)	2277.7	9.16

FIGURE 4•6 Half Tile Display showing annotation positioned below the graphical display.

☑ Draw 3D Planes

When ticked, two rectangular planes, beneath and at the rear of the spectra are filled with a colour defined on the Colour property page. If un-ticked the display appears as seen in Figure 4•7 compared to the display in Figure 4•5 on page 72. Note, Figure 4•7 is drawn using the Draw Graph Box un-ticked, thus only axes appear on the 3D plot.

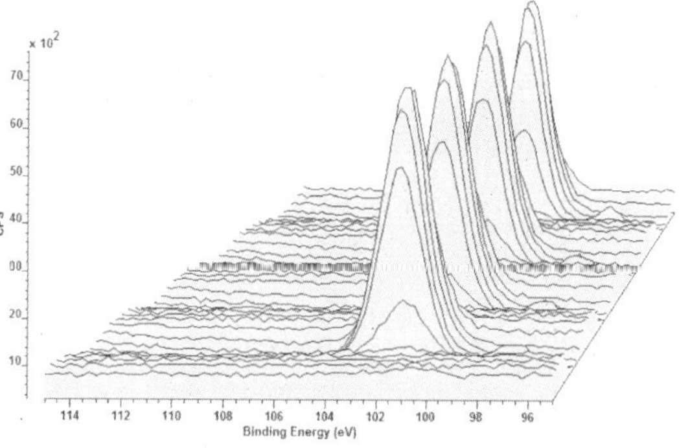

FIGURE 4•7 3D display mode where the Draw 3D Planes tick-box is un-ticked and the Draw Graph Box tick-box is un-ticked too. Compare Figure 4•5 on page 72.

Draw Lines

When ticked, a spectrum is drawn using the data points connected by line segments.

Draw Points

When ticked and **Draw Lines** is not ticked, the data points appear as symbols with connecting line segments, one symbol per data point. If ticked together with Draw Lines, a reduced number of symbols are drawn at regular intervals and the data points are connected using line segments.

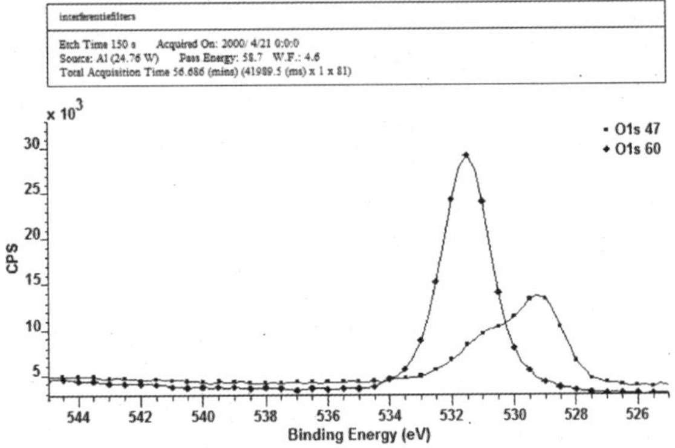

FIGURE 4•8 Spectra drawn using Draw Lines and Draw Points ticked. Header information is also added to the display using the Vamas Block Comment and Acquisition Parameters tick boxes. The key is added to the display using the Draw Key tick box.

Fill Below

The **Fill Below** tick box results in different plots depending on the state of the **Draw Lines** and **Draw Points** tick boxes. If the **Draw Lines** tick box is ticked at the same time as the **Fill Below** tick box, then the area between the graph and the X axis is filled with the colour determined from the 3D Fill button on the Colour property page. An alternative drawing mode results if the **Fill Below** tick box is ticked and the **Draw Points** tick is ticked whilst the **Draw Lines** tick box is not. Under these circumstances, the data points are used to draw sticks from the data point to the X axis as shown in the Inset Tile below.

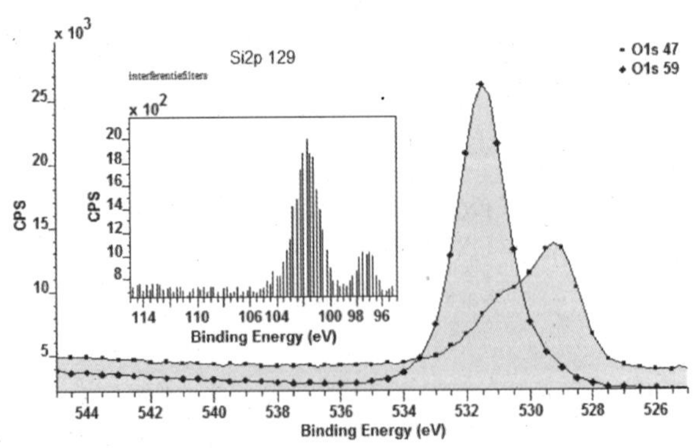

FIGURE 4•9 Fill Below ticked. The parent tile is displayed using both the Draw Lines and Draw Points tick boxes ticked, while the inset tile is plotted with the Draw Lines unticked and the Draw Points tick box ticked.

Draw Key

When ticked, a key or legend is added to the display of two or more spectra (Figure 4•9). The default source for the text is the Block Id for the block containing the data and is therefore the string used to display the data block in the right-hand-side of the Experiment Frame. The block identifier can be adjusted using the following toolbar buttons: Block Info or ? .

Key on Left

When ticked, the key or legend is positioned in the top-left side of the graphical area; otherwise the key appears in the top-right-hand side of the graphical area.

Description Key

When ticked, the text appearing in the key or legend derives from the first line of the block comment.

Sample Id Key

When ticked, the text appearing in the key or legend derives from the Sample Id block field. The Sample Id text may appear as an alternative to the Block Id or in conjunction with the Description Key tick box text.

Include Exp Variable

When ticked, the current experimental variable is appended to the text appearing in the key.

Display Comps from First Spectrum

When ticked, synthetic components defined on the first spectrum of several spectra displayed in a tile are displayed over the data. This tick box is used when fitting a set of linked components that span multiple acquisition regions.

Display Spectrum Index

When ticked, the index with respect to the ISO file displayed in the Experiment Frame appears in the top-left-hand side of the graphical area. Data in ISO files are specified by two indices, the block index and the corresponding variable index within a given block. Displaying data derived from the corresponding variables is achieved by holding the Control Key down and pressing the PageUp and PageDown buttons on the keyboard. Control Home and Control End move to the first and last corresponding variables in a block. The transmission function for a spectrum, if present in the block, appears as the second corresponding variable indexed 1, below.

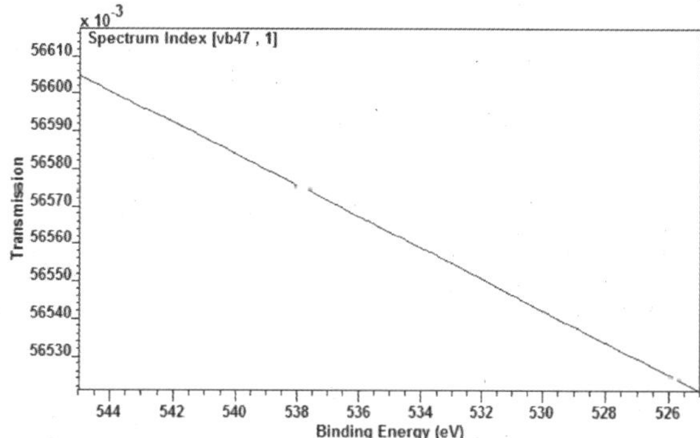

FIGURE 4•10 Transmission data plotted from block index 47 appearing in the corresponding variable with index 1.

When ticked and the line width is one or zero, the synthetic components or the spectra are drawn using a range of line styles.

Axes Font

The font used to display the numerical values associated with the X and Y axes are adjusted using the resulting Font dialog window. The font only takes affect following pressing the OK or the Apply button on the Tile Display Parameters dialog window.

Colour Property Page

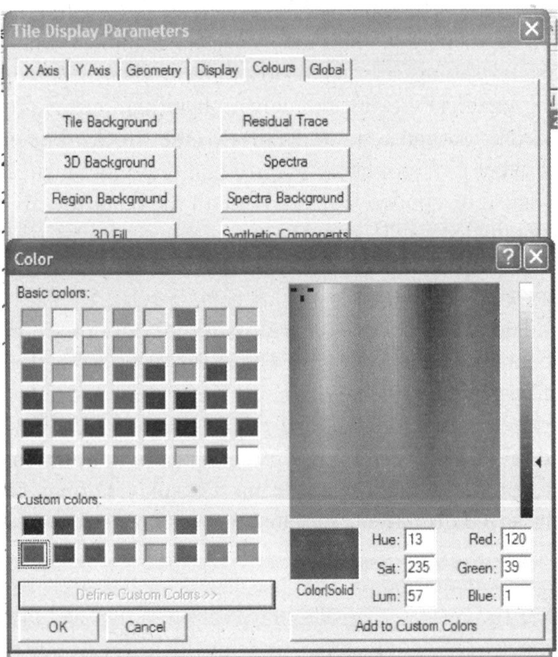

The Colour property page offers a means for adjusting the colours used to display spectra and related information in the display tiles. The set of buttons on the property page all use the standard Microsoft Color dialog window, however two modes of operation are used depending on whether the button selected adjusts one colour or a set of colours. If one colour is involved, for example the Tile Background colour, then simply selecting a colour from any of the colour squares on the Color dialog window and pressing the OK button will alter the parameter, however the colour displayed by the tile is only updated once the OK or Apply button is pressed on the Tile Display Parameter dialog window. The alternative mode involved sets of colours rather than individual colours. For Example, a range of colours are used to display overlaid spectra, therefore the important colours are those appearing in the Custom Color palette on the Color dialog window. There are sixteen colours available for definition via the extended section of the Color dialog window and these colours are used to display overlaid spectra in the order top-left to top-right, then bottom-left to bottom-right. The button labeled **Define Custom Colors**, when pressed, reveals a mechanism for making adjustments to the current palette of Custom Colors. Click a colour square from the set of Custom Colors, adjust the colour using the cursor over the large coloured square, then press the Add to Custom Colors button. Once all the changes to the custom color palette are made, press the OK button on the Color dialog window. Again, the tile display is only updated once the OK or Apply button is pressed on the Tile Display Parameters dialog window.

Global Property Page

The Tile Display Parameter dialog window alters the display settings for the Active Tile. If however, the settings prepared and applied to the Active Tile are also applicable to other tiles currently displayed in the left-hand-side of the Experiment Frame, then the Global property page can be used to apply these new settings to all the tiles in the scrolled list.

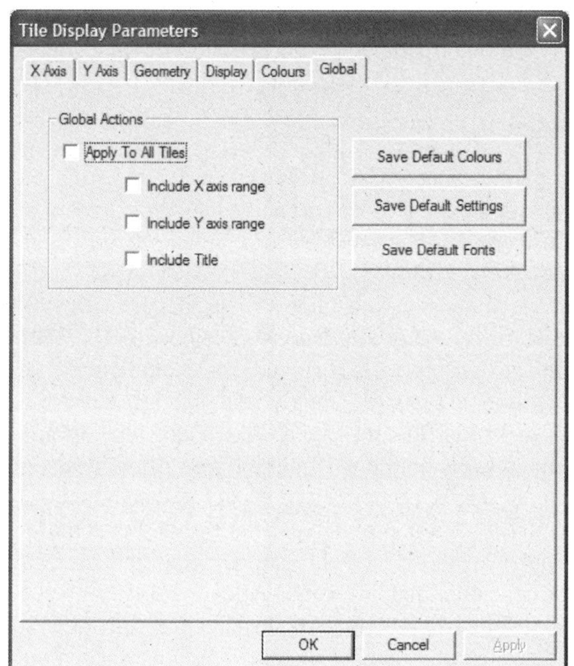

When applying certain display setting to a range of tile, not all the display settings are necessarily appropriate. For example, if two tiles exist displaying an O 1s and a C 1s peak, then although the desire to unify the spectrum colour used to display both spectra is valid, accepting a single range of energies for the X axis is not. Similarly, the intensity range may be inappropriate as well as unifying the title. These display settings are generally excluded when the display settings are applied to a range of tiles, however to accommodate the situation where these parameters are intended to be updated too, there are a set of tick-boxes which when tick, explicitly specify that these parameters should also be applied together with all the other tile display parameters.

The Global property page also offers a means of saving default settings for the tile display between CasaXPS sessions. The directory containing the CasaXPS.exe program also contains a directory called CasaXPS.DEF, in which a set of configuration files are stored. Once the tile display in terms of fonts, colours and display options is established for the Active Tile, then pressing the buttons on the Global property page will save most display settings to disk. On restarting CasaXPS, the saved display settings are read and used to display the data.

References

[1] Genet M. (private communication).
[2] Briggs D. (private communication).

CHAPTER 5

Qualitative Analysis and the Element Library

The Element Library dialog window provides a means of managing the information used to identify spectrum features and creating regions and components with predefined values. Strictly it is not an "element" library at all, but a list of spectroscopic lines which, while derived from characteristic transitions within the electronic structure of atoms of a particular element, can also reflect the exact chemical environment of those atoms - that is provide chemical compound specificity. Correct peak identification and quantification of XPS/AES spectra relies on the proper maintenance of the Element Library's line positions (transition energy values) and relative sensitivity factors. Three property pages within the module offer two views of the data held in the element library file, and a mechanism for loading and merging data into the current session.

The module is launched from the upper toolbar (See "Second Group" on page 55.) as shown below

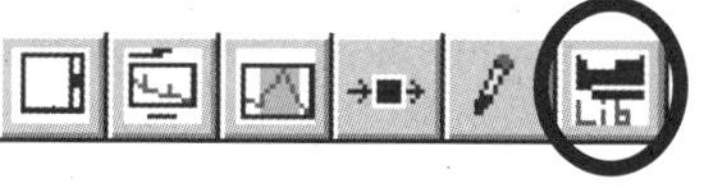

FIGURE 5•1 Launching the Library

In order to function correctly, CasaXPS should have an element library present in the system (although some basic features will function without one) - but for many commands and operations, particularly involving qualitative and quantitative analysis, a library file ***must*** be present. The file may be changed at any time using the controls described below (so that several library files may be held on disk), but only one library (of unlimited size) may be active at any time. The module recognises both CasaXPS and Kratos (proprietary) formats.

The default CasaXPS element library is compiled using rough peak positions and Scofield cross-sections for aluminium and magnesium X-ray anodes to represent the corresponding relative sensitivity of the photoelectric peaks relative to the C 1s transition. While appropriate for some purposes, these peak energies and relative sensitivity factors will not satisfy all requirements and so the CasaXPS system offers mechanisms for customized application and sample specific element library creation. This section describes the means for modifying the element library in CasaXPS.

The Element Library File Format

The CasaXPS element library is an ASCII file. The first line in the file is a version number (Figure 5•2), which may be '0' or '1' depending on whether the entries for a peak appear on separate lines or are separated by TAB characters, respectively. The latter, TAB spaced format is displayed in Figure 5•2 and supersedes the "Type 0" format with a view to improve the edit ability of the data in spreadsheet programs. For large number of changes, using a spreadsheet is the best means of constructing a CasaXPS library, however for small alterations the Element Library property page offers dialog window based adjustments of the library.

Each peak within an element library entry consists of the following fields:
1. Element
2. Transition
3. Label/Name
4. Mass (Daltons)
5. Energy Type (BE or KE)
6. Energy (eV)
7. F.W.H.M.
8. Line shape (e.g. GL(30))
9. Relative Sensitivity Factor
10. Excitation source string

Figure 5•2 illustrates a set of peak entries in version 1 format. Note how the library entries visible include two different excitation source strings, namely "Al" and "Mg". A single library file may contain any number of entries corresponding to different excitation source strings, however when used in CasaXPS, these strings are matched to the excitation source string in the displayed data block in the Active Tile and only those library entries for which a match occurs are displayed in the Element Table property page on the Element Library dialog window.

Microsoft Excel - casaXPS.lib

J10 Mg

	A	B	C	D	E	F	G	H	I	J	K
1	1										
2	Bi	6p	Bi 6p	0.00E+00	BE	3.00E+00	1.00E+00	GL(30)	8.41E-02	Al	
3	Po	6p	Po 6p	0.00E+00	BE	5.00E+00	1.00E+00	GL(30)	1.36E-01	Al	
4	At	6p	At 6p	0.00E+00	BE	8.00E+00	1.00E+00	GL(30)	1.98E-01	Al	
5	Rn	6p	Rn 6p	0.00E+00	BE	1.10E+01	1.00E+00	GL(30)	2.72E-01	Al	
6	Fr	6p	Fr 6p	0.00E+00	BE	1.50E+01	1.00E+00	GL(30)	3.37E-01	Al	
7	Ra	6p	Ra 6p	0.00E+00	BE	1.90E+01	1.00E+00	GL(30)	3.96E-01	Al	
8	Th	6p	Th 6p	0.00E+00	BE	4.30E+01	1.00E+00	GL(30)	4.99E-01	Al	
9	U	6p	U 6p	0.00E+00	BE	1.90E+01	1.00E+00	GL(30)	5.09E-01	Al	
10	Bi	6p3/2	Bi 6p3/2	0.00E+00	BE	3.00E+00	1.00E+00	GL(30)	4.56E-02	Mg	
11	Po	6p3/2	Po 6p3/2	0.00E+00	BE	5.00E+00	1.00E+00	GL(30)	7.45E-02	Mg	
12	At	6p3/2	At 6p3/2	0.00E+00	BE	8.00E+00	1.00E+00	GL(30)	1.10E-01	Mg	
13	Rn	6p3/2	Rn 6p3/2	0.00E+00	BE	1.10E+01	1.00E+00	GL(30)	1.52E-01	Mg	
14	Fr	6p3/2	Fr 6p3/2	0.00E+00	BE	1.50E+01	1.00E+00	GL(30)	1.93E-01	Mg	
15	Ra	6p3/2	Ra 6p3/2	0.00E+00	BE	1.90E+01	1.00E+00	GL(30)	2.30E-01	Mg	
16	Th	6p3/2	Th 6p3/2	0.00E+00	BE	4.30E+01	1.00E+00	GL(30)	2.93E-01	Mg	
17	U	6p3/2	U 6p3/2	0.00E+00	BE	1.90E+01	1.00E+00	GL(30)	3.01E-01	Mg	
18	Pb	6p	Pb 6p	0.00E+00	BE	1.00E+00	1.00E+00	GL(30)	3.48E-02	Mg	

casaXPS

Ready

FIGURE 5•2 The CasaXPS default element library loaded into a spreadsheet.

Although the version 1 library format is aimed at the creation of library files within a spreadsheet program, there exits a mechanism within CasaXPS for making occasional changes and additions to the currently loaded library file. Figure 5•3 on page 84 shows the dialog window aimed at making these individual changes to an entry. The Edit Entry dialog window is invoked by placing the cursor over the Name field for an entry in the Element Table scrolled-list, then right-clicking the mouse button. The current set of library fields for the selected peak are displayed on the dialog window, from which these fields can be adjusted, then either updated or a new entry created.

To update the current entry, the OK button should be pressed, however a new entry is created with the values taken from the state of the dialog window, when the Create button is pressed. Alternatively, the Delete button will remove the entry used to invoke the dialog window. Modifications to the element library, on exit, cause CasaXPS to offer to save the element library to a new file name.

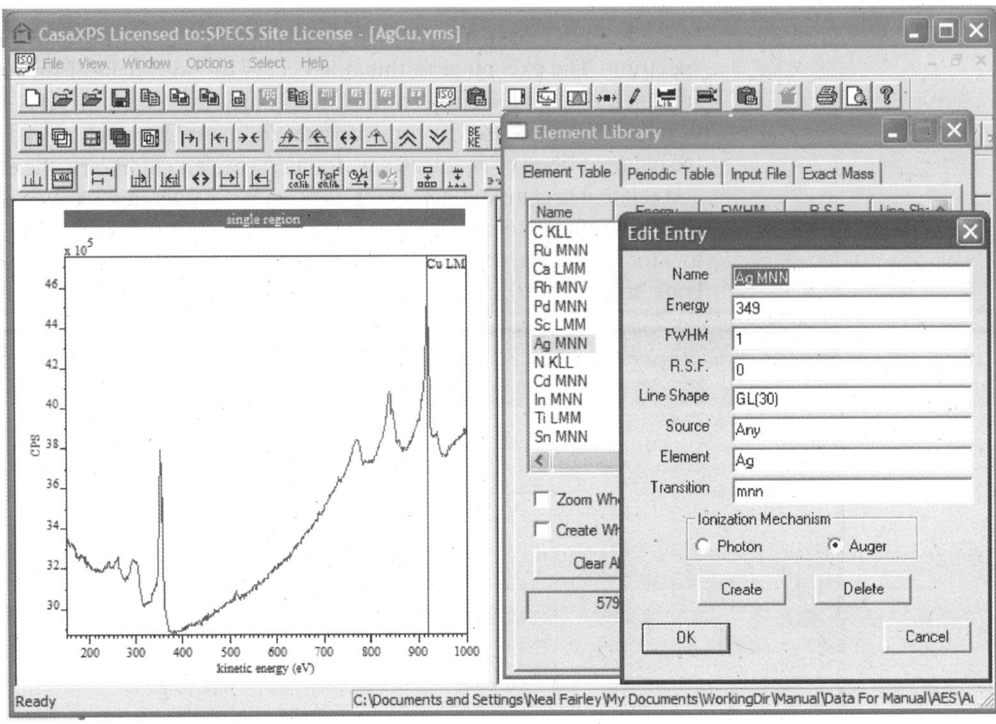

FIGURE 5•3 Edit Entry dialog window used to make changes to the fields for an element library entry.

Since the default CasaXPS element library does not contain any relative sensitivity factors for AES peaks, the greatest need to create an element library is, therefore, for those using Auger data. The concepts and procedure are identical for XPS entries; however there are Auger specific features within the library mechanism and so AES spectra deserve special attention in the context of the element library. The Auger spectrum in Figure 5•3 will be used to illustrate editing the element library.

The Element Library and Auger Spectra: a General Example

When CasaXPS is started, the executable file CasaXPS.exe reads the library file CasaXPS.lib located in the same directory as the executable file (the name of the executable file determines the name for the default library file). When a spectrum is displayed in the Active Tile, on invoking the Element library dialog window, the Element Table property page displays those entries for which the excitation source string from the entry matches the corresponding field in the data block holding the spectrum. The exception to this rule occurs for AES data. If the excitation source field in the data block is set to ElectronGunDummy, then for AES spectra the electron gun energy is used to construct the string used to match an excitation source in the element library. That is to say, for an AES element library, the excitation source and relative sensitivity factors for electron beam energies 3 keV, 5 keV and 10 keV should be entered using the strings S(3), S(5) and S(10) (Figure 5•4). Sensitivity factors for AES are given in a number of publications, perhaps most usefully in a recent book edited by Briggs and Grant[1].

Microsoft Excel - CasaXPS_aes_CuAg.lib

I13 fx 0

	A	B	C	D	E	F	G	H	I	J	K
1	1										
2	Cu	1	Cu 1	0	KE	916	1	GL(30)	0	S(3)	
3	Cu	1	Cu 1	0	KE	916	1	GL(30)	0.489	S(5)	
4	Cu	1	Cu 1	0	KE	916	1	GL(30)	0	S(10)	
5	Cu	2	Cu 2	0	KE	837	1	GL(30)	0	S(3)	
6	Cu	2	Cu 2	0	KE	837	1	GL(30)	0.135	S(5)	
7	Cu	2	Cu 2	0	KE	837	1	GL(30)	0	S(10)	
8	Cu	3	Cu 3	0	KE	766	1	GL(30)	0	S(3)	
9	Cu	3	Cu 3	0	KE	766	1	GL(30)	0.08	S(5)	
10	Cu	3	Cu 3	0	KE	766	1	GL(30)	0	S(10)	
11	Ag	1	Ag 1	0	KE	357	1	GL(30)	0	S(3)	
12	Ag	1	Ag 1	0	KE	357	1	GL(30)	1.385792	S(5)	
13	Ag	1	Ag 1	0	KE	357	1	GL(30)	0	S(10)	
14	Ag	2	Ag 2	0	KE	351	1	GL(30)	0	S(3)	
15	Ag	2	Ag 2	0	KE	351	1	GL(30)	0	S(5)	
16	Ag	2	Ag 2	0	KE	351	1	GL(30)	0	S(10)	
17	Ag	3	Ag 3	0	KE	289	1	GL(30)	0	S(3)	
18	Ag	3	Ag 3	0	KE	289	1	GL(30)	0	S(5)	

CasaXPS_aes_CuAg

Ready

FIGURE 5•4 Auger element library prepared for Cu and Ag transitions at electron gun energy of 5 keV.

Once an element library file is prepared for a particular application, the user may choose to make the library the default CasaXPS library or only load the library when appropriate. To make a CasaXPS formatted library file the default element library, the library file must be placed in the same directory as the CasaXPS executable and given the same base-name as the executable file. It is probably best to retain the

1. "Surface Analysis by Auger and X-ray Photoelectron Spectroscopy"
D. Briggs and J.T Grant (Editors), (ISBN: 1 901019 04 7)
IM Publications & Surface Spectra Limited (2003) (page 825)

standard naming conversion, namely, CasaXPS.exe for the executable and therefore CasaXPS.lib for the default library. Other configuration files and directories also use the base-name of the executable and therefore retaining the same base-name for all is important. Allowing the user to have several copies of the executable with different base-names permits custom behaviour. For example, it would be possible to set-up a copy of CasaXPS with the name CasaXPS_aes.exe, for which the library and configuration files would be named CasaXPS_aes.lib and CasaXPS_aes.DEF. These associated files are then free to be configured for use with Auger data. The alternative is to load a library file as and when required. To load a CasaXPS formatted library file, select the Input File property page (Figure 5•5) on the Element Library dialog window, use the Browse button to select the required library file, select the CasaXPS lib radio button and choose to either overwrite the existing library (Load button) or merge the new entries from the selected file with the current library (Merge button). Once a new file is either loaded or merged into the CasaXPS library, the entries can be examined via the Element Table property page and will be used when new quantification regions are created.

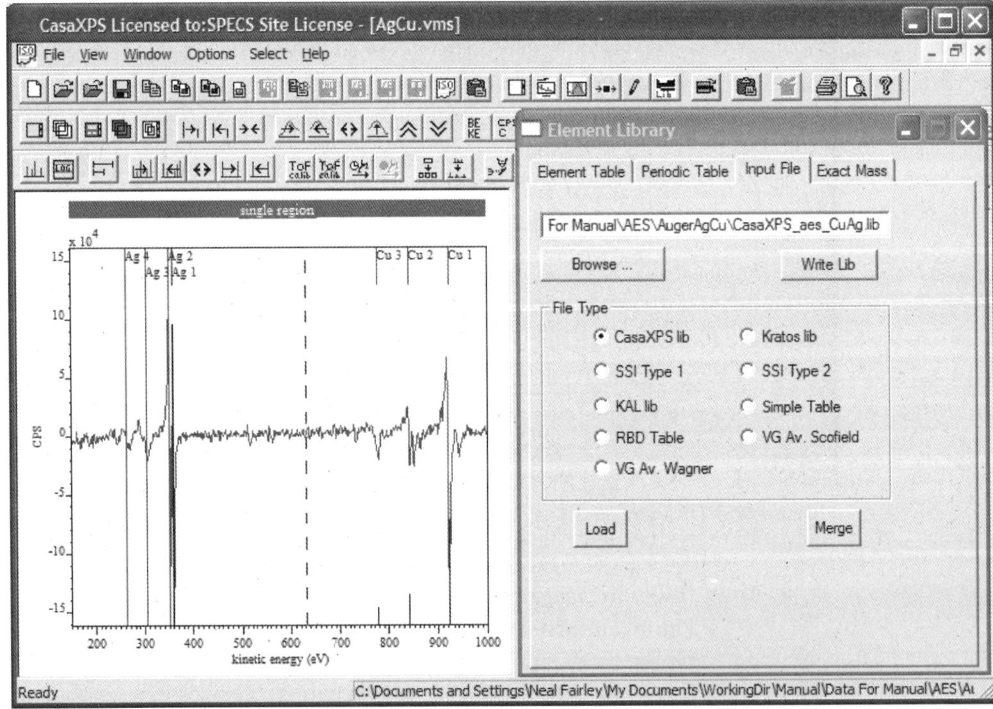

FIGURE 5•5 The Input File property page on the Element Library dialog window, where the derivative of the spectrum shown in Figure 5•3 is displayed using the element markers and relative sensitivity factors from the library in Figure 5•4.

The CasaXPS window in Figure 5•6 shows the differentiated Auger spectrum from Figure 5•3, where quantification regions are defined for each of the resonance structures from the Cu and Ag elements. Note that there are three regions defined for each of the Cu Auger lines and while regions for all three Cu lines are displayed on the spectrum, only the Cu1 peak has been assigned a non-zero relative sensitivity factor. Any quantification item, region or component, with zero RSF will be excluded from the resulting quantification table (also shown in Figure 5•6). Similarly, only one of the Ag peaks is assigned a non-zero RSF.

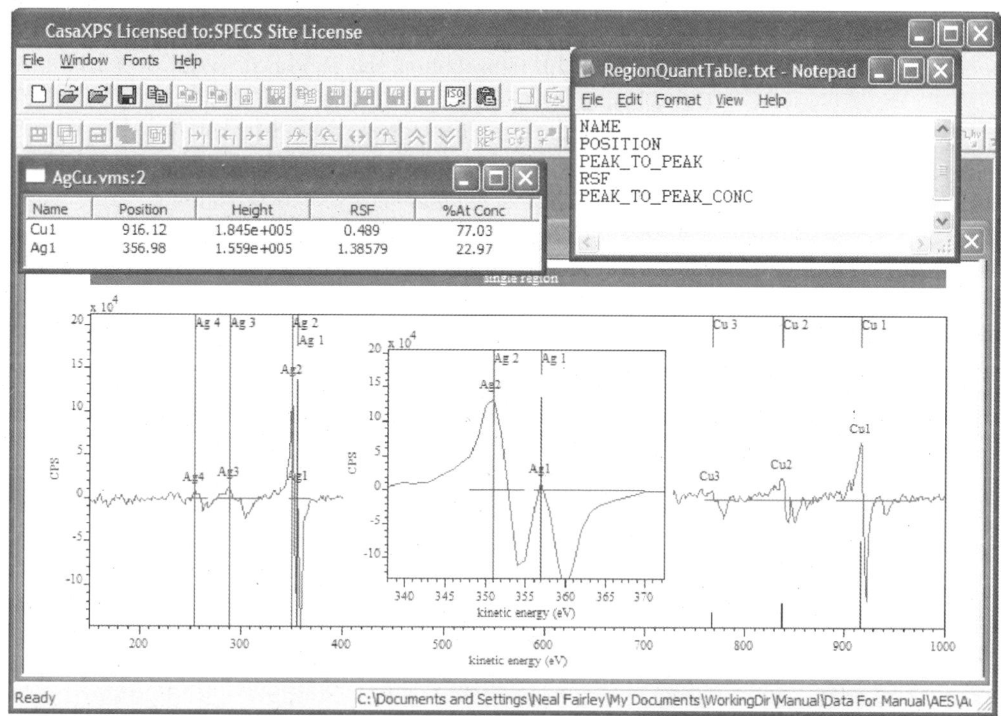

FIGURE 5•6 Differentiated Auger spectrum from an Ag Cu Alloy, where quantification regions are defined for every peak in the library file.

The quantification report in Figure 5•6 is generated using a configuration file defining the fields for inclusion in the table. The default Standard Report tabulates many fields not appropriate for AES data and therefore the configuration file RegionQuantTable.txt (located in the directory CasaXPS.DEF) is modified to extract only those values of interest for Auger data. The entries in the configuration file used to create the quantification report (Report Spec property page on the Quantification Pa-

rameters dialog window) are also included in the Notepad window shown in Figure 5•6 on page 87. Once a report is created, the Copy toolbar button invokes a Copy to Clipboard dialog window, from which the table can be transferred via the clipboard to a spreadsheet or saved to a text file.
Chemical state information can be extracted using quantification regions appropriately positioned to differentiate between shifted peaks. Figure 5•6 illustrates two closely positioned Ag peaks, which are plotted within the inset tile, where the region limits are defined to permit a measurement of the peak-to-peak intensity from each of the transitions. While these two peaks are not due to distinct chemical compounds in the conventional sense, for chemically shifted peaks a similar judicious positioning of regions offers a means of measuring chemical information.

Identifying Peaks Interactively

Peak identification for both XPS and AES is performed using the scrolled list of transitions and the "Periodic Table" property page. A spectrum may be viewed with markers added at energies where peaks might be expected (Figure 5•5 - note that the position line sizes (heights) are scaled to reflect the RSF values stored in the library) To add and remove these markers, select an element through either the periodic table interface or from the element table. The markers are toggled on and off by repeatedly selecting the same element.
An alternative method for examining a spectrum is to have the scrolled element table visible and then run the cursor over the tile display window with the left mouse button held down. The energy order list will scroll with the energy position of the cursor. Release the mouse button when over a peak of interest and then use the table to toggle the peak markers until a match is established.

A more detailed example of how to approach peak identification (in XPS) is given below.

Step 1: Activate the element library

The Element Library dialog window is invoked or brought to the front of any CasaXPS dialog windows by pressing the toolbar button indicated in Figure 5•7 on page 89. The first property page of the element library dialog window is a scrolled list of line positions where the list is ordered by the energy of the entries. If the mouse is left-clicked on the spectrum displayed in the left-hand-side of the subframe shown in Figure 5•7, then the scrolled list of element library entries will scroll to the energy indicated by the mouse click. Therefore, by pointing at the peaks in the spectrum via the mouse, the names for the various features can be brought into view and if the name field displayed in the scrolled list is selected, again using the mouse, a set of markers will appear on the spectrum positioned at the energies at which spectral lines should appear for the chosen element. Figure 5•7 shows the spectrum after the O 1s and C 1s lines have been activated via the element library scrolled list. If the name field of an element library entry is selected a second time, the element markers are removed from the display. The current set of active element markers can be view on the Periodic Table property page also located on the element library dialog window. The element markers may be activated and deactivated via the Periodic Table display as well as the scrolled element list.

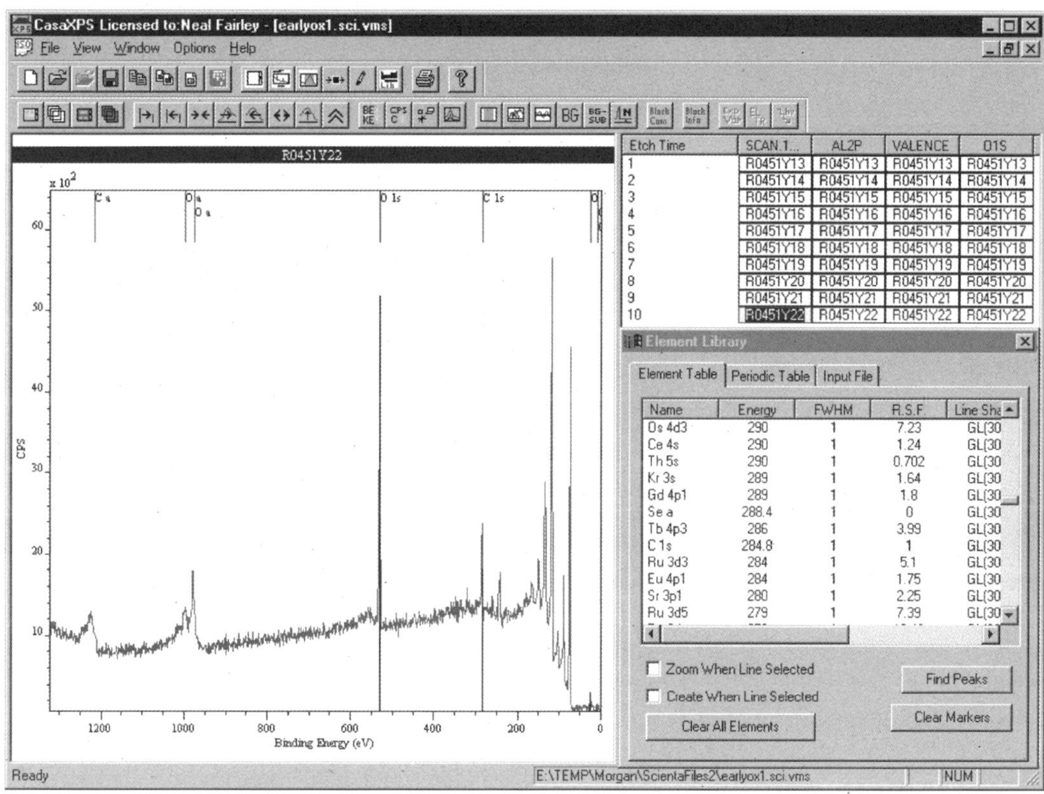

FIGURE 5•7 XPS peak identification via the element library.

Step 2: Annotate peaks using element markers

Once all the spectral lines are identified and markers have been placed on the displayed spectrum, the peaks can be annotated using the names stored in the element library. This step not only creates a labelled spectrum, but these annotation peak labels are used to create a set of quantification regions at the press of a button.

The annotation dialog window includes a property page entitled Peak Labels (Figure 5•8). All the peak markers active on a spectrum are listed on the Peak Labels property page and may be selected using the mouse and the control key (to add to the selection). When the Apply button is pressed those labels selected in the list of names on the Peak Labels property page become annotation labels on the spectrum. The peak labels are positioned on the spectrum according to the location of the nearest peak, so it is important that the spectrum is calibrated before this procedure is performed.

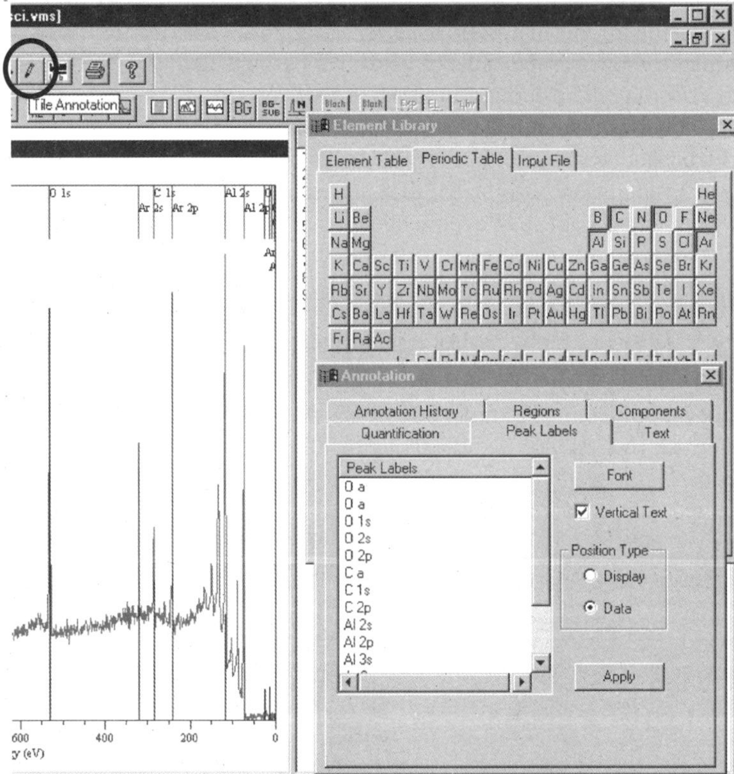

FIGURE 5•8 Peak Labels Property Page on the Annotation Dialog Window

Peak and element markers are maintained within a session on a spectrum by spectrum basis. Altering the markers on one spectrum does not change those previously chosen for a different spectrum (block) even within the same experiment. Simply select the tile used to display the data and the periodic table will update to reflect the state of the markers chosen for that spectrum.

CHAPTER 6

Quantification

Data can be quantified at a number of levels. The most direct form of quantification uses “regions” applied to a wide scan spectrum, the results being written on the displayed spectrum using an annotation option. For cases where peaks are not well resolved, both regions and synthetic components can be prepared over a set of narrow scan spectra and the results specified through a set of expressions involving the names used to label the objects. A report generated in this way can be viewed within CasaXPS or exported as a “tab” separated ASCII (plain text) file for use with a spreadsheet program such as Excel.

Figure 6•1 shows the main toolbar button for the quantification dialog window (See “Second Group” on page 55.) used to perform these two extreme procedures, plus a number of reporting mechanisms in between.

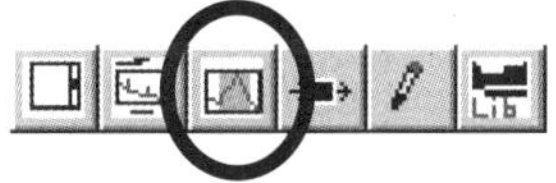

FIGURE 6•1 Launching the Quantification dialog

Three property pages are used in the quantification process, labelled “Regions”, “Components” and “Report Spec”.

Regions Property Page

Region definition is the key to quantification of XPS spectra. For each spectrum, quantification is performed using the information prepared on the associated Regions Property Page of the Quantification Parameters dialog window (Figure 6•2).

Regions can be created using:

- Create pushbutton (direct creation from energy range defined in current zoom state: other parameters must be manually entered).
- Create from Labels pushbutton (each annotation peak label is used to create regions: most parameters entered automatically. Background is set as the last "BG type" previously selected. "Linear" is the default). Note: "Create from Labels" assumes a well calibrated spectrum.
- A combination of selecting lines from the Element Library dialog window whilst still viewing the Regions Property Page.

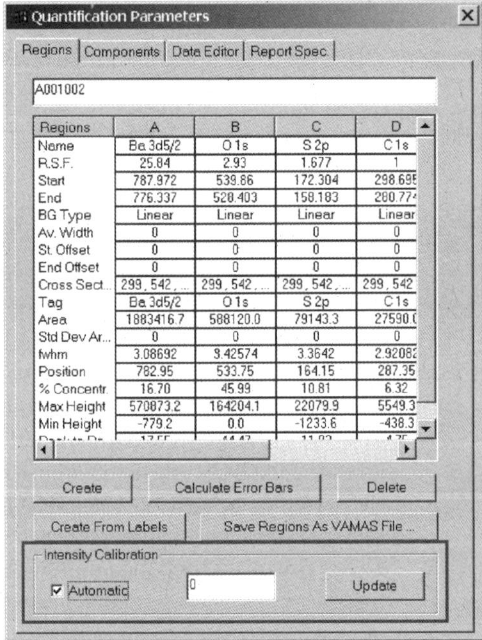

FIGURE 6•2 Regions Property Page

Integration regions define the parameters used to identify the signal associated with a peak in the spectrum. These parameters are tabulated on the Regions Property Page in columns where each item in a column is either an editable input or a read-only value determined from the data. Editable items, when selected with the left mouse button, become text entry fields and may be altered using keyboard entry. These text fields are listed in Table 6•1 on page 93.

TABLE 6•1. Editable items available on the Regions Property Page

Name	The name used to identify the region in quantification tables
RSF	Relative Sensitivity Factor for a transition.
Start	Start energy for the integration region.
End	End energy for the integration region.
BG Type	Background type used as a lower bound to the peak.
Av Width	Determines the number of points used to tie the background to the data. Actual number of points averaged = 2 * Av Width + 1.
Start Offset	Percentage offset applied to the start of the background after the initial tie point is determined from the Av Width.
End Offset	Percentage offset applied to the end of the background after the initial tie point is determined from the Av Width.
Cross-section	Parameters used to determine the background. Meaning depends on the chosen background type.
Tag	A string used to identify the region for computational purposes.

Background (BG Type)

Figure 6•3 shows the list of background types on offer in CasaXPS. The dialog window in Figure 6•3 is invoked by holding the Control Key down at the same time as selecting the BG Type field within a Region. The BG Type text-field is updated from the selected background when the OK pushbutton is pressed. The region is updated with the chosen background once the Return Key is pressed on the Regions Property Page. Values updated include:

Area

The reported area is determined from the background-subtracted data after normalisation with respect to the total acquisition time (dwell-time times number of sweeps), energy step size and is also adjusted for transmission correction using the Intensity Calibrations fields marked in Figure 6•2. If the data file contains transmission correction and the Automatic checkbox is ticked, then the reported area includes the transmission adjustment. There is also an energy dependent exponent text-field for compensating for mean free path and analyser variations. The measured area is multiplied by the kinetic energy of the ejected electron raised to the power of the value entered in this text-field. The energy dependent adjustments are typically required when Scofield cross-sections are used to compensate for the relative transition probabilities for the measure photoelectric lines and, for many instruments, a value of –0.5 would be typical (however this value is not necessarily universally applicable).

St Dev Area

The standard deviation in the Area parameter read-back is zero unless the Calculate Error Bars pushbutton (Figure 6•2) is pressed following any adjustments to the regions parameters. The standard deviation in the area is computed for each region defined on a spectrum using a Monte Carlo procedure for determining the error distribution for each of the computed areas. If the standard deviation in the Area is calculated, the standard deviation for the percentage concentration is also reported in the Standard Reports generated by the Report Spec Property Page also on the Quantification Parameters dialog window.

The Monte Carlo error analysis assumes noise in the data is consistent with the expected noise in XPS data. This requires raw counts within the original data file coupled with total acquisition time. The procedure estimates the error in the recorded peak intensities with respect to the measurement process and is therefore a measure of the precision with which the values are measured and is not a measure for the accuracy in a given concentration value.

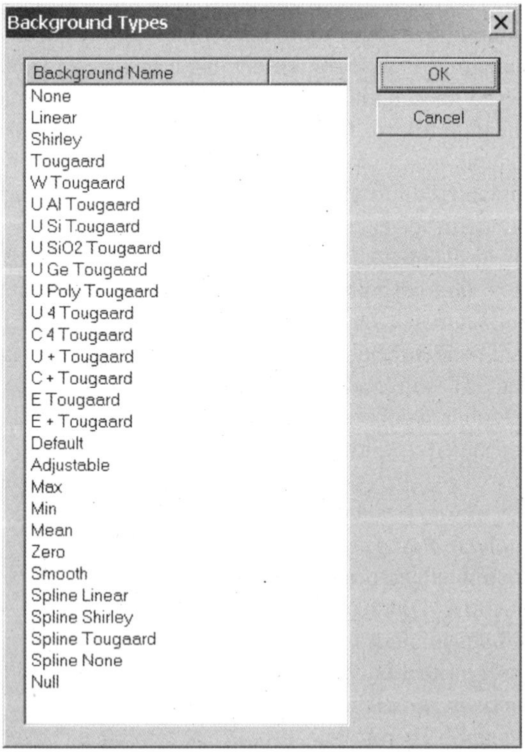

Background Types

OK
Cancel

Background Name
None
Linear
Shirley
Tougaard
W Tougaard
U Al Tougaard
U Si Tougaard
U SiO2 Tougaard
U Ge Tougaard
U Poly Tougaard
U 4 Tougaard
C 4 Tougaard
U + Tougaard
C + Tougaard
E Tougaard
E + Tougaard
Default
Adjustable
Max
Min
Mean
Zero
Smooth
Spline Linear
Spline Shirley
Spline Tougaard
Spline None
Null

FIGURE 6•3 Background Types offered on the Regions Property Page. To invoke this dialog window, hold the Control Key down when selecting the BG Type field.

Report Specification Property Page

Standard Reports represent comprehensive tabulations of quantification results, which may result from either:

1. Integration regions.
2. Intensities determined from synthetic peaks used to fit the data envelope.
3. A combination of 1 and 2.
4. Integration regions proportioned by synthetic peaks for only those quantification items (regions or components) for which Tag fields are defined.

Custom Reports, on the other hand, allow atomic concentrations to be computed from user-defined intensities. These intensities are calculated from formulae specified in terms of the quantification names associated with integration regions or synthetic components, and allow complex relationships between these basic quantification items to be exploited.

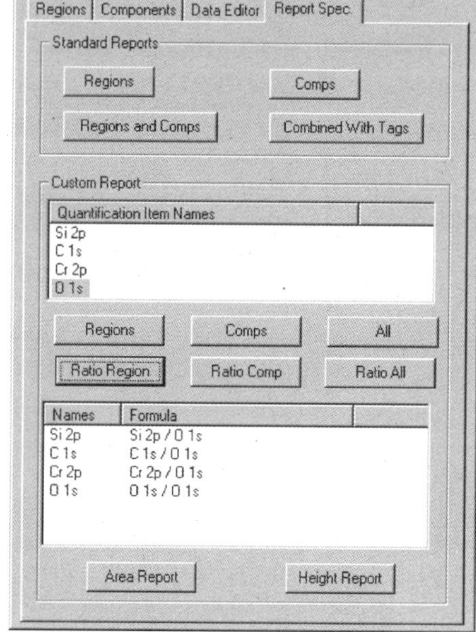

FIGURE 6•4 Report Specification Property Page.

Figure 6•4 shows the current set of available quantification items in the list. From this list the formulae can be constructed in the list headed "Formula" where in the case shown in Figure 6•4 the formulae are simply the ratio of the elemental intensities to that of the O 1s intensity.

Constructing Columns of Intensities from Regions and Components

The set of push buttons positioned between these two lists allow a quick definition for some common situations. The "Regions" button enters a set of Names/Formulae into the lower list that simply derives from quantification items corresponding to integrations regions. Similarly, only intensities from synthetic components are entered when the "Comps" button is pressed or alternatively all quantification items are entered when the "All" button is selected. Ratio formulae are similarly entered into the Name/Formulae list, but an item from the Quantification Item Names list must be selected before either of the buttons is pushed. The quantification item chosen is used as the denominator in the ratio formulae as shown in Figure 6•4.

Where do the Quantification Item Names come from?

The entries in the Quantification Item Names are determined from the data blocks selected in the browser of the active Experiment Frame. The custom report defined using these buttons allows a table of intensities to be generated, where each selected row in the Experiment frame results in a row in the quantification report and the intensities/Atomic Concentrations are gathered from the formula for a Name in the lower table.

WARNING: Intensities from quantification items located in the same row in the Experiment Frame with the same name are summed. Thus, if a peak fit is constructed and each of the synthetic peaks is given the same name then the intensity for the entire set of peaks will be summed into the common quantification item name. If however the region used to define the background is also given the same name as the synthetic peaks, the intensity from the data will effectively be doubled since the peaks and the integration region will both approximate the same intensity for an element. To avoid this erroneous situation, the RSF value for either the region or the set of components should be set to zero. A zero RSF will exclude the (potential) quantification item from the quantification calculation.

Adjusting the Formulae

Once an initial Name/Formula table is prepared, the entries can be edited. Right-click the mouse over the name field to invoke an edit dialog window (Figure 2). Here the Name and the Formula can be adjusted, deleted or a new entry added.

- To add a new entry, edit the fields and press the "Insert" button.
- To delete the current entry select the "Delete" button.
- To modify the current entry, edit the text fields and press the "OK" button.

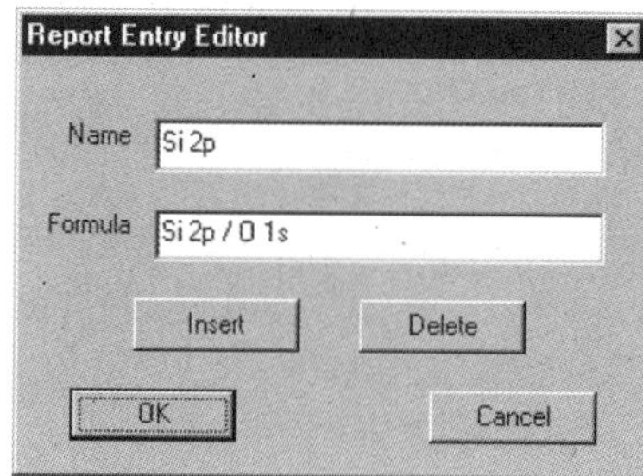

FIGURE 6•5 Name/Formula Edit Dialog Window.

The formula may be any arithmetic expression involving the quantification item names and numerical constants. For example Si 2p – O 1s * 0.5 would be a valid formula given the set of quantification items shown in Figure 6•4.

Preparing a quantification report

The CasaXPS interface for preparing a quantification report is the Quantification Parameters Dialog Window (see Figure 6•2 and Figure 6•4), available from the Quantify button on the Options Menu (page 91).

Quantification Regions
The intensity of a peak in an XPS spectrum is measured by integrating the recorded counts per second minus a computed background over an energy interval delimiting the peak. For the purposes of quantifying a sample, the integrated intensity must be adjusted for both relative sensitivity of the various photoelectric lines and energy dependent instrumental/measurement artefacts before computing the elemental composition for a surface. The primary purpose of quantification regions as defined in CasaXPS is to provide the information required for this calculation.

Synthetic Components
When photoelectric peaks overlap, the tool most often used to partition the contributions from the individual transitions to the recorded envelope is optimisation using predefined line-shapes for modelling the underlying transitions. The intensity for a peak is determined by fitting a set of synthetic components to the data by minimising the sum of squares residual, where the intensity for a peak is extracted as one of the optimisation parameters. A synthetic component is therefore analogous to a quantification region; however quantities such as position, area and full-width-half-maximum (FWHM) are determined from an optimization procedure using a given line-shape rather than, in the case of quantification regions, being computed directly from the data.

Quantification Items and Reports
Since quantification regions and synthetic components both result in apportioning

characteristic responses - quantification values - to XPS peaks, and since each may be used independently, both are referred to as quantification items. Quantification reports are constructed from the characteristic values determined from these quantification items.

A quantification report may be either a Standard Report or a Custom Report. The Standard Report option allows a tabulation of quantification items, grouped by experimental variable, where values taken from each quantification item appear as a line in the final report. Various options are available for Standard Reports to permit the user to define the format for each row of information, however each quantification item, if included, appears as a row in the table of results. Custom Reports, on the other hand are column orientated and one or more quantification items can contribute to the values reported in a column.

Essentially, a column in a Custom Report is defined in terms of arithmetic expressions involving the name fields given to the quantification items and are aimed at extracting trends within a data set defined by the variation of quantification items with respect to the experimental variable.

Quantification of Spectra

The following is a set of examples illustrating the various methods for preparing a quantification report. First, however, an overview to the possible analysis scenarios is presented before the mechanics of performing the tasks is detailed.

Quantification of a Survey Spectrum

The most common and basic analysis of a sample is probably the elemental quantification using quantification regions on a low resolution survey spectrum. Figure 6•6 shows a survey spectrum acquired from a PMMA sample using a Kratos Axis 165 operating in Hybrid mode at a pass-energy of 160 eV, where the elemental quantification in the form of an atomic concentration table appears as annotation on the tile displaying the data.

In CasaXPS, the method for creating the quantification table may follow several paths:

1. Automatic peak identification leading directly to annotating the display with a region quantification table.
2. Manual peak identification followed by manual labelling of peaks leading to region creation based on the annotation labels.
3. Region creation using the Element Library entries.
4. Direct region creation using the Regions Property page.

Quantification tables based on regions can be displayed over the data, as seen in

Figure 6•6, or alternatively separate quantification table text files are created from the information derived from the quantification regions. The route taken leading to the quantification table depends on the nature of the spectrum under analysis and also personal preference.

The easiest method, when it works, is the automatic peak identification and region creation route. The regions shown in Figure 6•6 were created using this procedure: PMMA is a very simple spectrum and thus lends itself to automatic analysis because of the good signal to noise available and peaks close to the expected energy position for the various photoelectric lines. Automatic peak identification generally works well provided that peaks can be distinguished from the noise and the spectrum is calibrated with respect to energy.

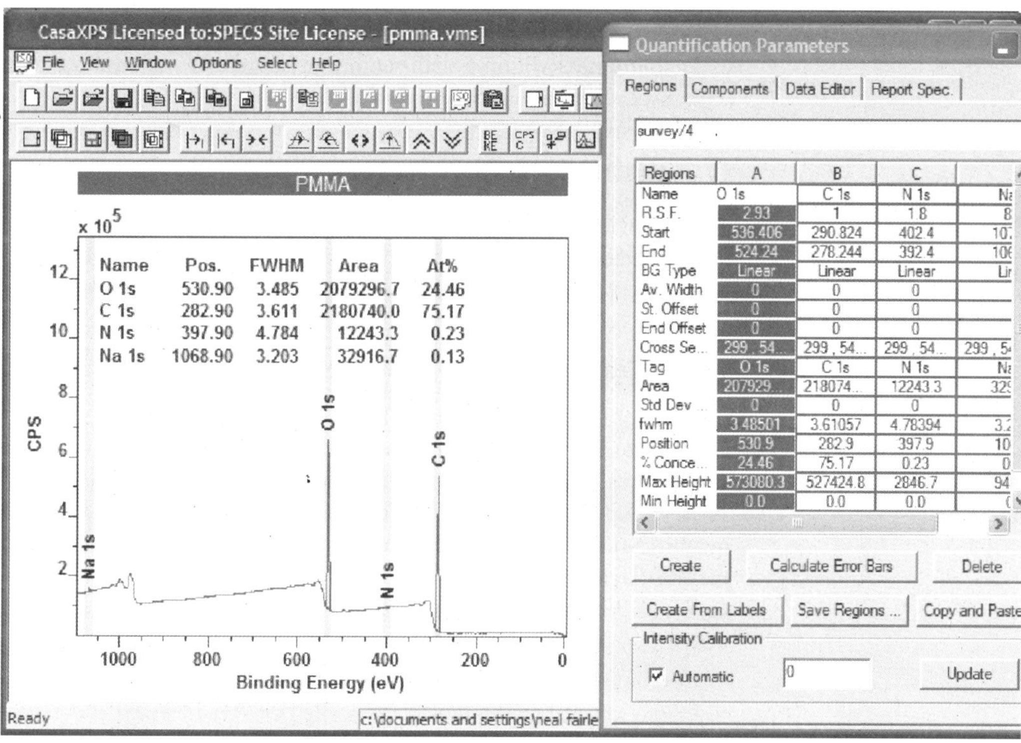

FIGURE 6•6 Quantification of a Survey Spectrum

The steps to perform an automatic quantification are as follows:

1. Display the survey spectrum in the Active Tile.
2. Calibrate the energy scale using the Processing dialog window.

3. Invoke the Element Library dialog window.
4. Press the Find Peaks button on either the Element Table property page or from the Periodic Table property page. Both buttons perform the same task.
5. Use the Periodic Table buttons to add or remove elements from those identified in the previous step until the set of element markers match the peaks in the data.
6. While still displaying the element markers, press the Create Regions pushbutton located on the Element Library dialog window. A set of quantification regions are created and the corresponding quantification table is displayed over the spectrum.
7. At this point it is worth checking the background start and end positions for each region and, if required, creating any additional regions. The need to manually create regions is sometimes required because the automatic creation procedure will only create a region when a peak can be distinguished from the background and occasionally minor peaks may be ignored by this process.

Visually reviewing regions: A quick means of checking the energy limits for a set of regions is to press the Reset toolbar button. The Reset toolbar button loads a set of energy intervals on to the zoom list determined from the current set of quantification regions. Once the zoom list is loaded in this way, pressing the Zoom Out toolbar button causes the display to step through the quantification regions.

The regions created via the Element Library dialog window are chosen from those elements identified using the Find Peaks button and subsequently adjusted using either the Element Table Scrolled List or the Periodic Table property pages on the Element Library dialog window to add or remove elements from those originally identified. For any given element, the transition with the largest RSF is selected when the quantification regions are created. There may be good reasons why the default transition is not appropriate and depending on whether the choice is specific to a given sample or preferred in general, the regions can be either modified manually or the default transition overridden using a configuration file. An example of an element for which the default transition is rarely used during quantification is aluminium. The Al 2p transition is generally preferred to the Al 2s when quantifying a sample, presumably due to the complex background structures resulting from metallic aluminium, and since the RSF for the Al 2s transition is larger than the Al 2p transition, it becomes necessary to force the creation mechanism to choose the Al 2p transition when automatic region creation is performed. A configuration file CasaXPS_quant.lib located in the same directory as the CasaXPS.exe executable file is used to override the default creation choice. The entries in the file CasaXPS_quant.lib are formatted identically to the standard CasaXPS element library, therefore to override a given transition simply copy the corresponding entry from the element library file into the configuration file CasaXPS_quant.lib. The sig-

nificant information in the CasaXPS_quant.lib file is the species and transition columns; all other values in this file are ignored, but nevertheless must be included. The second means of creating quantification regions is based on annotating the peaks prior to creating the regions. A typical spectrum will require the labelling of peaks for presentation purposes and so as part of the labelling process quantification regions can be created *en route* to labelling the spectrum. Peak labels themselves are created in conjunction with the Element Library dialog window and since one problem associated with the quantification process is the correct extraction of the relative sensitivity factors for the peak in question, this link between the peak label and the element library is helpful when creating regions.
The methodology is as follows:

1. First annotate the peaks used to quantify the sample. The Peak Labels property page on the Annotation dialog window offers those lines for which the Element Library has been used to display element markers and when created via the Peak Label property page, the labels are assigned the text used in the Element Library.
2. Once the peaks intended for use in the quantification are labelled using the text from the Element Library, the Regions property page is used to create the quantification regions. That is to say, the Create from Label pushbutton causes the creation of a quantification region for each peak label.
3. The remaining peak labels for those peaks not used in the quantification can now be created using the same mechanism as that used in step 1

A survey spectrum data block does not initially contain information identifying peaks in the data, hence the importance of linking the Element Library text strings with quantification region creation via the annotation peak labels. A further means of specifying the transition associated with a peak, but without the need to create peak label annotation, is via the Element Table scrolled-list (Figure 6•7). The creation of quantification regions is performed by selecting the Regions property page on the Quantification Parameters dialog window and, on the Element Library dialog window, ticking the two tick-boxes below the scrolled list in Figure 6•7 labelled **Zoom When Line Selected** and **Create When Line Selected** followed by selecting a line via the name field in the scrolled list. These tick boxes modify the behaviour of the scrolled list on the Element Table property page. The **Zoom When Line Selected** tick-box causes the action of picking a transition in the Element Table to zoom the display so that the energy scale visible in the Active Tile matches a small interval around the line indicated via the name field of the scrolled-list. The **Create When Line Selected** tick box similarly modifies the same action by creating a quantification region using the energy interval currently used to display the spectrum. The act of creating a quantification region by selecting a line in the Element Table ensures the correct relative sensitivity factor is used in the newly created region.

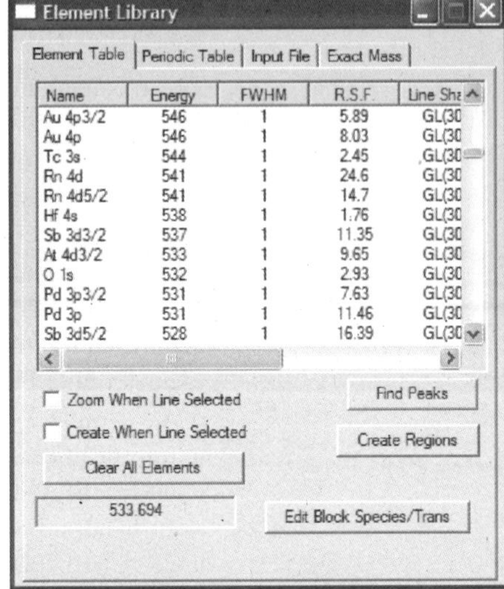

FIGURE 6•7 Element Library Property Page

The fourth method for creating a quantification region is to simply press the Create button on the Region property page of the Quantification Parameters dialog window. The current zoom state determines the start and end energies for the newly created region, however all other parameters are merely a repeat of the parameters previously used to create a quantification region. If the species/transition fields are consistent with an entry in the Element Library, then the relative sensitivity factor is retrieved from the library and used in the quantification region. For survey spectra, owing to the lack of species/transition information, the basic creation mechanism is of little use, however if narrow scan spectra are involved, the species/transition fields should be correctly assigned, in which case the Create pushbutton will indeed result in the correct assignment for the relative sensitivity of the line.

Although the Create pushbutton, when applied to a survey spectrum, creates a region without retrieving the RSF from the library, the RSF from the library can be entered using the name field for the region. The default name field in the created region will not match any element library entry, however if the region name is re-entered using a name from the library preceded by the # character, then on pressing the return key the name field will be entered without the # character and the element library field matching the string is used to update the RSF in the region. In the case of a survey spectrum, an atomic concentration table may be displayed over the spectrum as seen in Figure 6•6 or alternatively a quantification report is generated using the Report Spec property page on the Quantification Parameters dialog window. When using the Report Spec property page, the data block containing the survey spectrum must be selected in the right-hand-side of the Experiment Frame prior to pressing the button used to generate the report table.

Quantification of High Resolution Spectra

The mechanisms described for the creation of quantification regions on a survey spectrum are also applicable to high resolution data from a sample typically recorded as a sequence of narrow scan spectra. The key difference is that each narrow scan spectrum is assigned one or more quantification regions and since more than one spectrum is included when quantifying a sample, the quantification table generated from these separate narrow scan spectra requires the use of the browser in the right-hand- side of the Experiment Frame to perform the task. Those spectra for which a quantification report is desired must be selected in the right-hand-side of the Experiment Frame. Further more, quantification using multiple acquisition regions for a given sample requires all corresponding spectrum blocks used in the quantification to appear in the same row of the browser pane in the right-hand-side of the Experiment Frame. If the data file does not result in the correct alignment of blocks in the right-hand-side, then the values for the species/transition and the experimental variable must be adjusted until those blocks containing the spectra and quantification regions for each sample all appear in a single row. A quantification report is generated from only those quantification regions for which the data blocks are selected in the right-hand-side and a separate row will appear in the quantification table for each quantification item in the row of selected blocks in the right-hand-side.

Quantification with Tags

A problem when quantifying samples, particularly for older instruments, is the variation of an analyser's transmission characteristic with pass energy, lens modes and any other acquisition variables. A consequence of these operating mode dependent variations in the measured peak areas is that, without transmission correction, quantification must be performed using the same acquisition settings. Transmission correction usually takes the form of, so called, transmission functions and in modern instruments these are typically saved along with the spectra for use during quantification. Older instruments rarely include transmission correction so, when more than one operating mode is used, an alternative strategy is offered in CasaXPS for quantification.

Quantification using TAGS is designed to allow for the situations where elemental concentrations determined from the survey operating mode are combined with high resolution data to reveal chemical state information unavailable in the lower resolution survey spectrum. Peak models determined from the high resolution spectra are associated with quantification regions defined on the survey spectrum and the relative intensity of the synthetic peaks from the high resolution spectrum are used to proportion the elemental concentration from the survey data. Thus, the chemical state information present in the high resolution spectra is referenced back to the survey spectrum. The result is a combined quantification report in which the transmission variations are eliminated from the quantification report. Both quantification regions and synthetic components include as part of their definition a TAG field. These TAG fields are text strings, which when the same string is used for both a region and a corresponding synthetic component, the region intensity is further tabulated using the relative intensities of synthetic peaks with the same string in the TAG fields as the string entered for TAG field in the region. Quantification using TAG fields is one of the options on the Report Spec property page of the Quantification Parameters dialog window.

Quantification of Depth Profiles

The general philosophy in CasaXPS is to handle larger data sets by identical mechanisms to those used for smaller set of data. The same quantification items used to quantify survey and narrow scan regions are used to reduce XPS depth profiles to traces for each chemical state monitored as a function of depth. The structure of a profile experiment is essentially a sequence of survey or narrow scan spectra recorded following an etch cycle or stage tilt; thus the data blocks appear in columns corresponding to the assignments for the species/transition where each row is aligned with respect to a common etch time or angle. The route to tabulating a set of intensities as a function of experimental variable is via defining regions and/or synthetic variables for a spectrum and then propagating these quantification items to spectra with similar requirements. There is no requirement to propagate the quantification items through an entire set of spectra; rather the propagation is performed only on those data blocks selected in the right-hand-side of an Experiment Frame. Thus, a peak model, for example, can be adjusted at intervals during a depth profile in order to account for new or absent peaks resulting from a change of matrix material or interfacial anomalies. It should be emphasized, however, that each spectrum used to profile a sample must have quantification items assigned and so regions or synthetic components must be created directly or propagated to each spectrum in the profile.

Survey Spectra: Creating Quantification Regions, Relative Sensitivity Factors and Transmission Correction

As discussed above, the easiest method for defining quantification regions on a survey spectrum is using the automatic peak identification options on the Element Library dialog window followed by pressing the Create Regions button on the same dialog window. For well behaved spectra, such as the data shown in Figure 6•6 on page 99, the results are generally achieved with minimal user intervention, however not all samples are as accommodating and some form of adjustments to the automatically created regions becomes necessary. The primary aim of the following is to describe the options available for the manual creation of quantification regions. The example chosen to illustrate the manual creation of quantification regions is also useful in understanding some of the issues associated with the quantification of a sample. The sample is pure gold, so the only significant peaks derive from the same material at different kinetic energies and from a range of symmetries (4s, 4p1/2, 4p 3/2, 4d, 4f and 5p3/2), which makes it possible to investigate the factors influencing quantification in a known regime. Figure 6•8 is a gold spectrum taken from a Kratos Axis Ultra at UMIST in Manchester UK. The initial problem is to define the appropriate quantification regions for this unusual situation in which every transition for which relative sensitivity factors are available are used to measure the gold signal. If all is well, the atomic concentration table should report the same value regardless of the transition chosen. Figure 6•8 does indeed offer a quantification report demonstrating this property; however the route to such a report involves several steps worthy of discussion.

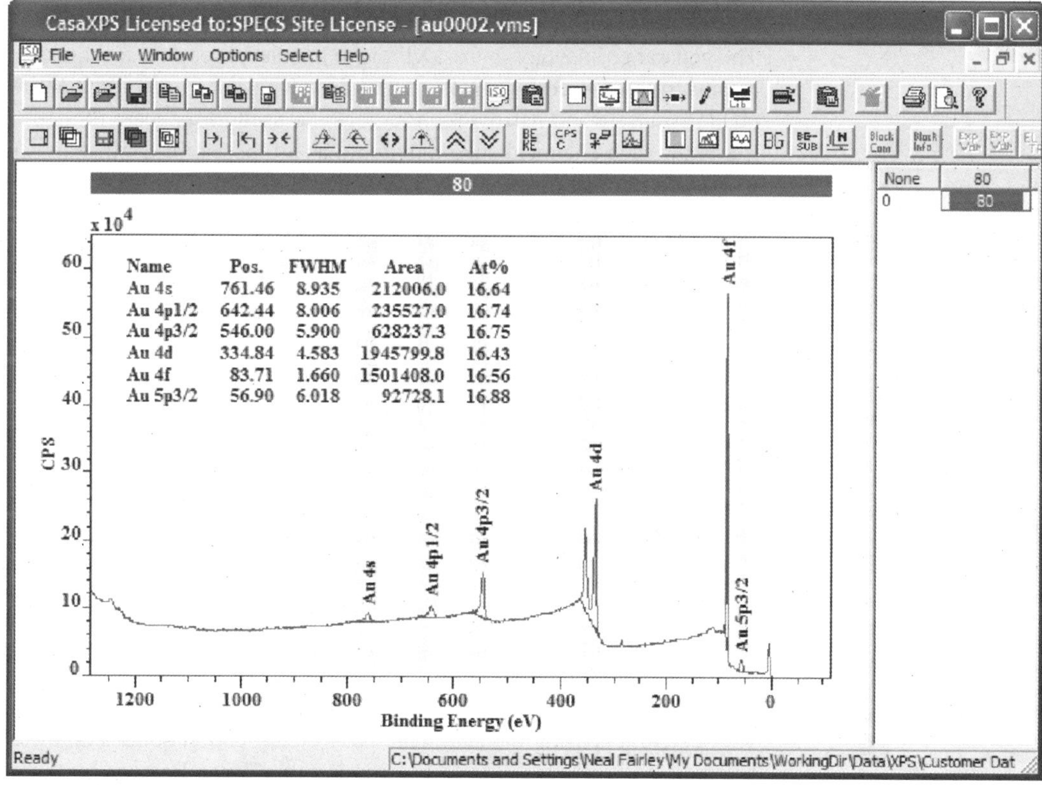

FIGURE 6•8 Gold Spectrum taken from a Kratos Axis Ultra

Manual Creation of Quantification Regions

The basic means of creating a quantification region is the Create button on the Quan-

tification Parameters dialog window shown in Figure 6•2. On pressing the Create button, a new quantification region is created on the first spectrum displayed in the Active Tile, where the start and end energy values for the region are determined from the interval displayed in the Active Tile.
Additional information from the element library (such as the appropriate RSF, for example) is entered into the region, *provided* that the name field in the newly created region matches the element/transition fields for an entry in the current element library in use by CasaXPS.
Creation of the regions' name fields by this route derives directly from the species/transition fields for the data in the Active Tile, so the RSF retrieval will only work for spectra in which the species/transition fields have been correctly assigned in the ISO 14976 file. In the case of a survey spectrum where multiple species/transitions are present, the RSF will not be retrieved and therefore must be entered manually.

When creating regions on a survey spectrum such as the gold spectrum in Figure 6•8 on page 105, the steps to follow are:

1. Display the spectrum in the Active Tile.
2. Zoom into the peak for which the region must be defined.
3. Press the Create button on the Regions property page on the Quantification Parameters dialog window.
4. Set the name for the region just created by entering the element/transition strings for the name preceded by the # character and press the enter key on the keyboard.

To display a spectrum in the Active Tile: Using the mouse left-hand button, double click on the data block in Experiment Browser. Alternatively, select the block using the left-hand mouse button and press the toolbar button.

To zoom into a section of a spectrum: Use the left-hand mouse button to drag out a box over the display area of interest and once a box is displayed, click inside the box. The display will rescale to show the rectangular region previously bounded by the box.
Once the enter key on the keyboard is pressed, the # character is removed from the string entered for the name and the RSF is retrieved from the matching entry in the element library. The region name field can subsequently be altered to any string not starting with the # character without further updates to the RSF

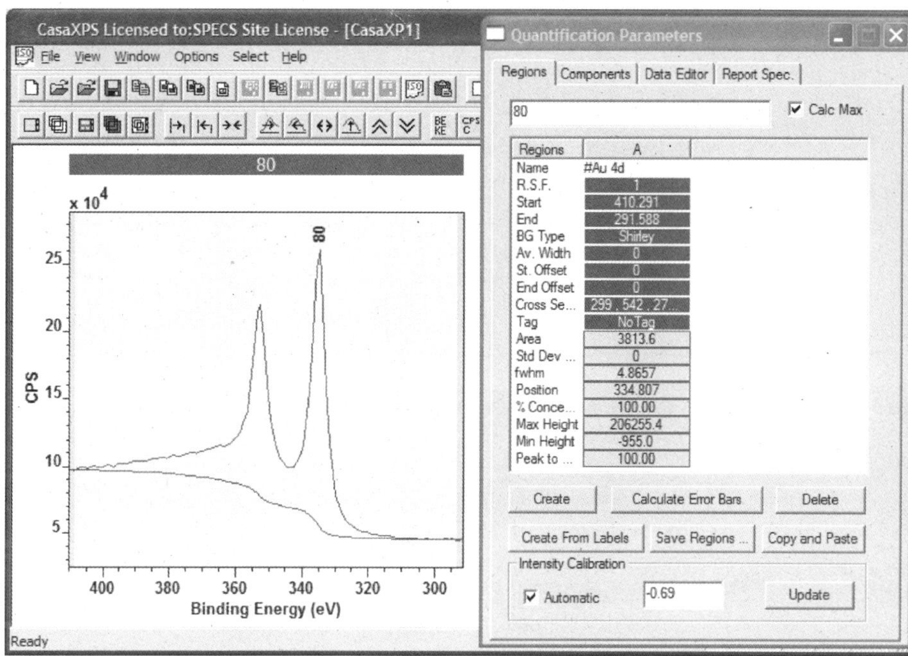

FIGURE 6•9 Zoomed energy interval showing a newly created region prior to the use of the region name field to trigger the update of the RSF and TAG fields.

Figure 6•9 illustrates the use of the # character to force the region to retrieve the RSF from the element library. The element library name field for the doublet peaks displayed in the Active Tile in Figure 6•9 is Au 4d, therefore to retrieve the RSF for the region create for the displayed energy interval, the name field for the region is entered using the string #Au 4d. When the Enter Key is pressed, the region name is modified to Au 4d and this new string is used to match an entry in the current element library. If a match is found, both the RSF and the TAG string are updated within the region parameters.

Relative Sensitivity Factors and Other Corrections

If information is retrieved from the element library, the name entered for the region must be an exact match for an entry in the element library. In the case of doublet states, spectral features derive from two peaks; these two peaks are labelled with the same principal and orbital quantum numbers n and l, but differentiated by the total angular momentum quantum number j. For example, in a gold spectrum, the doublet pair at about binding energy 83 eV are assigned to the two transitions Au $4f_{5/2}$ and Au $4f_{7/2}$ (Figure 6•10). The RSF for these individual peaks are different and so if

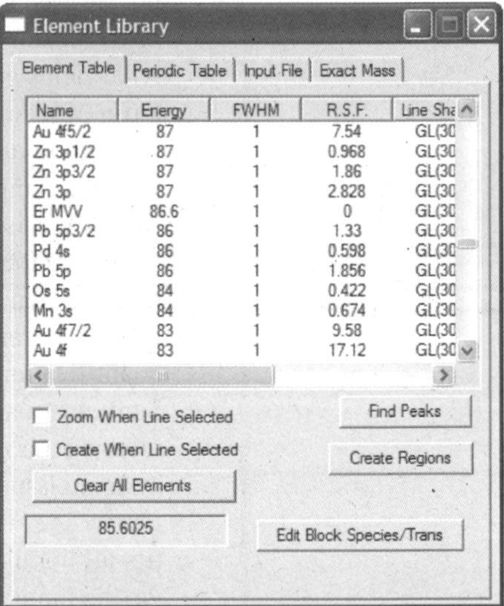

FIGURE 6•10 Element Library dialog window showing the three possible choices for the Au 4f transition

only one peak from a doublet pair is used to measure the intensity for an element, the name used for the region should be identical to the name field in the element library for the chosen transition. In the case of the gold spectrum shown in Figure 6•8 on page 105, the Au 4f doublet pair overlap to the point where it is necessary to use the total counts associated with the two peaks, in which case the RSF corresponding to the name Au 4f is used to scale the intensity prior to comparison with other transitions. This is in contrast to the $4p_{1/2}$ and $4p_{3/2}$ transitions, where the peaks are easily resolved and so the individual RSF entries for the two peaks are included in the quantification table in Figure 6•8. Ideally, the use of the combined intensity of the doublet pair is preferred from a statistical perspective; however in practice the influences of time, background and interfering peaks often force the use on one peak when quantifying a material. The default CasaXPS library therefore includes three entries per doublet pair, one for each individual peak in the doublet and one for the total intensity for the two peaks. The rule for using doublet peaks is to use the entry without a subscript for situations where both peaks are included in the measurement, that is both peaks should have the same RSF, and to use only the subscripted entries for cases where one peak from a doublet pair is included in the quantification table. The default CasaXPS library contains Scofield cross-sections for the relative sensitivity factors. These Scofield cross-sections were computed using Hartree-Fock theory to determine the relative intensity of the transitions throughout the periodic table for a given photon energy. The cross-sections are relative to the C 1s transition and the default CasaXPS library includes entries for both Mg and Al x-ray anodes. The theory used to calculate the cross-sections does not, however, include corrections for instrumental and sample dependent intensity variations. Without these corrections, XPS yields good relative measurements but poor absolute quantities. That is to say, a run of identical samples, when quantified using Scofield cross-sections, will produce good agreement between the resulting tables, however the numerical values will not compare favourable to the true proportions of the materials comprising the samples. It is therefore important to account for these additional intensity variations as part of the quantification procedure. The principal corrections are: instrumental transmission correction, instrument dependent angular distribution corrections to the measured transition and escape depth variations as a function of kinetic energy. The quantification table shown in Figure 6•8 on page 105 has been computed using Scofield cross-sections corrected for all three intensity adjustments. Once these corrections are made to the intensity of the peaks used in quantification, the only remaining issues are that of background definition and matrix dependent effects. The former is at the user's discretion, while the latter is beyond the scope of a one-size-fits-all library approach; however multiple CasaXPS libraries are easily maintained via the Input property page on the Element Library dialog window and so matrix dependent RSF tables can be accommodated within CasaXPS.

Most modern instruments offer transmission correction to account for the variation in performance of a spectrometer as a function of kinetic energy for the ejected electrons when measured using the different operating modes of the instrument. Various mathematical models are employed to describe these transmission functions; however the one that best fits the Scofield library is the transmission function correction procedure developed by the National Physical Laboratory (NPL) in the UK. For each operating mode of an instrument, a set of survey spectra from clean copper, silver and gold samples are acquired and these are reference back to a spectrum determined on a well characterised spectrometer at NPL. The result of the procedure is a transmission curve for each operating mode of the instrument and these transmission curves are added to the data whenever an acquisition is performed. Regardless of the source for the transmission curve, when added to the ISO data file as a corresponding variable, CasaXPS automatically detects the presence of the transmission curve and performs the correction at the time a quantification table is computed.

Once the instrumental transmission is accounted for, there is still an energy depend correction required originating from the sample. The depth below the sample surface from which electrons are collected by the spectrometer varies as a function of kinetic energy of the ejected electrons, sometimes referred to as the escape-depth correction, this adjustment to the peak intensities takes the form of dividing the area of a peak by the kinetic energy of the peak maximum raised to a constant power n, where n is typically a value between 0.5 and 1.0. In CasaXPS, the intensity is adjusted by multiplying by the energy raise to an exponent, and therefore the value used for the exponent should be a negative number.

Both transmission correction and escape-depth correction are applied by default, assuming the information is made available to CasaXPS. The transmission curve must be present in the data block as a second corresponding variable; however the value for the escape-depth correction is either present in the data block containing the spectrum or entered on the Regions property page in the text-fields within the Intensity Calibration section (Figure 6•2 on page 92). A configuration file entry in the ParameterFile.txt file located in the CasaXPS.DEF directory allows the default value for a new block to be specified different from zero. The values used to correct the peak intensities for transmission and escape depth are computed at the energy of the peak maximum. There is, however, a subtle difference between the meanings of "the energy at the peak maximum" for these two quantities. The transmission curve relates to the peak maximum with respect to the instrument and therefore relates to the kinetic energy of the electrons in flight through the analyser. In contrast, the escape-depth is considered by CasaXPS to relate to the kinetic energy of the ejected electron within the sample. For these reasons, the transmission correction is computed using the energy at the peak maximum before any energy calibration is performed, while the escape depth correction is performed using the kinetic energy for the calibrated peak position.

Transmission and escape-depth corrections are energy dependent adjustments applied to the measured intensities; however there is a further correction related to the instrumental geometry and the symmetry of the electronic states used in quantification. The so called angular distribution correction can be viewed as a correction, which must be applied to the Scofield cross-sections, to account for intensity variation resulting from the angle of the x-ray source to the angle of the ejected electrons (analyser axis). If the angle in question is different from the, so called, magic angle of 54° 44' for which correction is unnecessary, the values for the Scofield cross-sections must be adjusted to account for the angular distribution of photoelectrons ionized by unpolarized x-rays impinging on a core sub-shell. The instrument used to acquire the gold spectrum in Figure 6•8 on page 105 (a Kratos Axis Ultra) had an angle of 60° between the x-ray source and the analyser axis, therefore a direct application of uncorrected Scofield cross-sections would not yield the quantification shown in Figure 6•8. In fact, the RSFs used to characterize the set of gold transitions where determined by correcting the Scofield cross-sections for both the angular distribution correction factors and the transmission function determined using the NPL instrument calibration procedure. The relative percentage concentration values were also computed using an escape-depth correction where the value for the exponent was -0.69. The table in Figure 6•8 illustrates the possibility of measuring the gold signal from any one of the six transitions resolvable in the gold spectrum. Given the range of kinetic energies for the peak positions and the variation in the electronic states, the consistent nature of the relative intensities is very good. The problem with offering this table as a paradigm for all quantification lies in the choice of background type and the quantification region limits. In the case of the gold spectrum in Figure 6•8, all the quantification regions are defined with linear background types and the energy intervals over which integration is performed are chosen to be reasonable subject to the constraint that a good result is achieved. There is greater uncertainty associated with real samples and the choice of background type and integration limits are without doubt the greatest source of error with respect to the precision of an XPS measurement. Nevertheless, without these corrections to the intensity calculation, it was not possible to reproduce the quantification in Figure 6•8 with the same consistency.

Angular distribution corrected Scofield cross-sections are calculated in CasaXPS for properly defined ISO 14976 files. The required information for the use of automatically corrected RSFs is a comment line placed in the file comment section of the ISO file and an appropriate entry in the ParametersFile.txt configuration file. At present only files converted from PHI MULTIPAK data files contain the required information, but as an alternative, CasaXPS libraries of corrected Scofield cross-sections are possible to construct. The library used to quantify the gold spectrum was created for a specific instrument, but the corrected Scofield cross-sections are such that spectrometers characterised relative to the gold spectrum in Figure 6•8, could write relative transmission functions to the data enabling the same modified Scofield library to be used in general. Since the gold spectrum was acquired on a Kratos Axis Ultra, the use of this gold spectrum and associated library is most appropriate for other Kratos Axis instruments.

Atomic Concentration Report

The quantification report displayed over the gold survey spectrum in Figure 6•8 on page 105 is typical of the types of information used to characterise a sample. Various quantities are computed from the data, such as peak positions, FWHM and peak area, however the column reported under the heading At%, is in general, the atomic concentration calculated from the intensities following the application of all the corrections discussed above. Clearly, in the case of a pure gold sample the atomic concentration is 100% gold and the quantification table in Figure 6•8 is merely establishing the relative accuracy of measuring the gold intensity using any one of the six regions on display. When quantifying a real sample, the quantification table derives from a selection of regions, where one transition is chosen for each element identified as contributing to the peaks in the data. Given that m elements are so defined and therefore m regions defined on the spectrum, the calculation for the relative proportions of the sample surface or percentage atomic concentration is given by the formula:

$$X_i = 100\frac{A_i}{\sum_{j=1}^{m} A_j}$$

The percentage atomic concentration for the *i*th element X_i is defined by the adjusted intensity A_i as follows:

$$A_i = \frac{I_i}{T(E_i)R_iE_i^n}$$

The terms contributing to the adjusted intensity are: the measured intensity for a peak I_i (either integrated peak area or peak height), the transmission function evaluated at the peak position $T(E_i)$, the relative sensitivity factor R_i, kinetic energy E_i and escape-depth exponent n.

Quantification Examples in Detail

The general overview presented above is now expanded using specific examples to illustrate the practical aspects of using CasaXPS to quantify spectra.

Quantification using High Resolution Spectra

Although elemental quantification is commonly performed on low resolution survey spectra, circumstances occasionally require the use of high resolution spectra for this purpose. A common reason for using narrow scan spectra is poor signal-to-noise in the data. The greatest source of error when computing the area of a peak results from noise-levels within the data channels used to tie the calculated background to the data. For weak peaks it may be necessary to increase the dwell-time compared to stronger peaks; the aim of most analysts is to acquire the data in the minimum time subject to the constraint of sufficient precision. The experiment shown in Figure 6•11 is an example of the case in point. A small amount of chromium is barely evident on the survey spectrum in Figure 6•11 however, using an increased dwell-time for a narrow energy interval containing the Cr 2p doublet allows efficient use of time when measuring the Cr signal to the required precision.

Data acquired using different dwell-times, number of scans and energy step sizes are accounted for correctly by integrating the peak areas using the units of counts per sec electron Volts (CPSeV), thus eliminating these parameters from the quantification calculation. Switching lens modes or pass energy when acquiring the different narrow scan spectra will require the proper use of transmission functions to allow quantification to proceed.

An analysis of the experimental data shown in Figure 6•11 follows: a more detailed discussion of other aspects of this analysis is given in "Background Subtraction: an example" on page 193.

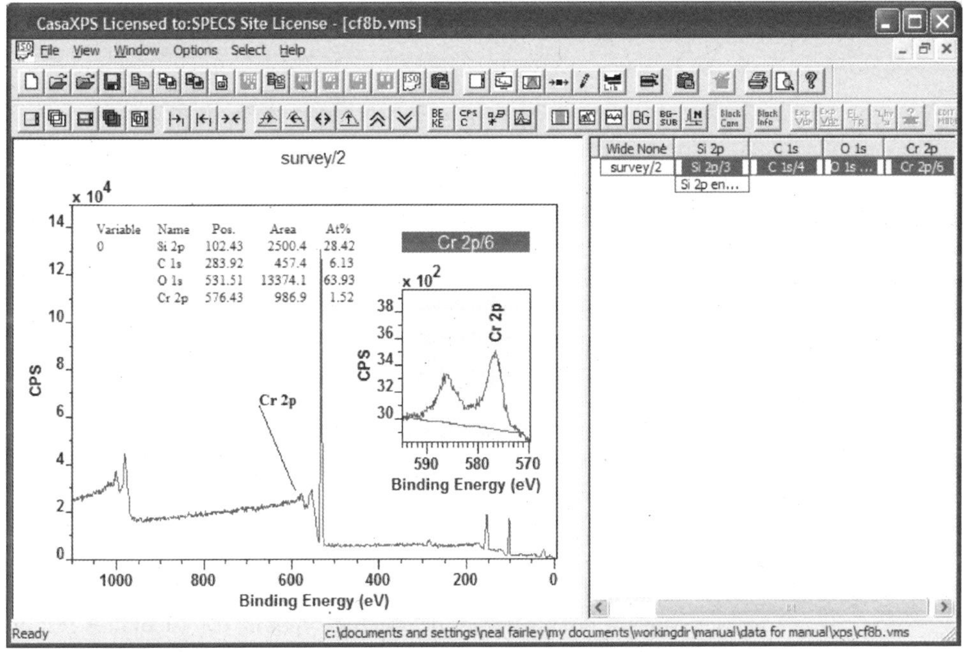

FIGURE 6•11

Charge correction: The first step is usually to correct the data for energy shifts due to charging of the sample under the influence of the x-rays. The Calibration property page of the Processing dialog window offers a range of procedures for charge correcting spectra (Figure 6•12 on page 114). When the experiment consists of a set of narrow scan spectra, most likely the set of spectra all require an identical energy shift, where the value for the energy shift is determined by locating the peak on one spectrum.

There are several ways of performing this task, however in the example shown in Figure 6•11 the calibration of the file was performed using the following approach:

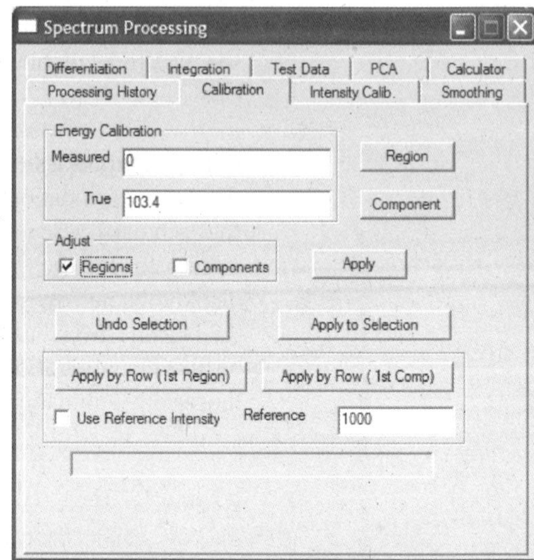

FIGURE 6•12 Charge correction

1. Define a quantification region on the Si 2p spectrum (described below).
2. Select the data block containing the Si 2p data.
3. In the True text-field on the Calibration property page, enter the require energy for the peak position defined by the peak maximum in the quantification region previously defined on the Si 2p spectrum.
4. Un-tick the box labelled Use Reference Intensity and press the button labelled Apply by Row (1st Region).

The result of these steps is to calibrate the energy for each spectrum in the same row as the Si 2p by shifting the energy scale by the amount required to place the maximum of the Si 2p peak at the energy entered in the True field on the Calibration property page. The position of the Si 2p peak is determined from the first region defined on the Si 2p spectrum. In this case there is only one quantification region defined on the Si 2p spectrum, but if more than one were defined, then the region in column A of the Regions property page of the Quantification Parameters dialog window is the one used in the shift operation. It is also worth noting that each row of spectra in the Browser, when calibrated using the Apply by Row (1st Region), is independently adjusted. That is to say, each row in which the appropriate Si 2p data block is selected will be treated as a separate calculation to determine the energy shifts. Thus multiple samples loaded into the same ISO 14976 file, which appear in separate rows, can be treated in this manner. The restriction, however, is that every spectrum in a row is shifted without exception. For situations where for whatever reasons only a subset of a row should be calibrated identically, other means of calibrating selected spectra are available on the Calibration property page of the Processing dialog window.

Quantification issues: Quantification using narrow scan spectra requires the creation of quantification regions for each significant peak in the data. The data block for each narrow scan spectrum used in the quantification maintains the region information for the data held within the block. It is therefore necessary to individually display the narrow scan spectra and create appropriate quantification regions. Once the regions are created, quantification proceeds via the Report Spec property page on the Quantification dialog window, where quantification is performed on only those data within the blocks selected in the right-hand-side of the Experiment Frame.

Region creation: For the data in Figure 6•11 on page 113, the blocks are assigned the correct species/transition fields, so creating quantification regions is as simple as displaying the data in the left-hand-side of the Experiment Frame, invoking the Quantification Parameters dialog window and pressing the Create button on the Regions property page. The RSF for each region is retrieved from the current element library; this is because the name field within the element library matches the strings assigned for the species/transition in the data block. Once regions are created for each narrow scan spectrum, switching to the Report Spec property page on the Quantification Parameters dialog window and selecting the data blocks as seen in Figure 6•13 allows quantification to proceed based on the users choice.

Quantification via the Standard Report: The data shown highlighted in Figure 6•13 represents four narrow scan spectra in which a single quantification region has been created on each spectrum, namely, the Si 2p, O 1s, C 1s and Cr 2p peaks. When a region is initially created using the Create pushbutton on the Regions property page, the name field for the region is automatically entered using a concatenation of the element/transition strings for which a match occurred with an element library. Although the regions are assigned to individual spectra, there is a feedback mechanism for checking which regions are active with respect to the Report Spec option as a consequence of the current browser selection: all quantification items defined on the selected block will have their names entered into the scrolled list entitled Quantification Items Names in the Custom Report section of the Report Spec property page It is therefore worth glancing through the list following making the selection of those data blocks for inclusion in the quantification table.

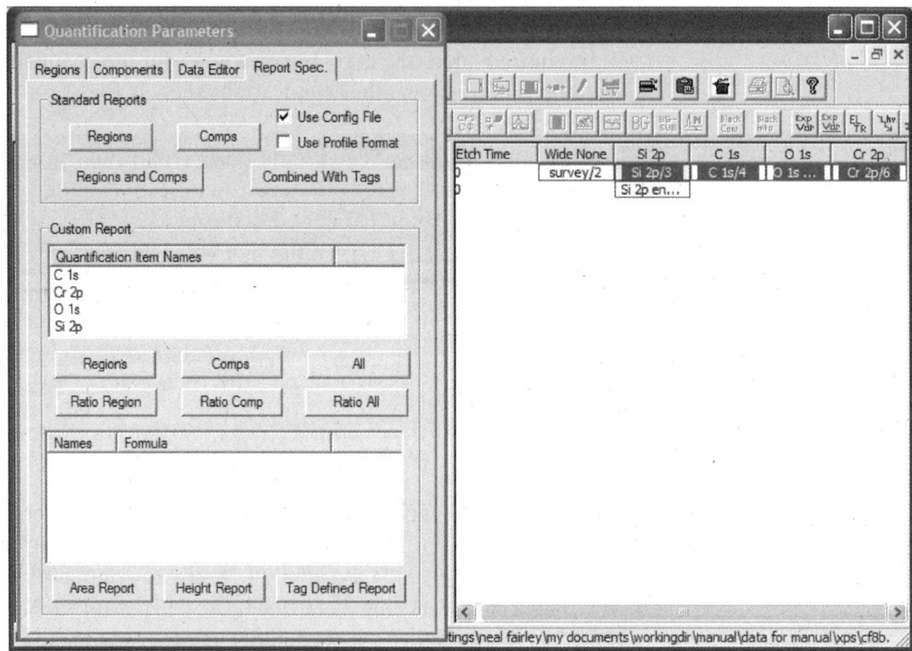

FIGURE 6•13 Report Spec Property Page on the Quantification Parameters Dialog Window.

Note the Quantification Item Names scrolled-list only includes distinct strings in the list. That is to say, if two regions are assigned the same name, then only one entry will appear in the Quantification Item Names scrolled-list, however, a Standard Report will create an entry in the final table, one for each region or component defined on the selected data, subject to which Standard Report button in pressed. This is in contrast to the Custom Report, in which any combination of regions and components defined on the selected data are always included in the calculation of the quantification table. Further, if two or more regions and/or components are assigned the same name fields, then only one entry will appear in the Custom Report table for identically named quantification items and the intensity from all items with the same name are summed to produce the intensity used in the resulting table.

The steps to generate a Standard Report are as follows:

1. Select the blocks containing the required data in the browser (right-hand-side of the Experiment Frame).

2. Press the Regions pushbutton on the Standard Report Section of the Report Spec property page.
3. If required, press the copy button to transfer the quantification tables onto the clipboard for use in other applications such as Excel.

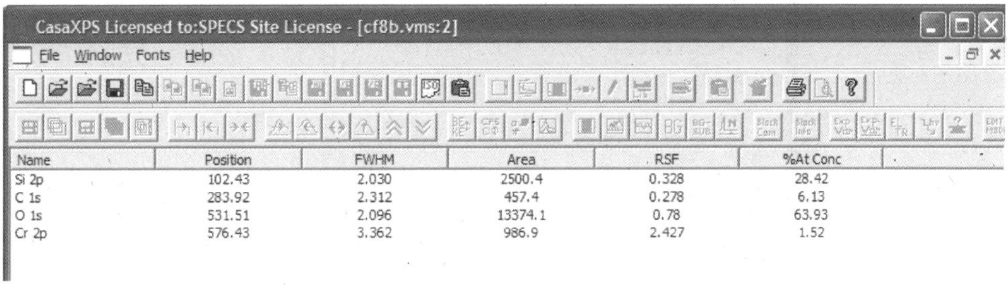

CasaXPS Licensed to:SPECS Site License - [cf8b.vms:2]

File Window Fonts Help

Name	Position	FWHM	Area	RSF	%At Conc
Si 2p	102.43	2.030	2500.4	0.328	28.42
C 1s	283.92	2.312	457.4	0.278	6.13
O 1s	531.51	2.096	13374.1	0.78	63.93
Cr 2p	576.43	3.362	986.9	2.427	1.52

FIGURE 6•14 Standard Report Table.

The quantification report shown in Figure 6•14 was generated using the Regions pushbutton, where the Use Config File tick-box was ticked. A much wider range of quantification table fields are possible, however for many only a subset of the full table is desirable and using a configuration file named RegionQuantTable.txt in the CasaXPS.DEF directory allows the specification of which fields are displayed when the table is generated. If the tick box for the use of the configuration file is not ticked, a much more verbose from of the table is created. It is also worth noting, the configuration file definition file also allows the user to display additional fields not visible in the default quantification table.

Quantification of a survey scan using Tags

The data file shown in Figure 6•15 represents an example where a combination of a low resolution survey spectrum and a high resolution C 1s narrow scan spectrum results in a more complete quantification table than if the two spectra are treated separately. Figure 6•15 shows data taken from a Kratos Axis 165 and, although the Kratos system does include a method for transmission correction, the data shown was exported without the original transmission curves, so a direct comparison of the intensities measured from the survey data and the peak model shown for the C 1s narrow scan region is not possible. Further, the ISO file containing these spectra include neither the N 1s nor O 1s data acquired using the same operating modes as the C 1s region, therefore the only way to exploit the information from the C 1s data envelope is to use the Tag mechanism in CasaXPS; that is, to proportion the elemental concentration for C 1s measured using the survey spectrum in accordance with the relative intensities from the high resolution peak model.

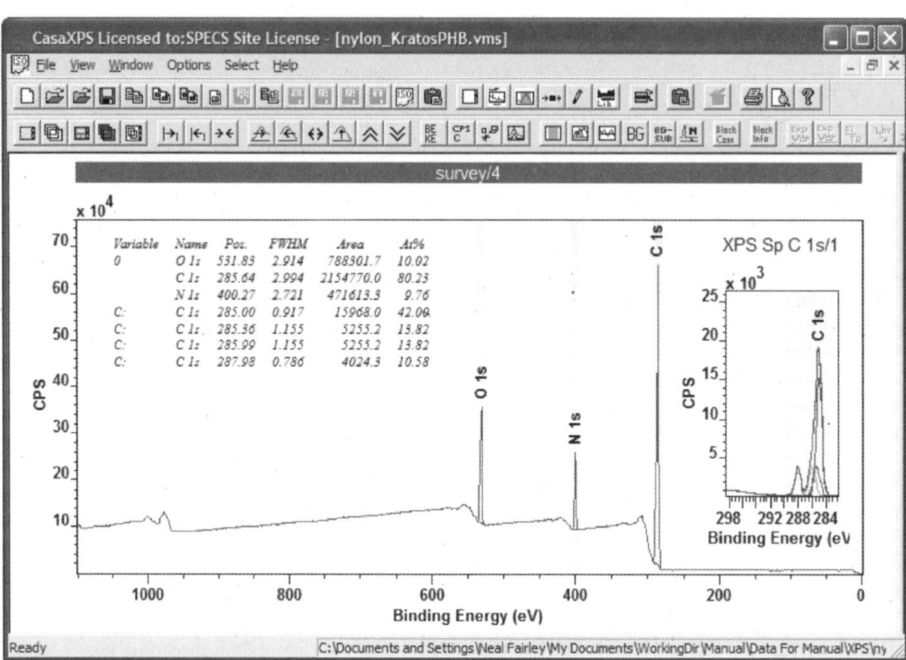

FIGURE 6•15 Quantification using Tag fields.

The Tag mechanism in CasaXPS relies on the proper assignment of strings for each region or component in the calculation. That is to say, the Tag strings are identical for only those components and quantification region for which the concentration, as measured by the region, is to be broken down in the proportions calculated from the relative intensities of synthetic peaks. In the example of nylon seen in Figure 6•15, there are three peaks used in the elemental concentration. When the regions defining the O 1s, C 1s and N 1s intensities are created, the Tag field must be entered with a value other than NoTag and must also be distinct from one another. The default Tag field, when a region is created using information from the element library, is the same string found in the Name field for the region. Figure 6•16 shows the table from the Regions property page where the Tag fields are entered using the element/transitions for the regions in question, although any other strings would have been equally as good provided the three strings are all different. However, whatever the Tag string used for the C 1s region must also be used for the Tag fields entered in the C 1s components seen in the insert tile in Figure 6•15. Since the peak model for the C 1s

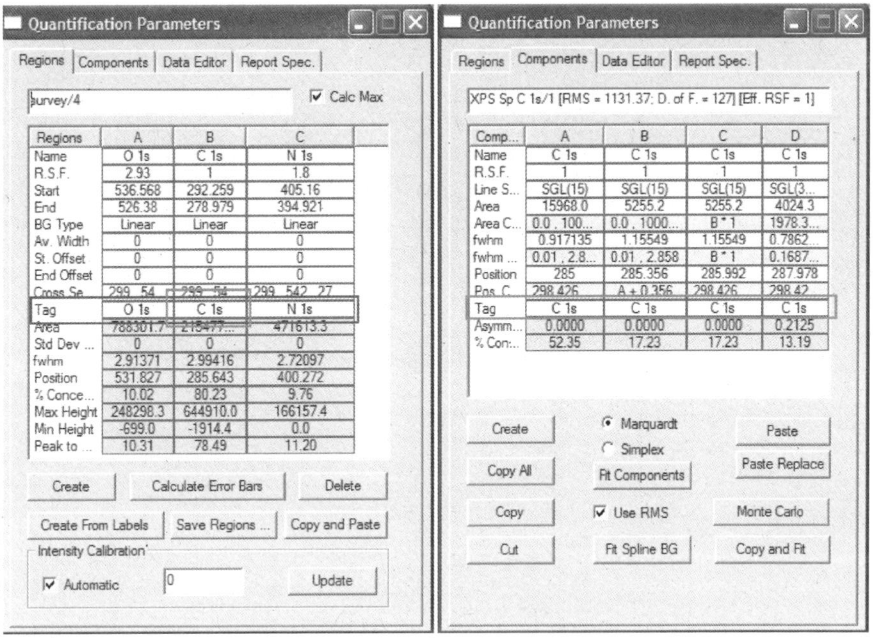

FIGURE 6•16 Tables from the Regions and Components Property Page on the Quantification Parameters dialog window.

region also requires the creation of a region, it is as equally important to ensure the Tag field for the C 1s region defined on the narrow scan spectrum is assigned the string NoTag. The significance of the string NoTag is that only quantification items for which the Tag field is different from NoTag are included in a Tag based quantification report. Since the area of the peak determined from the region on the narrow scan C 1s data is of no use when compared to the O 1s and N 1s areas measured from the survey spectrum (owing to the different instrumental operating modes used to acquire the data), it is important the region from the C 1s narrow scan spectrum is excluded from the quantification calculation by this means. A simple way to set a Tag field to NoTag is to delete all characters from the text-field and press return. The Tag field will automatically load the NoTag string.

The quantification report displayed over the survey spectrum in Figure 6•15 on page 118 is a form of annotation. A table of information added to the display using the Quantification property page on the Annotation dialog window (Figure 6•17) is both very useful, but also very dangerous if misused. The information in the table does not come directly from the ISO block maintaining the data display in the Active Tile, but rather is dependent on the blocks selected in the right-hand-side of the Experiment Frame. The danger is clearly that the table shown on the display may derive

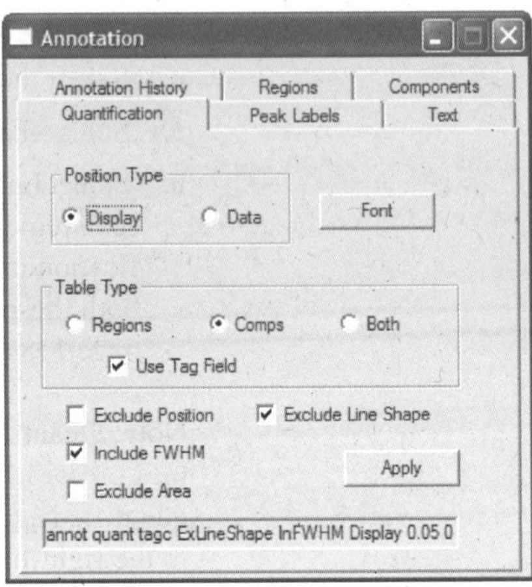

FIGURE 6•17 Annotation dialog window displaying the Quantification property page.

from a completely different spectrum in the same file; however the advantage of these quantification tables is that related spectra can be combined to produce information as useful as seen in Figure 6•15. Namely, the survey spectrum provides the elemental quantification, while the C 1s narrow scan data block provides the relative intensities of the four carbon peaks. The mechanism by which the quantification table is created for the annotation is in fact identical to the procedure used to generate general quantification tables using the Standard Report on the Report Spec property page from the Quantification Parameters dialog window. If the desire is to annotate a spectrum using information only derived from the spectrum displayed in a tile, then either the Regions or Component property pages on the Annotation dialog window are appropriate. One consequence of using the quantification table annotation is also seen in Figure 6•15, where the peak model for the C 1s envelope must also conform to the chemistry of the sample, namely the intensity of the O 1s peak must relate to the highest binding energy C 1s synthetic peak. It is therefore possible to adjust the C 1s peak model whilst monitoring the affect of the relative intensities on the O 1s and C 1s peaks. Further, Figure 6•15 illustrates a use for the insert tile, where the parent tile displays the survey spectrum together with the quantification table, whilst the insert tile displays the C 1s narrow scan region, thus the components for the C 1s spectrum can be changed whilst still viewing the quantification table. A display such as the one shown in Figure 6•15 is created as follows:

1. Select the survey spectrum block in the right-hand-side of the Experiment Frame.
2. Press the Display tool bar button.

3. Draw a zoom box in a vacant area of the tile sufficiently large to display the C 1s spectrum and press the Insert keyboard key.
4. Select the data block containing the C 1s spectrum.
5. Select the newly created insert tile and press the Overlay tool bar button.
6. Select both the data blocks containing the survey spectrum and the C 1s narrow scan region in the right-hand-side of the Experiment Frame. Given that a quantification table annotation is defined on the survey spectrum, the act of selecting both these blocks will result in the display of the table shown in Figure 6•15 on page 118.

Note: Quantification annotation behaves in the same way as the Standard Report tables created from the Report Spec property page. That is, data blocks can only be quantified against each other provided they appear in the same row of blocks as seen in the right-hand-side of the Experiment Frame.

A table of quantification results identical to the quantification annotation table in Figure 6•15 is generated using the Standard Report button (see Figure 6•13 on page 116) on the Report Spec property page of the Quantification Parameters dialog window. Again, the C 1s and survey data blocks must be selected in the right-hand-side of the Experiment Frame prior to pressing the button. Once the quantification table window is displayed in CasaXPS, the values from the table are saved to file via the File menu or alternatively placed on the clipboard via the Copy toolbar button.

A Multi-layer Depth Profile Example

The data set shown in Figure 6•18 is a profile measured using a PHI Quantum 2000 where the sample was a multi-layer material.
Not all depth profiles have so many interfaces, however even the presence of one boundary within the profile may require specific adjustments to both quantification regions and synthetic components to permit an accurate description of the chemistry as the environment alters across an interface. The ability to create quantification items on a spectrum-by-spectrum basis allows variations due to matrix effects to be easily accommodated within CasaXPS and results in profiles representative of the material rather than the limitations imposed by requiring one model to fit all. The key to analysing depth profiles is to realise that:

1. Each spectrum requires the definition of quantification regions and components.
2. The browser on the right-hand-side of the Experiment Frame permits the propagation of a specific model to only those blocks currently selected.

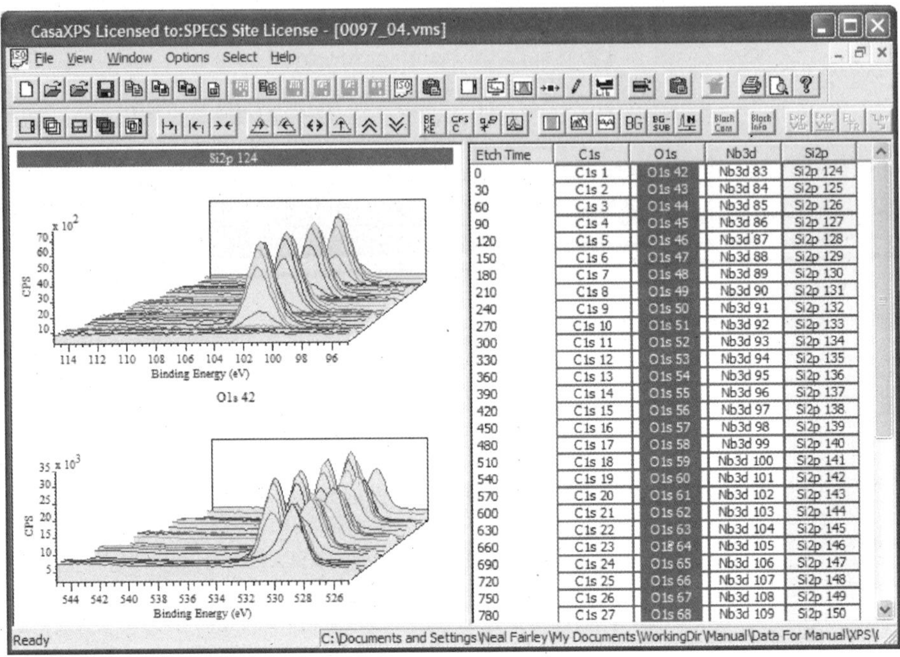

FIGURE 6•18 Multi-layer Depth Profile Data Set (PHI Quantum 2000)

If a sequence of spectra can all be modelled with the same set of region and components, the task is therefore to define the required region and components on one representative spectrum, then using the browser, select the remaining spectra for which the model is appropriate and propagate the model from the one spectrum to all those so indicated.

The importance of maintaining the quantification items as individual parameters for each spectrum is illustrated in Figure 6•19 on page 123, where the niobium spectra are displayed together with the quantification regions corresponding to the initial surface before ion-gun action and following the first ion-gun etch cycle. The action of the ion-gun transforms a well defined doublet pair into an envelope derived from a number of such doublet pairs. It is clearly preferable to select spectrum specific start and end energies for the integration regions; the consequence of only using one set of region parameters applied to all the Nb 3d spectra is a more noisy profile trace in the final result. By way of example, consider the Si 2p spectra in the data set shown in Figure 6•18. The silicon signal is effectively reduced to noise when the etch cycle is within the niobium layer, however at the interface between the silicon and niobium oxides a small Si 2p peak appears to the right of the silicon oxide peak.

In order to minimise the influence of noise and accommodate the intermittent appearance of the minor peak in the silicon profile, it is necessary to adjust the region limits on a spectrum by spectrum basis.

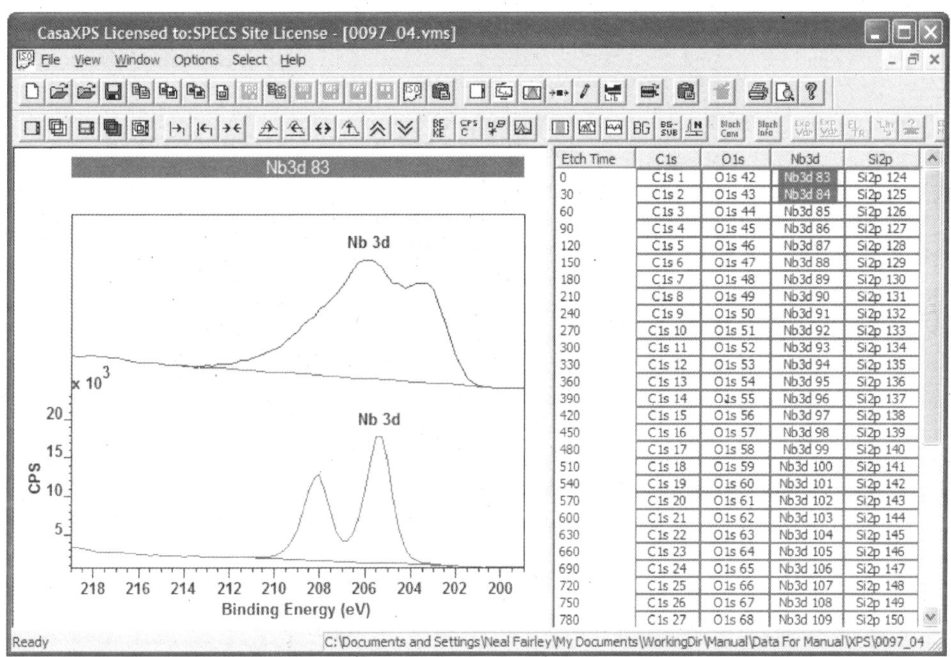

FIGURE 6•19 Niobium spectra illustrating the variation in the data envelope before and after the first ion-gun etch cycle.

One method of creating the required quantification regions throughout the profile is as follows:

1. Select all the Si 2p spectra in the right-hand-side of the Experiment Frame. If the heading button for the column of Si 2p block in the right-side browser is pressed once, every block below the header button will become selected.
2. Press the Overlay tool bar button. The entire set of Si 2p spectra will appear overlaid in the Active Tile.
3. Using the Regions property page on the Quantification Parameters dialog window, create a quantification region by pressing the Create button. The element/transition fields for these data are correctly assigned and agree with the desired entries in the current element library; therefore the RSF is retrieved from the library when the region is created.
   Note that the region is only created on the first spectrum in the Active Tile,

namely, the data block selected first when the browser selection was made prior to pressing the Overlay tool bar button.

4. Adjust the quantification region parameters, for example the Av Width field might be set to "1" and the start and end parameters adjusted under mouse control such that the region looks plausible for all the spectra display in the Active Tile.
5. Propagate the region to all the Si 2p spectra: ensure the Si 2p column of spectrum blocks are all selected in the Experiment Browser, right-click the mouse over the tile displaying the Si 2p spectrum for which the region was defined, ensure the Region tick-box in the propagate section of the Browser Actions dialog window is ticked and none others, confirm the list of destination data blocks agrees with the intended Si 2p list and press the OK button. Since the Si 2p spectra are overlaid in the Active Tile, the newly created quantification regions together with the resulting backgrounds will appear in the display showing all the silicon spectra. Note, however, the fact that the spectra are displayed in the Active Tile is a convenient way of visualising the result of the propagation and it is the Browser selection which dictates the destination for the propagation action, not the set of spectra in the Active Tile.
6. Use the arrow keyboard keys to step through the set of Si 2p spectra making adjustments to the region parameters as required. If the Quantification Parameters dialog window is open and the Region property page is the front-most page, then the mouse can be used to adjust region limits.

The depth profile in Figure 6•18 is a relatively small profile and the interfaces are close together, however for larger sequences of etch-cycles where sub-ranges of spectra are accommodated by the same set of region parameters, the propagation mechanism can be used to selectively transfer a region to a subset of spectra. That is, the propagation operation is performed on only those blocks selected in the right-hand-side of the Experiment Frame and there is no requirement to select the entire column of Si 2p spectra as suggested above. Similarly, regions and synthetic components can be propagated using the identical mechanism when the use of subsets within a profile sequence becomes more important; optimised peaks require more attention than regions typically do. Small adjustments to peak constraints can often make a big difference to the quality of the final depth profile, particular when these small alterations to the peak model are applied to the appropriate part of the data set. Now, continuing with the remaining element/transitions in the profile, regions and/or components must be defined for each of the four columns of blocks seen in Figure 6•18. Only the O 1s spectra are quantified using synthetic components, where two O 1s Gaussian-Lorentzian line-shapes are used to fit the data. These two peaks are labelled O 1s 1 and O 1s 2, principally to emphasize the importance, when using the Custom Report to generate the quantification table, of distinct names for quantification items. If two quantification items (regions or components) are assigned the same name, then the Custom Report (Figure 6•20) will add the intensities from these identically named quantification items together before calculating the concentration table. If, for example, the region for one of these O 1s spectra had been named O 1s and the same name O 1s given to both the synthetic peaks, the result from the Custom Report would be in error by a factor of two. That is, the intensity from the region and the two synthetic peaks would be summed together and used to compute the oxygen concentration. It is therefore important to assign distinct names to quantification items unless it is desired to sum the intensities involved. There are

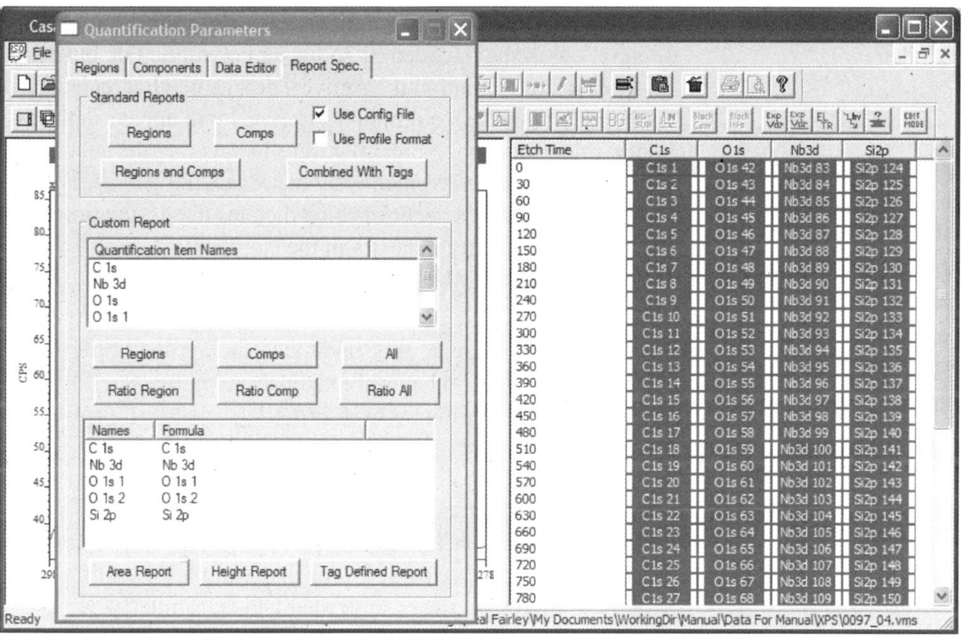

FIGURE 6•20 Custom Report section prepared for generating a profile from the data selected in the right-hand-side of the Experiment Frame.

many situations where two or more regions might be appropriately assigned the same name with the view to improving the statistic in a measurement. For example, Figure 6•8 on page 105 shows the Au 4p doublet pair widely separated such that a single region spanning the energy interval containing both peaks would struggle to define a plausible background. On the other hand, two regions (as seen in Figure 6•8), easily define the background for the individual peaks and by assigning the same name to these regions, the intensity from both would be summed by the Custom Report.

A key point to note is that the Custom Report always works with all the quantification items defined on the selected data blocks, regardless of whether the Region or

Component button is pressed to initialise the Name/Formula table seen in Figure 6•20 on page 125. It is the Name/Formula table that defines the report and the buttons between the Quantification Items table and the Name/Formula table are only a quick way of initialising the Name/Formula table. Unwanted entries in the Name/Formula table can be removed by right-click over the name field of the Name/Formula list; a dialog window appears offering the option to modify, insert or delete a field in the Name/Formula table. The Name/Formula table shown in Figure 6•20 was prepared by first selecting the entire data set, then pressing the button labelled All between the two scrolled lists on the Custom Report section. Since the O 1s spectra have both a region named O 1s and two components named O 1s 1 and O 1s 2, it is necessary to exclude the region from the calculation. A consequence of calling these quantification items by different names is that each has a distinct entry in the Name/Formula table and therefore the O 1s entry corresponding to the region can be deleted. The strings used in the Name column of the table is at the user's discretion, however the Formula entries must be constructed using exactly the strings used to name the quantification items.
Thus the formula to add the intensity from the two components should be entered as follows: O 1s 1 + O 1s 2, where both the case and the spaces in the names O 1s 1 and O 1s 2 are important. Since formulae in the Name/Formula table can be any arithmetic expression using constants such as 1 and 2, and the quantification item names also include these digits, allowing such terms as 2 * O 1s 2 + 1 within a formula forces the use of exact strings for quantification item names.

The goal of processing a depth profile is to monitor the trend of each element as a function of depth. The data set seen in Figure 6•20 lists the data blocks in rows corresponding to etch-time; however the quantification table displayed in Figure 6•21 on page 127 lists the intensity and concentration determined from the blocks as a function of depth. The conversion from etch time to depth, in this example, is achieved by specifying the depth of the crater corresponding to the total etch-time.

The toolbar button, when pressed, offers a dialog window in which the interval for the crater depth dimension can be entered corresponding to the sequence of etch-times, and selecting the Depth radio button prior to pressing the OK button results in the new abscissa values and labels. The change of experimental variable from etch-time to depth is made under the assumption of a constant sputter-rate and a uniform etch-time during the course of the experiment. If the etch-time is none uniform, the Depth (Interpolate) radio button should be used rather than the Depth option. If more complex relationships exist between the etch-time and depth, then the quantification table must be exported to a spreadsheet for further processing.

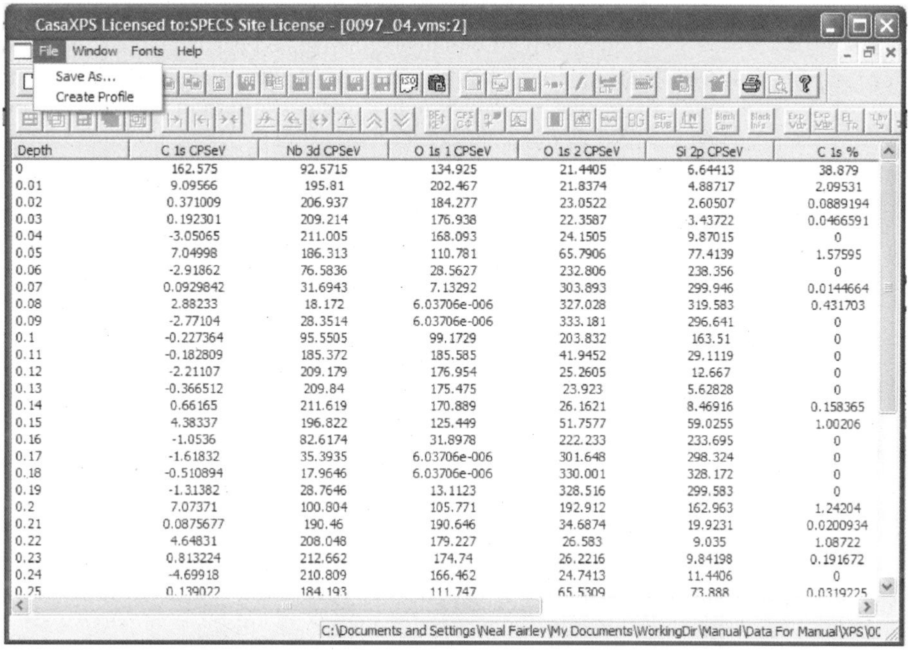

CasaXPS Licensed to:SPECS Site License - [0097_04.vms:2]

File Window Fonts Help

Save As...
Create Profile

Depth	C 1s CPSeV	Nb 3d CPSeV	O 1s 1 CPSeV	O 1s 2 CPSeV	Si 2p CPSeV	C 1s %
0	162.575	92.5715	134.925	21.4405	6.64413	38.879
0.01	9.09566	195.81	202.467	21.8374	4.88717	2.09531
0.02	0.371009	206.937	184.277	23.0522	2.60507	0.0889194
0.03	0.192301	209.214	176.938	22.3587	3.43722	0.0466591
0.04	-3.05065	211.005	168.093	24.1505	9.87015	0
0.05	7.04998	186.313	110.781	65.7906	77.4139	1.57595
0.06	-2.91862	76.5836	28.5627	232.806	238.356	0
0.07	0.0929842	31.6943	7.13292	303.893	299.946	0.0144664
0.08	2.88233	18.172	6.03706e-006	327.028	319.583	0.431703
0.09	-2.77104	28.3514	6.03706e-006	333.181	296.641	0
0.1	-0.227364	95.5505	99.1729	203.832	163.51	0
0.11	-0.182809	185.372	185.585	41.9452	29.1119	0
0.12	-2.21107	209.179	176.954	25.2605	12.667	0
0.13	-0.366512	209.84	175.475	23.923	5.62828	0
0.14	0.66165	211.619	170.889	26.1621	8.46916	0.158365
0.15	4.38337	196.822	125.449	51.7577	59.0255	1.00206
0.16	-1.0536	82.6174	31.8978	222.233	233.695	0
0.17	-1.61832	35.3935	6.03706e-006	301.648	298.324	0
0.18	-0.510894	17.9646	6.03706e-006	330.001	328.172	0
0.19	-1.31382	28.7646	13.1123	328.516	299.583	0
0.2	7.07371	100.804	105.771	192.912	162.963	1.24204
0.21	0.0875677	190.46	190.646	34.6874	19.9231	0.0200934
0.22	4.64831	208.048	179.227	26.583	9.035	1.08722
0.23	0.813224	212.662	174.74	26.2216	9.84198	0.191672
0.24	-4.69918	210.809	166.462	24.7413	11.4406	0
0.25	0.139022	184.193	111.747	65.5309	73.888	0.0319225

C:\Documents and Settings\Neal Fairley\My Documents\WorkingDir\Manual\Data For Manual\XPS\0C

FIGURE 6•21 Custom Report Table.

The quantification report shown in Figure 6•21 is generated by preparing the Name/ Formula table as seen in Figure 6•20 and then pressing the Area Report button within the Custom Report section. The resulting quantification report only includes values determined from the intensity parameters computed using either the integrated quantification region area or the area parameter from synthetic components. The raw intensities are adjusted for relative sensitivity of the transitions using the RSF entered into the quantification items, transmission correction, if available and selected via the Regions property page, and escape-depth correction using the value also entered on the Regions property page. The first column in Figure 6•21 is the experimental variable and is followed by the adjusted integrated intensities for each entry in the Name/Formula list, which in turn is followed by the percentage atomic concentration computed from the intensity entries.

The Custom Report is designed for depth-profiles or any other trend experiments. The File Menu, also displayed in Figure 6•21, offers a choice of saving the table to an ASCII file or creating an ISO 14976 file containing one data block for each column in the quantification table. Once selected, the Create Profile menu option results in a new Experiment Frame appearing within the CasaXPS main window. The trends within the profile can be overlaid and annotated with a key as seen in Figure 6•22 using the Tile Display dialog window.

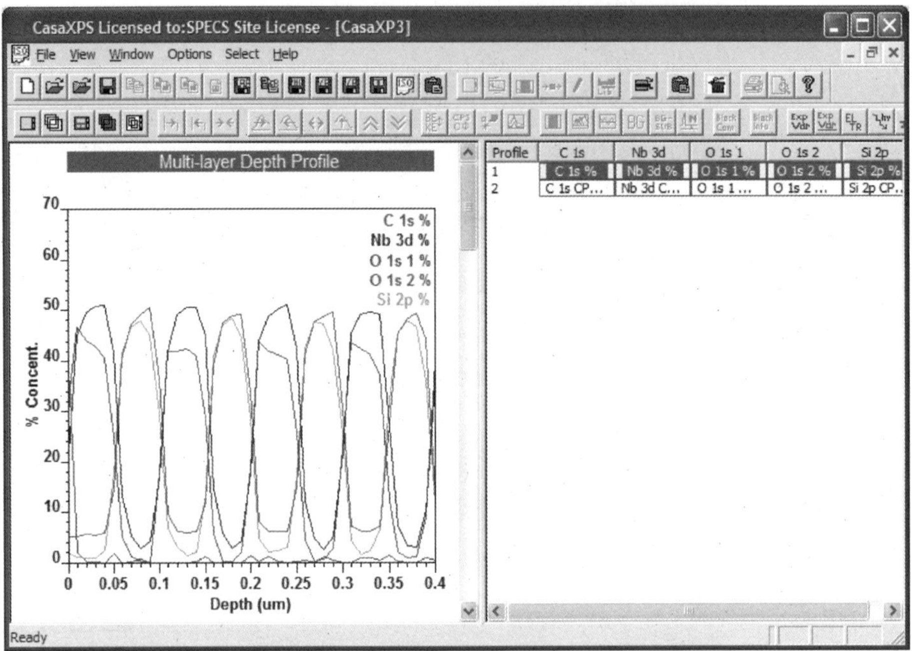

FIGURE 6•22 Depth Profile created from the table shown in Figure 6•21.

Apart from offering many configurable display options, in particular the Display property page on the Tile Display dialog window allows a key to be added to the display. A further option on the File Menu available for spectrum and profile display allows a prepared tile format to be saved to a separate file from the ISO file.
These ASCII tile format files default to a file extension of .tff and can be retrieved at a later time to return the display of a spectrum file to an earlier state. Saving and reloading a tile format file is performed using two options on the File Menu.

Quantification for Depth Profiles including Curve Fitting

For some analysts, depth profiles are the most common use of XPS while for others such experiments are rarely performed. Whether frequently used or otherwise, the size of these data sets requires an efficient means for processing the numerous spectra they contain. Figure 6•23 shows a typical data set from an XPS depth profile. Three hundred and eighty seven spectra were recorded in the course of this experiment and, to measure the chemical changes the spectra require analysis using both peak-fitting as well as simple integration regions.

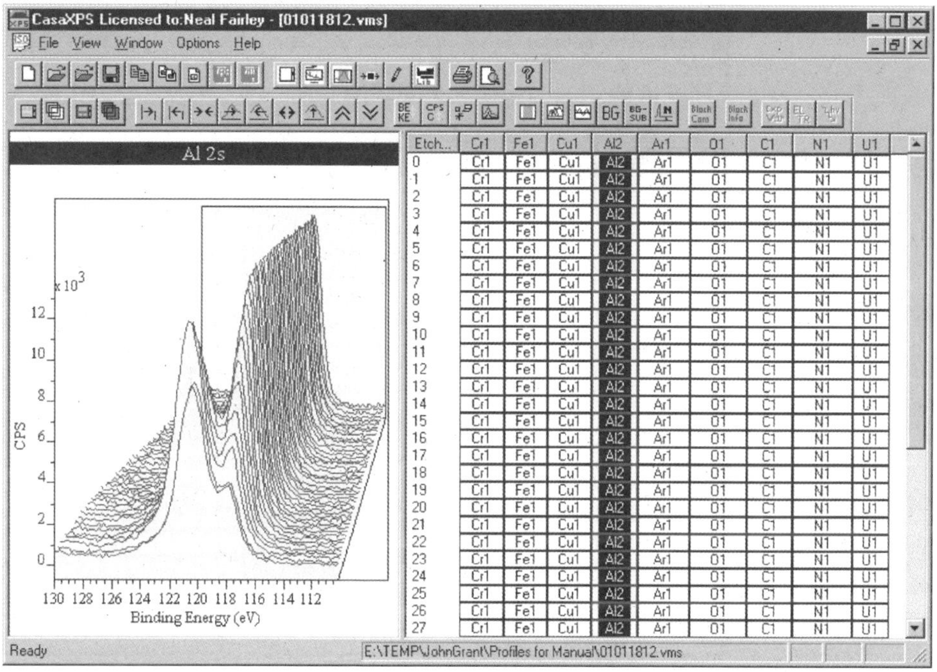

FIGURE 6•23 Depth profile acquired on a PHI 5700 using a monochromatic X-ray source and an argon ion-beam to etch the sample.

Analyzing the Profile Using CasaXPS

Eight narrow energy intervals were monitored during the course of the depth profile, some of which merely require a background and an integration region to measure the intensity for that species. Others are more involved. Aluminium metal and oxide are present in this particular material and the presence of copper also complicates the analysis since the Cu 3s peak lies in close proximity to the Al 2s lines and similarly Cu 3p interferes with the Al 2p doublet; hence the need for a synthetic peak model. The first step in generating a depth profile from the sequence of spectra is to identify the chemical states of interest. Consider the Al 2s region. To extract the information about the metal and oxide components it is necessary to define a synthetic model for the peaks located in the binding energy region [128,112] eV. Possible peaks found in this interval are Al 2s Oxide, Al 2s Metal, and Cu 3s (Figure 6•24). The spectra displayed in Figure 6•24 represent a sample of Al 2s regions taken form different depths in the material. Viewing the data in this way provides a means of assessing the variations in the peak proportions as the material is etched and allows a peak model to be built up which is suitable for the entire Al 2s profile.

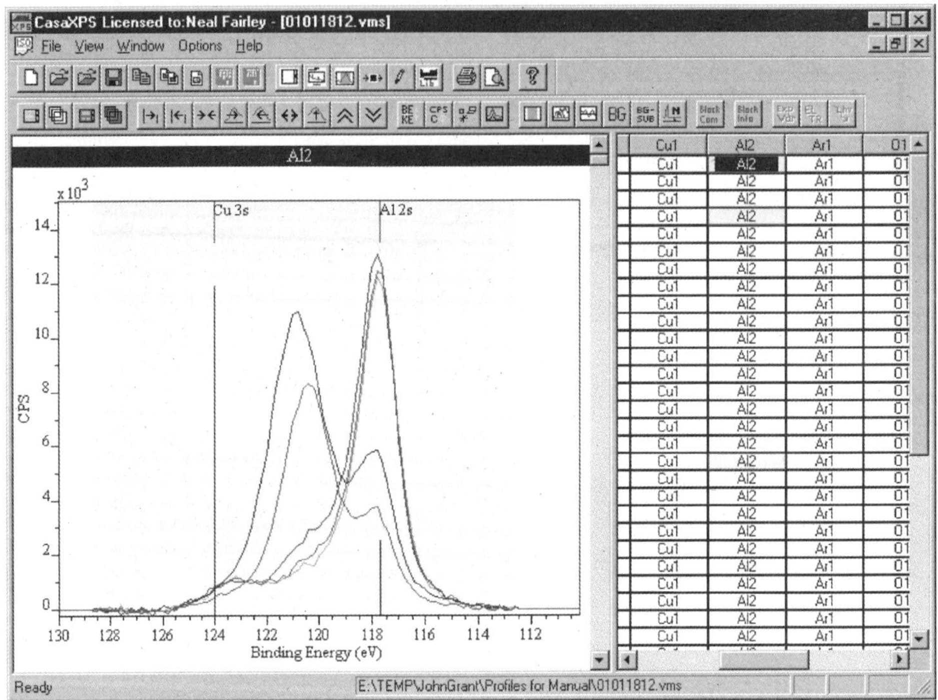

FIGURE 6•24 A sample of Al 2s spectra taken from various depths within the profile.

Display the spectra by first selecting the data blocks using the Experiment Browser and pressing the Overlay toolbar button .

Useful mechanisms for making a selection in the right-hand-side of the Experiment Frame are: Double-click the first block with the left-hand mouse button (this will both select the block and display the spectrum in the left-hand-side of the Experiment frame), then holding the Ctrl Key down left-click further data blocks (the control key adds to the current selection). Alternatively, left-click the first block in a range, then holding the Shift Key down left-click the last block in the range. The result of using the Shift Key is to select all the data blocks between the two mouse clicks.

A Shirley background and synthetic peaks are constructed for an Al 2s spectrum using the options on the Quantification Parameters dialog window (Figure 6•25). The tools for creating peak models and processing the profile all appear as property pages on the Quantification Parameters dialog.

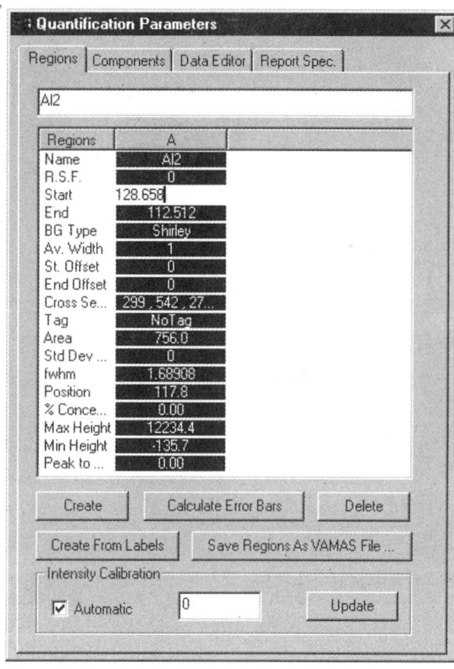

FIGURE 6•25 Regions property page for Al 2s

The first property page is labeled Regions and allows a background to be defined for the spectrum displayed in the active tile of the active Experiment Frame (now referred to as the active spectrum). To create the initial region it is sufficient to press the Create button on the Regions property page (Figure 6•25) and adjust the Start and End fields so that the background ties to the spectrum at appropriate points. It is often necessary to average the intensities either side of the tie points in order to reduce the influence of noise on the calculated background. The field labeled Av. Width specifies the number of data channels used to determine the point at which the background should tie to the data. Figure 6•25 shows the Av. Width set to "1" and results in 2 * Av. Width + 1 data channels determining the tie points, i.e. three data channels, the original center plus one point either side of the center.

The start and end points for an integration region may be entered using the edit text-field. Left click on the item for which a modification is required. This will enable the edit text-field, then enter a new value and press Return. If the Return Key is not pressed the edited value is left unaltered.

A background to an integration region is entered in the field labeled BG Type. Possible backgrounds are: Linear (l), Shirley (s), Tougaard (t), W Tougaard (w), Default (d), Adjust (a), Min (mi), Max (ma), Mean (me), Zero (z) and others relating to the three parameter Tougaard cross sections described elsewhere.

Note that the Relative Sensitivity Factor (RSF) for the Al 2s region shown in Figure 6•25 is set to zero. A zero RSF value will exclude the integrated intensity from any quantification calculation. For this example, the synthetic peaks will determine the intensities for the Al 2s profile and therefore RSF values for these items must be set appropriately.

Synthetic components (peaks) are constructed via the Components property page (Figure 6•26). Each time the Create button is pressed a new component is added to the active spectrum.

NB Species/transitions labels are important in CasaXPS. If the species/transition ISO fields are entered with the same text strings that are also entered in the Element Library, then when a region or component is created the predefined value for the RSF associated with the species/transition string are entered into the region or component. If the species/transition fields do not match then the energy-ordered scrolled list on the Element Library dialog window can be used to force the Element Library entries to be entered into the newly created regions/components. The selected line in the energy-ordered scrolled-list will be used when a region or component is created.

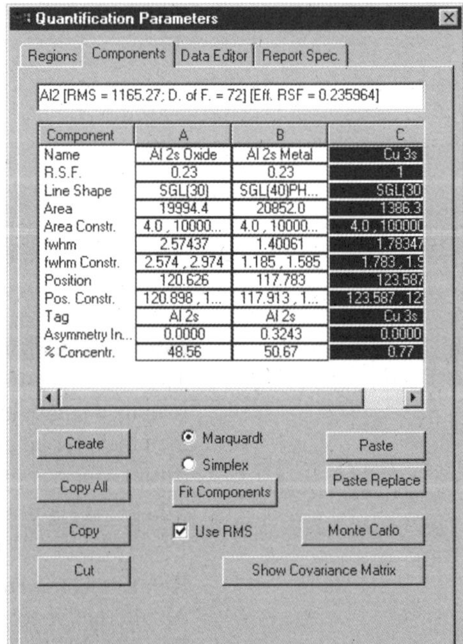

FIGURE 6•26 Components Property Page.

Figure 6•27 shows a set of fitted peaks for one Al 2s region within the depth profile. These peaks have been created using the Components property page, where the initial settings have been adjusted so as to be appropriate for the current problem.

Asymmetric Line Shape

An asymmetric line-shape is used for the metallic Al 2s peak. The string "SGL(40)PHI(1,10)" specifies the type of function used to introduce asymmetry into the line-shape. The first string "SGL(40)" defines the shape of the symmetric portion as a sum approximation to the Voigt function, where the Voigt approximation is constructed from 0.4 * Lorentzian + 0.6 * Gaussian. An asymmetric tail is defined in terms of an exponential function (Briggs and Seah) where the degree of asymmetry is specified using scale and length parameters. The string appended to the "SGL(40)"string , namely, "PHI(scale, length)" is defined by the Physical Electronics MultiPak software and is identical to the Briggs and Seah tail function when the scale parameter is equal to 1.

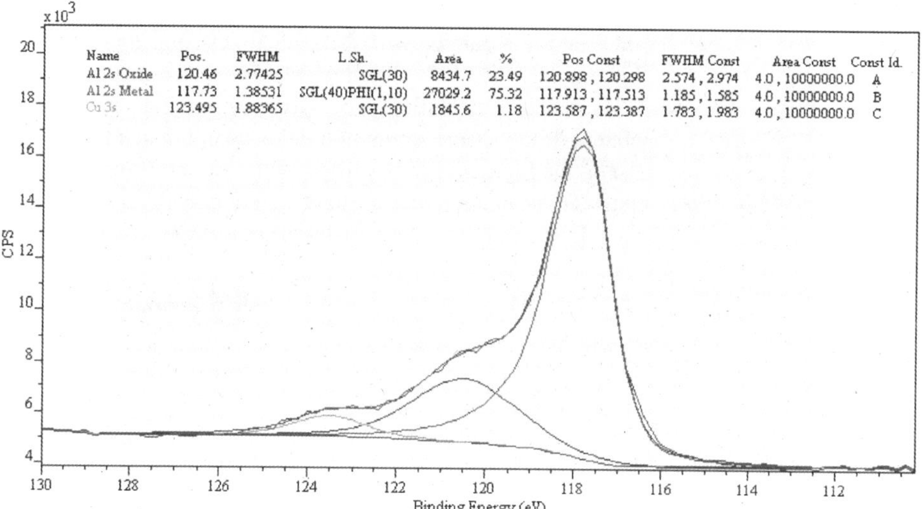

FIGURE 6•27 Peak parameters showing interval constraints.

Interval Constraints

The constraints in Figure 6•27 are simply interval constraints. These can be easily managed using the "#" constraint shortcut. If the constraint is entered as a "#" sign, then the interval constraint is set to be identical to the current value for the parameter (thus fixing the parameter). If, however, the constraint field is entered using "#0.2", say, the constraint interval is set to the current value for the parameter plus and minus the value entered after the "#" sign.

Relative Constraints

An alternative to constraint intervals is constraints with respect to other components. Figure 6•28 shows the same peaks but where the positions of the peaks in columns A and C are constrained to the position of the peak in column B by an offset and the FWHM for the peak in column A is constrained to the FWHM of the peak in column

B using a factor constraint. The column characters entered under the heading “Const Id” refer to the labels used to head up the parameter columns shown in Figure 6•26.

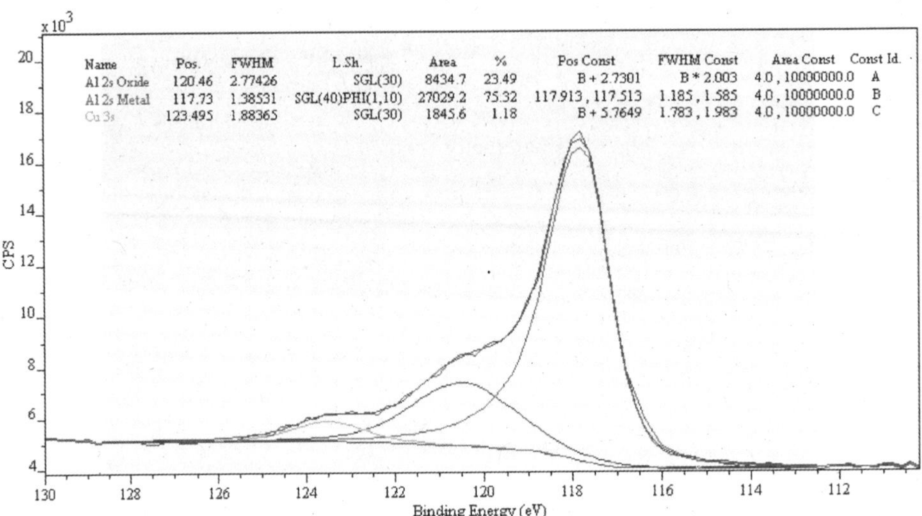

Name	Pos.	FWHM	L.Sh.	Area	%	Pos Const	FWHM Const	Area Const	Const Id.
Al 2s Oxide	120.46	2.77426	SGL(30)	8434.7	23.49	B + 2.7301	B * 2.003	4.0 , 10000000.0	A
Al 2s Metal	117.73	1.38531	SGL(40)PHI(1,10)	27029.2	75.32	117.913 , 117.513	1.185 , 1.585	4.0 , 10000000.0	B
Cu 3s	123.495	1.88365	SGL(30)	1845.6	1.18	B + 5.7649	1.783 , 1.983	4.0 , 10000000.0	C

FIGURE 6•28 Peak Parameters showing relational (relative) constraints.

Viewing Tables in CasaXPS

The component information shown in the tables within Figure 6•27 and Figure 6•28 are forms of annotation. These can be added to a spectrum display using the Component property page on the Annotation dialog window. Similarly, regions and tables based on multiple spectra quantification can also be added to the display using the Region and Quantification property pages of the Annotation dialog window. Once a suitable peak model has been prepared for the active spectrum, the next task is to transfer (or propagate) the quantification items to other Al 2s regions. This is achieved using the Browser Operations dialog window.

- Display the spectrum for which the peak model has been constructed.
- In the right-hand-side of the Experiment Frame (see Figure 6•23 on page 129), select the set of spectra for which the same peak model is appropriate. In this case all the Al 2s regions were selected although subsets may be selected if appropriate.
- Move the cursor over the spectrum displaying the peak model in the left-hand-side of the Experiment Frame and right-click the mouse. TheBrowser Operations dialog window appears (Figure 6•29).

- Select the Propagate operations required. That is, tick the Regions and Components check boxes and confirm that the list below includes only those spectra for which a peak model is required.
- Press the OK button and wait while the operations are copied to each of the spectra listed on the Browser Operations window. A progress meter offers a Stop button, which when pressed terminates the copy operation on completion of the current action.

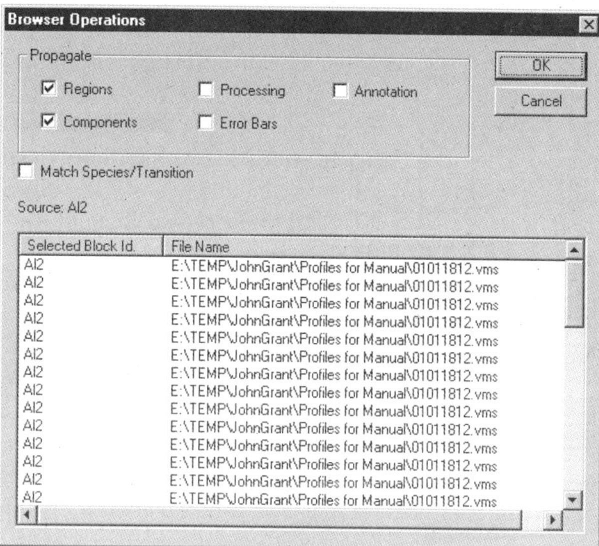

FIGURE 6•29 Right-click over the displayed spectrum to invoke the Browser Operation dialog.

Intensities for the regions other than the Al 2s region can be estimated using integration of the background subtracted data. Figure 6•28 on page 134 shows a representative sample from one layer of the profile, where integrations regions have been defined for all the spectral regions used in the analysis. Note that the Al 2 spectrum is the only one with synthetic peaks fitted. Once regions have been prepared for individual spectra, the propagate method described for the Al 2s synthetic peaks can be repeated for each of the regions in the profile.

- Create a region on one of the spectra.
- Select the set of spectra for which similar regions are required in the right-hand-side of the Experiment Frame.
- Right-click the mouse over the spectrum displaying the region in the left-hand-side of the Experiment Frame.
- Select the propagate option on the Browser Operations dialog window (Figure 6•29).
- Check that the appropriate spectra are listed on the Browser Operations dialog window and press OK if correct. NB if the Match Species/Transition check box is ticked, then only those selected regions with the same Species/ Transition ISO fields, as the source spectrum will receive the propagation operations.

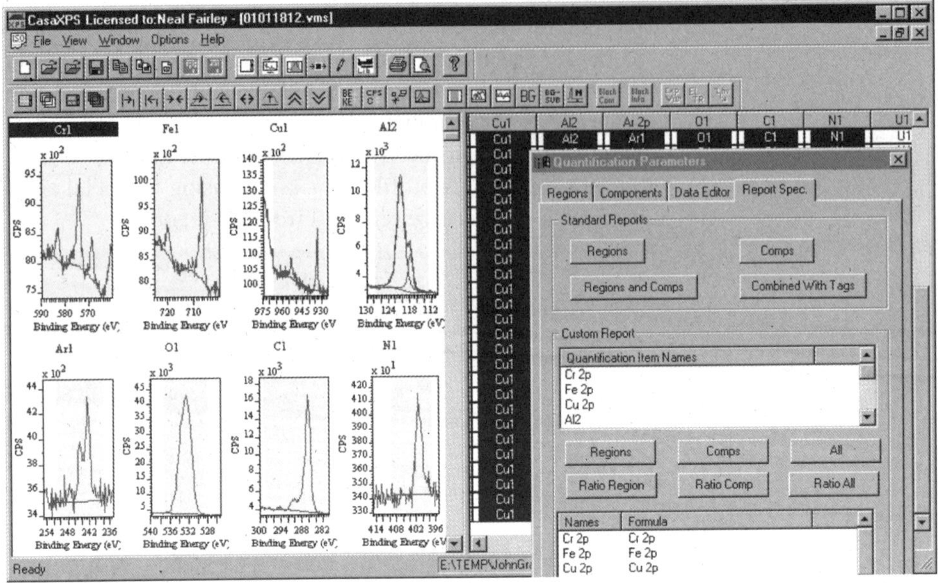

FIGURE 6•30 CasaXPS Program Frame showing the selected data blocks on the right-hand-side of the Experiment Frame and a representative set of spectra showing the regions and components used to quantify the profile. The Quantification Parameters Dialog window is offering the Report Spec property page.

The names associated with the regions and components are important when generating profiles using the Custom Report. The names form the basis for defining formulae in the Custom report. All intensities labeled using the same name are treated as contributing to the same species and are therefore summed together when a formula in a Custom Report uses that name. Figure 6•31 shows the state of the Custom Report tables when the Area Report button was pressed. The table behind the Quantification Dialog window shows the result of pressing the button. Each column within the report corresponds to the entries in the Name/Formula table shown in the Custom Report section.

Initializing the Custom Report

The Custom Report is initialized using the buttons between the two tables headed Quantification Item Names and Name/Formula. The Quantification Item Names list shows the list of region and component names defined for the spectra selected in the right-hand-side of the Experiment Frame (Figure 6•30). The top three buttons initialize the Name/Formula table using the items shown in the table above (the bottom three buttons initialize the Name/Formula using ratio formulae). If the Regions button is pressed, the Names/Formula are entered using the items in the above list corresponding to regions. If however, as is the case in this example, the formulae need to be defined using both regions and components, the button labeled All, when pressed, will fill the Name/Formula table using both region and component labels. After press a button there may be more entries in the Name/Formula table than is required for the profile. The current example includes an entry for the Al 2 regions, and since the Al 2s profile is measured using the metal and oxide components the regions item should be removed from the report.

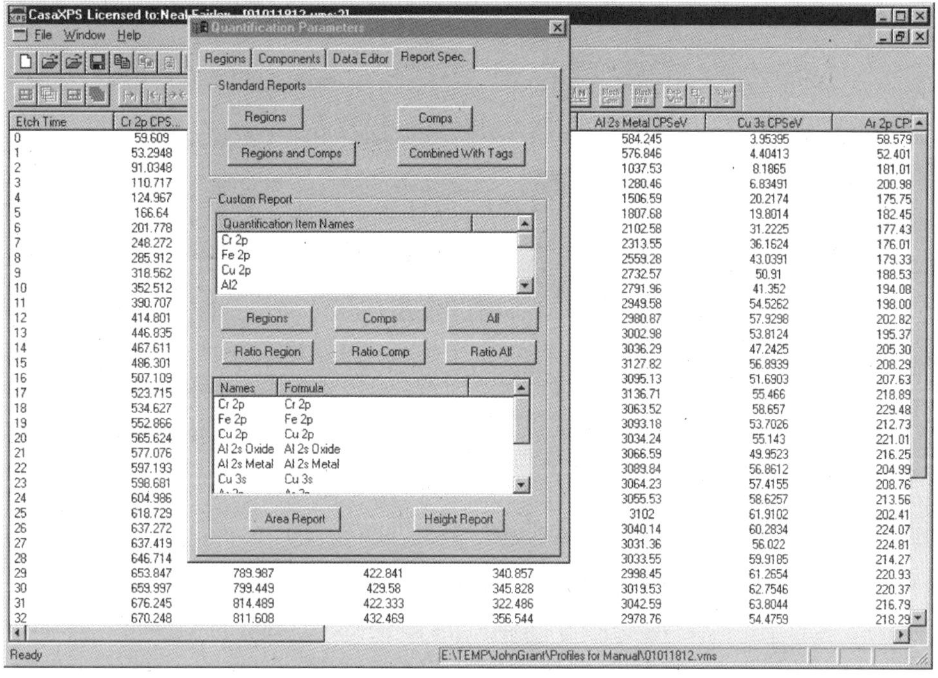

FIGURE 6•31 Report Spec property page together with the Custom Report generated from the formulae listed on the Report Spec.

To edit the Name/Formula list:

- Move the cursor over the Name field you wish to alter.
- Right-click the mouse over the Name field to invoke an edit dialog window (Figure 6•32). The Name and Formula fields are entered into the corresponding text-fields.
- To delete the chosen entry, press the Delete pushbutton. Similarly a new entry in the table is created from this dialog window by pressing the Insert pushbutton. If the OK button is pressed, any changes made to the text-fields will become the current values for the original table entry.

The report shown in Figure 6•31 is created when the button labeled Area Report is pressed. The result of pressing the Area Report button is to calculate the atomic concentration using intensities measured from area in counts-per-second x eV. If Auger data were used to profile a material then peak-to-peak measurement may be more appropriate and in this case the button labeled Height Report should be used instead. The sensitivity factors determine what type of intensity measurement should be used. In this example XPS sensitivity factors suitable for area measurements were used, hence the use of the area report.

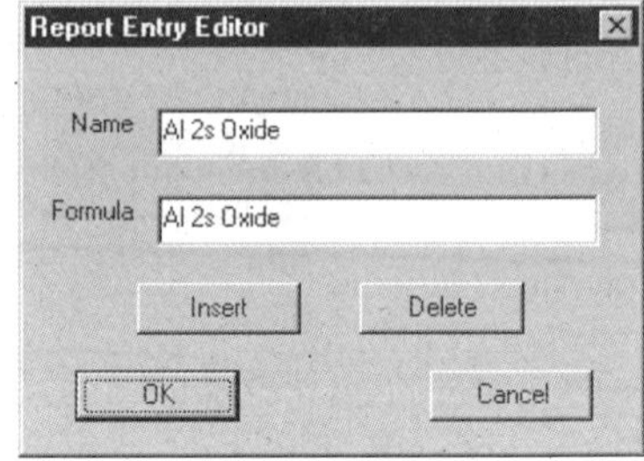

FIGURE 6•32 Dialog invoked by right-clicking the mouse over the name field in the Custom Report section.

Once a report has been generated (Figure 6•31), the resulting table may be saved to disk or taken through the clipboard to other graph plotting programs such as Excel or Sigma Plot. Figure 6•33 displays the profile tabulated in Figure 6•31 after formatting within Excel. The File menu offers a "Save As ..." option. This allows results to be written as a TAB spaced ASCII file or alternatively if the file is saved with a .vms file extension a VAMAS file is written to disk.

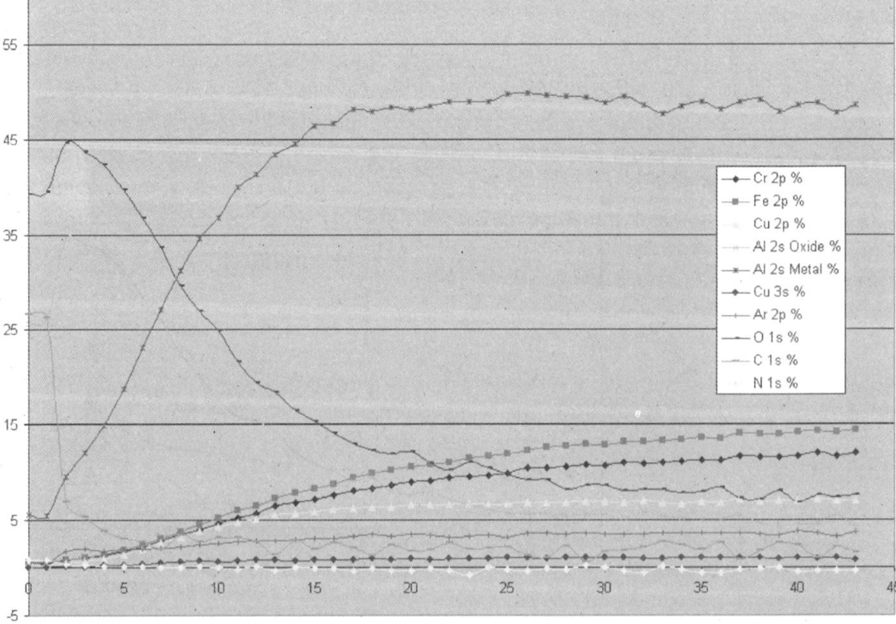

FIGURE 6•33 Depth Profile generated by CasaXPS and plotted using Excelä

(see also Converting the PHI MultiPak Data to ISO-14976 Format, on page 293)

Adjusting Spectra for Comparative Quantification

A common analysis scenario is comparing spectra from a batch of similar samples, where the principal difference between the recorded spectra is a set of minor peaks typically located near some dominant structure in the spectrum. If the set of samples are the result of a modification to a standard material, then manipulating the spectra using a systematic procedure may produce comparative results which otherwise would be difficult to achieve.

Consider the set of spectra shown in Figure 6•34. These data[1] correspond to a base material, referred to as the blank sample, and three spectra acquired following surface modification, designated samples 20107a1, 20107b1 and 20107c1. The data are acquired using different acquisition settings and is also subject to varying charging conditions. The feature of interest is the small Cr 2p doublet; variations in the intensity of the Cr 2p characterises the experiment. Two problems present themselves, namely, the Cr 2p lines are distorted by the relatively large O 1s plasmon loss structure and the variation in the acquisition procedure means that the data cannot be directly compared to one another. A solution is to shift and normalise the spectra with respect to the Si 2p lines and then subtract the normalised blank sample data from the modified sample spectra.

FIGURE 6•34

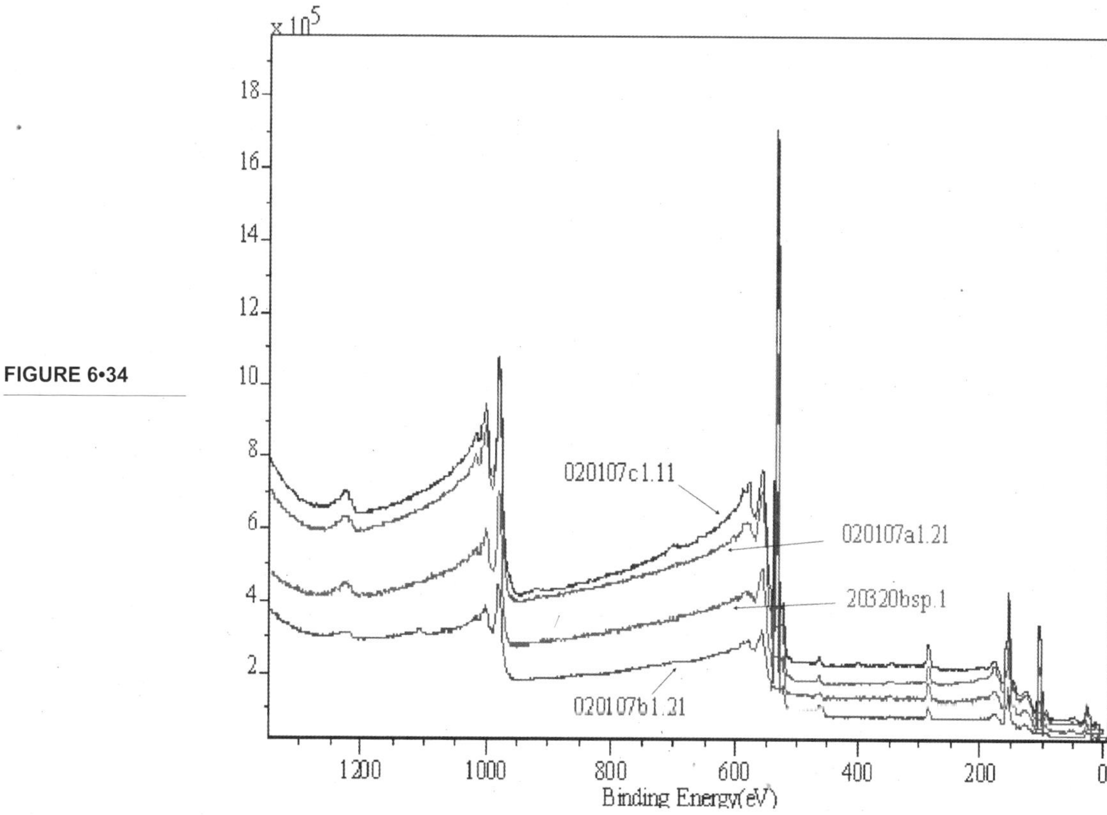

1. We are indebted to Dr. P.C. Thüne, Dutch Polymer Institute, Eindhoven University, for this example.

The ISO 14976 file shown in the Experiment Frame in Figure 6•35 includes the spectra displayed in Figure 6•34. Spectra acquired from each of four samples are collected into a single file where each sample is identified by a different index number (entered as the experimental variable). The resulting Experiment Frame displays the logical file as a set of spectral regions organised as rows of data blocks. CasaXPS provides a means of energy calibrating and intensity normalising the rows of such a file with respect to a particular peak within a spectral region on a row-by-row basis.

The steps required to both energy calibrate and intensity scale a file of spectra are as follows:

1. Choose a spectral region on which to calibrate the data. In the example shown in Figure 6•35 the Si 2p line has been chosen to provide the reference for each row of data.
2. Create a region and/or component on the reference spectra. Note that the first integration region or the first synthetic component will determine the position and/or the intensity used to calibrate the data. The term "first" means the region or component that appears first in the scrolled lists displayed on the Quantification Dialog Window.
3. In the right-hand-side browser, select the set of data blocks for which the calibration region or component is defined. In Figure 6•35, all the Si 2p data blocks are selected since these contain an integration region from which the position and scale factors will be determined.
4. On the Spectrum Processing dialog window, select the Calibration property page. Enter the energy for the selected spectral region consistent with the chosen region or component.
5. If the spectral intensities are to be normalised, the Use Reference Intensity checkbox should be ticked and an appropriate scale factor entered into the Reference text-field. The value entered into the text-field is simply to allow the scaled data bins to have values greater than one (if desired).
6. Press the appropriate button Apply by Row (1st Region) or Apply by Row (1st Comp) to perform the calibration for each row for which a corresponding data block is selected. If the Use Reference Intensity checkbox is ticked, a new Experiment Frame is created in the image of the original file, however the data is adjusted with respect to acquisition time instead of maintaining the dwell-time and number of scans as ISO fields. The data is adjusted under these circumstances to ensure that the spectra are normalised with respect to counts per second and removes the need to perform further pre-processing steps as part of the Calculator operations. This does mean, however, that once performed, the

data file should be saved to a different file name otherwise information previously held within the original file will be lost.

FIGURE 6•35

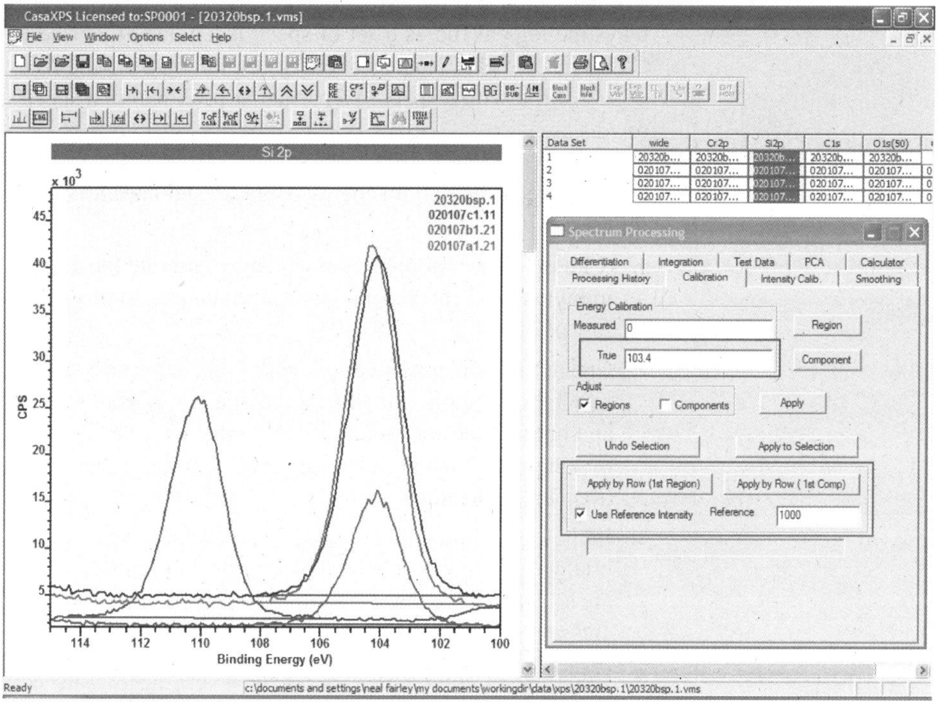

Once data have been normalised and calibrated, the Calculator Property Page can be used to subtract the blank sample spectra from the modified spectra.

Steps required to subtract the "blank" Cr 2p spectrum from the modified spectra are as follows:

1. Display the blank sample Cr 2p spectrum in the Active Tile.
2. On the Calculator Property Page, press the Set Operand pushbutton (see Figure 6•36 on page 142). The Active Spectrum in the Active Tile is loaded into the Calculator when the Set Operand pushbutton is pressed.
3. Press the Offset Adjust pushbutton. The spectrum loaded into the Calculator is shifted in intensity so that the maximum value for the data is zero. This ensures that the operand used in the calculation is below the spectra to which the operation is performed, thus resulting in positive values in the resultant (following the subtraction of the operand loaded into the Calculator).

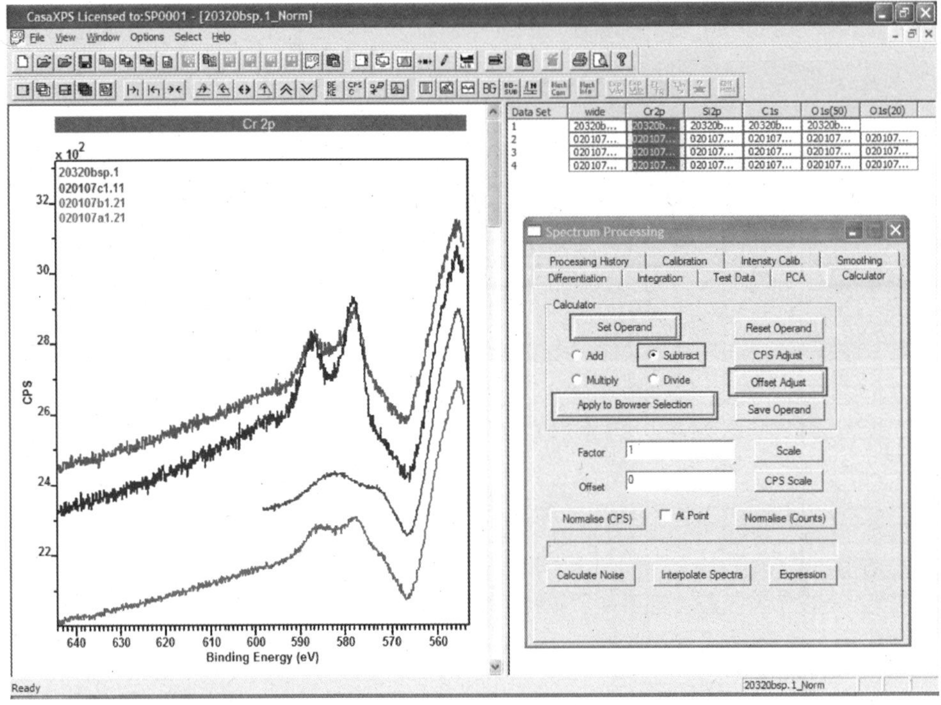

FIGURE 6•36

4. Select the spectra in the Experiment Browser for which the subtraction is required.
5. Select the Subtract radio button.
6. Press the Apply to Browser Selection pushbutton.
7. View the selected spectra in the left-hand side.

Figure 6•37 on page 143 shows the Cr 2p spectra after subtraction of the equivalent data recorded from the blank sample. The peak to the right of the Cr 2p doublet is a Ti 2s line, which is consistent with the Ti 2p features shown in Figure 6•34 on page 139 but is difficult to see in the data before subtracting the blank sample (Figure 6•36). The key point that has made this analysis possible is the consistent nature of the Si 2p lines, which offer a reference for both energy and intensity calibration.

FIGURE 6•37

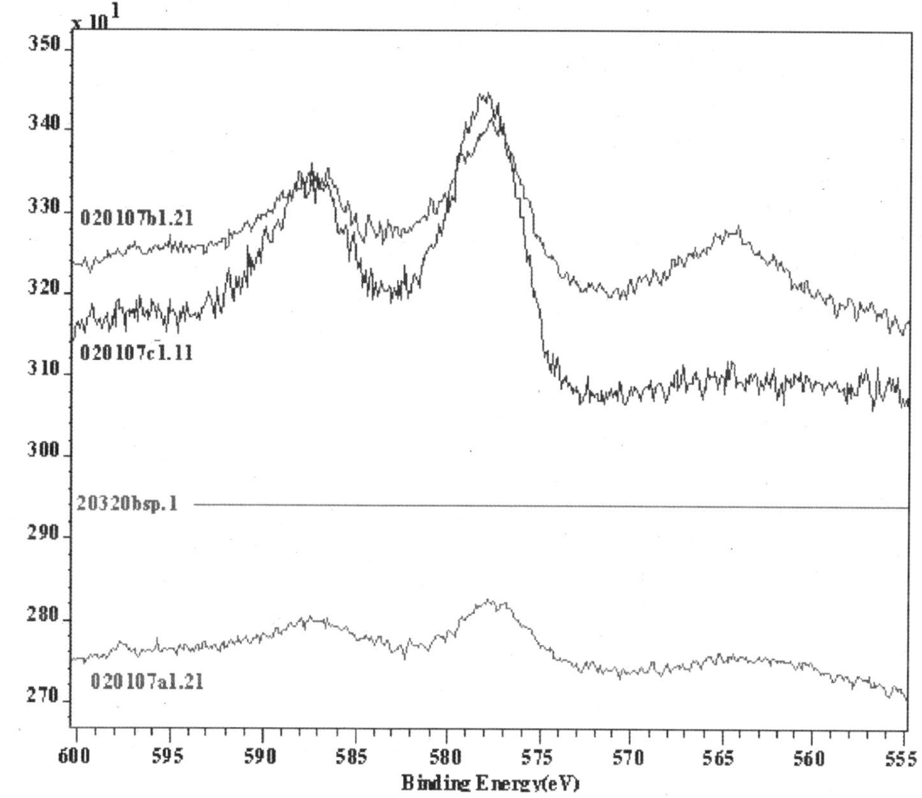

Customizing Columns in the Standard Report

A Standard Report is a collection of information generated from integration regions and/or synthetic components. The format of the Standard Report generated by the Regions or Comps push-button (see Figure 6•4 on page 95) can be defined using two configuration files located in the CasaXPS.DEF directory, RegionQuantTable.txt and ComponentQuantTable.txt. The order and number of columns generated is determined by the fields in these configuration files, where each line in the file determines the nature of a column within the final report. The number of valid entries in the configuration file determines the number of columns in the report. Table 6•2 on page 145 lists the set of strings used to define the report columns.

Figure 6•38 shows a report in terms of percentage mass concentration where the corresponding configuration file is displayed over the top of the report in a (Microsoft) Notepad window. Also note how the Experimental Variable shows labels associated with the ISO file rows rather than the numeric experimental value. This is achieved by generating the report whilst the Browser in the Experiment Frame is in Edit Mode.

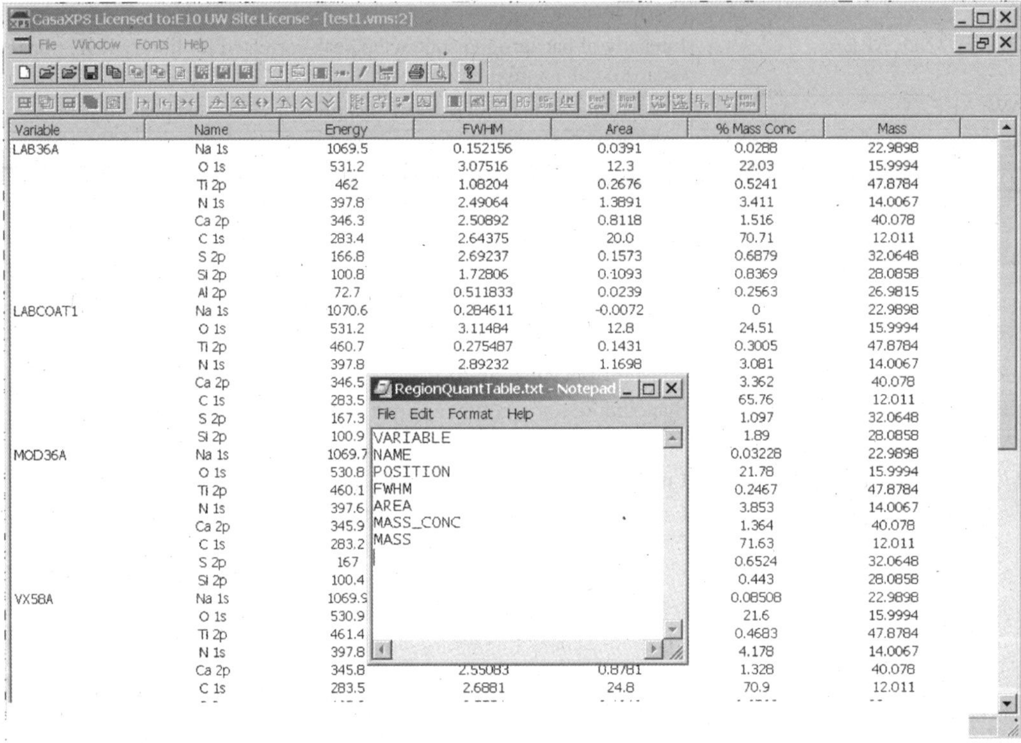

Variable	Name	Energy	FWHM	Area	% Mass Conc	Mass
LAB36A	Na 1s	1069.5	0.152156	0.0391	0.0288	22.9898
	O 1s	531.2	3.07516	12.3	22.03	15.9994
	Ti 2p	462	1.08204	0.2676	0.5241	47.8784
	N 1s	397.8	2.49064	1.3891	3.411	14.0067
	Ca 2p	346.3	2.50892	0.8118	1.516	40.078
	C 1s	283.4	2.64375	20.0	70.71	12.011
	S 2p	166.8	2.69237	0.1573	0.6879	32.0648
	Si 2p	100.8	1.72806	0.1093	0.8369	28.0858
	Al 2p	72.7	0.511833	0.0239	0.2563	26.9815
LABCOAT1	Na 1s	1070.6	0.284611	-0.0072	0	22.9898
	O 1s	531.2	3.11484	12.8	24.51	15.9994
	Ti 2p	460.7	0.275487	0.1431	0.3005	47.8784
	N 1s	397.8	2.89232	1.1698	3.081	14.0067
	Ca 2p	346.5			3.362	40.078
	C 1s	283.5			65.76	12.011
	S 2p	167.3			1.097	32.0648
	Si 2p	100.9			1.89	28.0858
MOD36A	Na 1s	1069.7			0.03228	22.9898
	O 1s	530.8			21.78	15.9994
	Ti 2p	460.1			0.2467	47.8784
	N 1s	397.6			3.853	14.0067
	Ca 2p	345.9			1.364	40.078
	C 1s	283.2			71.63	12.011
	S 2p	167			0.6524	32.0648
	Si 2p	100.4			0.443	28.0858
VX58A	Na 1s	1069.5			0.08508	22.9898
	O 1s	530.9			21.6	15.9994
	Ti 2p	461.4			0.4683	47.8784
	N 1s	397.8			4.178	14.0067
	Ca 2p	345.8	2.55083	0.8781	1.328	40.078
	C 1s	283.5	2.6881	24.8	70.9	12.011

FIGURE 6•38

TABLE 6•2. Keywords used to define the Standard Report configuration file.

NAME	Quantification item name (name of Region or Component)
POSITION	Energy location of the peak maximum
FWHM	Full Width Half Maximum
AREA	Intensity of background subtracted peak corrected for transmission and escape depth approximation
CORRECTED_AREA	Intensity of background subtracted peak corrected for library RSF, transmission and escape depth approximation (MFP)
RAW_AREA	Uncorrected or raw Intensity of background subtracted peak in CPSeV
RSF	Library RSF
TOTAL_SENSITIVITY_FACTOR	Combined correction factor of Library RSF * Transmission * MFP
TRANSMISSION	Value of the transmission function correction function at the kinetic energy of the peak maximum
MEAN_FREE_PATH	Value of the mean free path correction function at the kinetic energy of the peak maximum
CONCENTRATION	Percentage Atomic Concentration computed from the corrected area intensity
ERROR_BAR	Standard Deviation of the Percentage Atomic Concentration
AREA_ERROR_BAR	Standard Deviation of the Corrected Area Intensity
START	Start Energy for the definition of a quantification region
END	End Energy for the definition of a quantification region
PEAK_TO_PEAK	For differentiate AES data, the difference between the maximum and the minimum intensities within a quantification region
PEAK_TO_PEAK_CONC	Percentage Atomic Concentration computed from the peak to peak intensity after correction by the RSF.
LINE_SHAPE	The line-shape define for a synthetic component
TRANSITION_TAG	Text string used to identify and link components with quantification regions
POSITION_CONST	String defining the constraint placed on the energy position parameter for a synthetic components
AREA_CONST	String defining the constraint placed on the peak area parameter for a synthetic components
FWHM_CONST	String defining the constraint placed on the peak FWHM parameter for a synthetic components
COLUMN_LABEL	Alphabetic character heading to the synthetic component table used to identify the relative constraints in a peak model
DEGREES_OF_FREEDOM	The number of data channels used to optimize a peak model
VAMAS_BLOCK_NAME	Block Id corresponding to the data block from which a quantification item derives
CENTROID	Peak position from an integration region measured using the centroid
MASS	Elemental mass of the transition (Daltons)
MASS_CONC	Percentage mass concentration
SIGNAL_TO_NOISE	Standard deviation in the noise distribution assuming Poisson counting statistics

CHAPTER 7

Line Shapes and Backgrounds

A range of physically possible line profiles in core-level XPS is possible, and simple Gaussian or Lorentzian functions are very rarely adequate to describe them. In the case of metal samples it has been shown that asymmetric profiles should be expected on theoretical grounds [1], however recorded spectra exhibit deviations from idealized profiles due to a range of instrumental and physical effects:

1. The response function of the electron analyzer (which may be asymmetric),
2. The profile of the x-ray line-shape (predicted to be asymmetric for unmonochromatized metal anodes [1]),
3. Intrinsic life-time broadening of the core-level hole state (usually assumed to be Lorentzian in nature),
4. Phonon broadening,
5. Differential surface charging of the sample.

Peak fits and background choice

In addition to these instrumental considerations the shape of a synthetic peak is also influenced by the choice of background algorithm used to remove, so called, extrinsic electrons from the data. Figure 7•1 and Figure 7•2 are two examples of peak fits where the difference between the two outcomes lies in the choice of background. Doniach Sunjic profiles have been used to model the recorded doublet, where it has become necessary to adjust the asymmetry parameter as well as the width of the

Gaussian used to broaden the underlying Doniach Sunjic shape. The most obvious difference between the two peak fits is the reduction in the amount of asymmetry required in the line-shape when a Shirley [2] background is used. This is best measured by the asymmetry index [1], which changes from 0.49 in the case of a linear background to 0.008 when a Shirley background is employed.

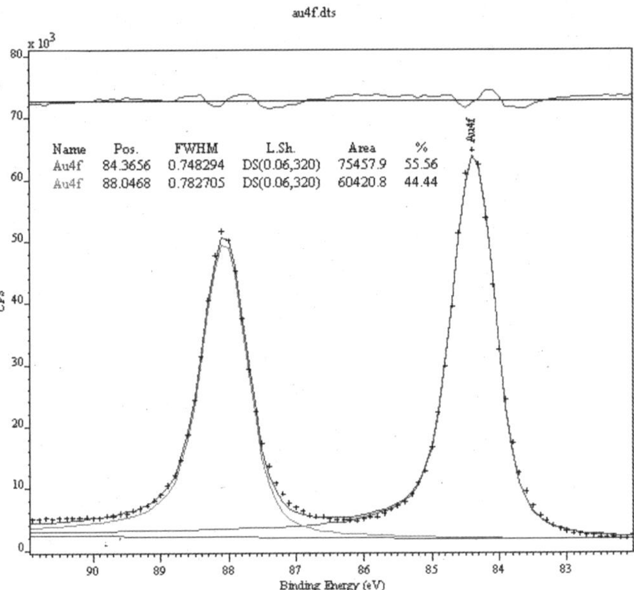

FIGURE 7•1 Au 4f doublet fitted using an offset linear background. The line-shape used in the fit is a Doniach Sunjic form convoluted with a Gaussian that results in an asymmetry index [1] of 0.49 for each peak.

Although theoretically based, the Doniach Sunjic profile suffers from ill-defined areas. The integral of the Doniach Sunjic function is infinite for non-zero values of the asymmetry parameter [3] and therefore any direct use of this line-shape requires a somewhat arbitrary use of cut-offs to allow finite peak areas to be reported. Changing the shape of the peaks through adjustments to the asymmetry parameter and the width of the broadening function has altered the relative proportion of the spin-orbit split peaks in the Au 4f doublet (Figure 7•1 and Figure 7•2). The background choice has altered the reported intensity as well as FWHM for the individual peaks. These statements emphasize the importance of comparing like-with-like when quantification results are used in practice.

The choice of line-shape is determined by the nature of the problem. Obtaining the best fit is not always as important as producing quantification results that can be compared with historical results. On the other hand if the results represent a sequence of experiments for which a trend is more important than the absolute values then a good model may take precedence over the need to supply numerical values consistent with the past. Provided the models are self consistent the use of Doniach Sunjic line-shapes may be acceptable.

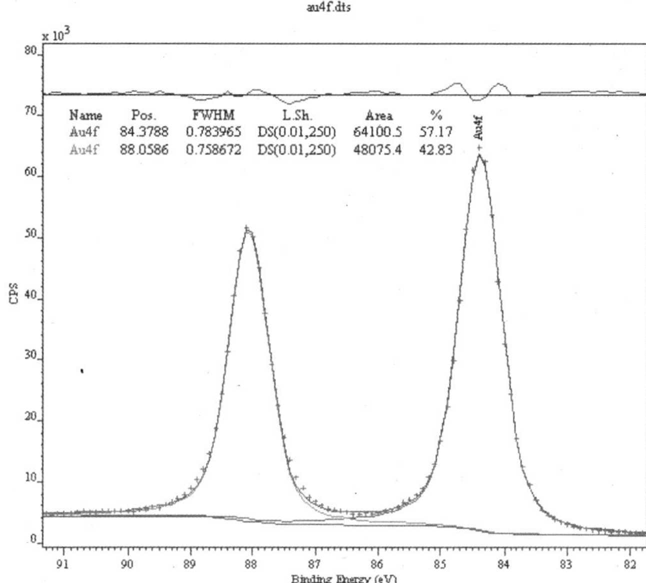

FIGURE 7•2 Au 4f doublet fitted using an iterated Shirley background. The line-shape used in the fit is a Doniach Sunjic form convoluted with a Gaussian that results in an asymmetry index [1] of 0.008 for each peak.

Analysis driven by precise quantification lie at the root of many *ad hoc* procedures used in XPS. Shirley backgrounds were introduced precisely [4] to remove as much asymmetry as possible from recorded data (Figure 7•2) in a well-prescribed fashion, so that near symmetric synthetic models can be used to characterize the intensity under a peak. A Gaussian/Lorentzian line-shape is finite and, with the appropriate relative sensitivity factors (RSF), can be used to compare intensities from fitted peaks to those calculated from integration regions. The key factor that has popularized Shirley backgrounds is the availability of well-characterized RSF values coupled with a specific background and line-shapes.

Some of the line-shapes introduced in CasaXPS have been constructed to allow Doniach Sunjic asymmetric behaviour to be associated with an underlying Gaussian/Lorentzian shape. The Voigt approximation is used to characterize the area, position and FWHM, while the asymmetric form approximates the rise in the signal much in the same way that the Shirley background is used to reduce the data to symmetric shape. The advantage of retaining the asymmetry in the synthetic model is apparent when a Tougaard background is used to remove the extrinsic contribution to a metal spectrum.

To facilitate both trend analysis and basic quantification a wide range of line-shapes is required.

A List of Line-Shapes

The line-shapes offered in CasaXPS are based around the following fundamental functional forms.

The **Voigt** functional form has been the basis for most quantitative analysis of XPS spectra. Unfortunately, an analytic form for the convolution of a Gaussian with a Lorentzian is not available [3] and so practical systems have adopted two approximations to the true Voigt function.

Gaussian/Lorentzian Product Form

$$GL(x, F, E, m) = \frac{\exp\left((-4\ln 2)(1-m)\frac{(x-E)^2}{F^2}\right)}{1+4m\frac{(x-E)^2}{F^2}}$$

Gaussian/Lorentzian Sum Form

$$SGL(x, F, E, m) = \exp\left(-4\ln 2\frac{(x-E)^2}{F^2}\right)(1-m) + \frac{m}{1+4\frac{(x-E)^2}{F^2}}$$

Exponential Asymmetric Blend Based upon Voigt-type Line-Shapes

Given either of the above Gaussian/Lorentzian symmetric line-shapes, an asymmetric profile is obtained from a blend function as follows.

$$Y(x) = GL(x) + (1 - GL(x)) \times T(x)$$

Where

$$T(x, k, F, E) = \begin{cases} \exp\left(-k\frac{(x-E)}{F}\right) & x \le E \\ 1 & otherwise \end{cases}$$

Alternative Asymmetric Line-Shapes

An asymmetric line-shape due to Ulrik Gelius (Uppsala, Sweden) offers a class of profiles by modifying the Voigt function via an *ad hoc* adjustment. The profile is given by

$$A(x, a, b, F, E) = \begin{cases} GL(x, F, E, m) + w(a, b)[AW(x, a, F, E) - G(x, F, E)] & x \le E \\ GL(x, F, E, m) & x > E \end{cases}$$

Where

$$AW(x, a, F, E) = \exp\left[-\left(\frac{2\sqrt{\ln 2}(x-E)}{F - a2\sqrt{\ln 2}(x-E)}\right)^2\right]$$

and

$$w(a, b) = b\left[0.7 + \frac{0.3}{(a + 0.01)}\right]$$

The parameters a and b determine the shape of the asymmetric portion of the curve.

Doniach Sunjic: A Theoretically Based Asymmetric Line-Shape

Doniach and Sunjic [1] performed an analysis for both photoemission and X-ray line-shapes, both of which result in an underlying profile given by the expression below. The formula includes an asymmetry parameter α that characterizes the asymmetry for a particular metal-like material. F is related to the FWHM and the position E is again related to but not equal to the position of the maximum intensity for the line-shape. It is therefore difficult to relate optimization parameters determined from the Doniach-Sunjic profile to similar quantities determined from Voigt-type line-shapes.

$$DS(x, \alpha, F, E) = \frac{\cos\left(\frac{\pi\alpha}{2} + (1-\alpha)\operatorname{atan}\left(\frac{x-E}{F}\right)\right)}{(F^2 + (x-E)^2)^{(1-\alpha)/2}}$$

Nevertheless, the Doniach-Sunjic profile offers an asymmetric shape that is particularly appropriate for non-monochromatic X-ray induced transitions; the profile is potentially present in both the photoemission process as well as the excitation source.

Line-Shapes Based upon Backgrounds

The above profiles are assumed to be entirely due to intrinsic electron energy variations, where a background subtraction algorithm is required before these line-shapes can be used to model the spectra. An alternative approach is to include the background shape as part of the model [4]. An analytic form for the Shirley background can be determined for each of the line-shapes and the sum of these backgrounds plus line-shapes is used to approximate the variation in the spectrum. A simple constant background is all that is required for this procedure although other forms for the background are still an option.

Castle *et al* [4] have developed a Shirley-type adjustment to a Voigt line-shape. The Shirley approximation is calculated from the current Gaussian/Lorentzian shape and a polynomial b0+b1(x-E) is used to scale the background in order to provide a fit to the observed spectra. The procedure yields a "Kappa" parameter (given by b0) that characterizes the "intrinsic" step in the spectrum observed for a particular sample.

Line-Shapes Available in CasaXPS

Curve fitting in CasaXPS is performed via the Quantification Parameters dialog window where the synthetic line-shapes are defined from the scrolled-list on the Components property page (Figure 7•3).

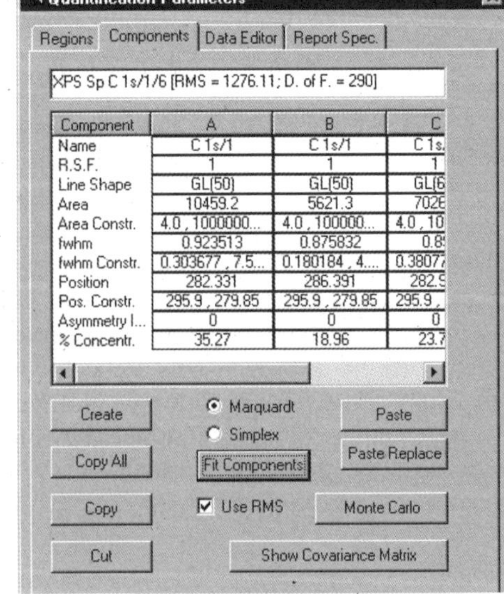

FIGURE 7•3 Property Page used to define synthetic line-shapes

The line-shape used to describe a photoelectric transition is entered in the row labelled "Line Shape" and takes the form of a text string.
GL(p) : Gaussian/Lorentzian product formula where the mixing is determined by m = p/100, GL(100) is a pure Lorentzian while GL(0) is pure Gaussian.
SGL(p) : Gaussian/Lorentzian sum formula where the mixing is determined by m = p/100, SGL(100) is a pure Lorentzian while SGL(0) is pure Gaussian.
GL(p)T(k) : Gaussian/Lorentzian product formula modified by the exponential blend.
SGL(p)T(k) : Gaussian/Lorentzian sum formula modified by the exponential blend
GL(p)K(b0,b1) : Gaussian/Lorentzian product formula modified by a Shirley-type background prescribed by Castle et al. A linear polynomial determined from b0 and b1 adjusts the step in the Shirley background.
SGL(p)K(b0,b1) : Gaussian/Lorentzian sum formula modified by a Shirley-type background prescribed by Castle et al. A linear polynomial determined from b0 and b1 adjusts the step in the Shirley background.

Further adjustments to the basic shapes

The basic shapes that result from the various functional forms can be further modified by numerically convoluting the profile with a Gaussian. For example, GL(100,50) will convolute a pure Lorentzian with a Gaussian characterized by a width of 50. The value for the width is the number of digital nodes used to describe the Gaussian and is therefore an arbitrary unit characteristic of the numerical form for the Gaussian.
N.B. Any line-shape that requires a digital convolution will cause slower performance than a shape that does not involve this procedure.

Asymmetric Line-Shapes

The exponential tail described by Sherwood in the book "Practical Surface Analysis" edited by Briggs and Seah, attempts to approximate asymmetric line-shapes commonly found in photoelectric peaks. Several researchers have offered alternative deformations of the symmetric Voigt profile however none but Doniach Sunjic backs their approaches with a theoretical basis. Nevertheless, practical surface analysis requires practical solutions and the need for a rigid line-shape model that matches the observed profiles makes *ad hoc* forms acceptable tools. Philosophically, theoretically based solutions are preferred and hence, at least for metals, Doniach Sunjic profiles ought to offer "better" fits.
A practical solution offered by Ulrik Gelius (QSA 7) [5] uses a Voigt-like function as the underlying shape and modifies the lower kinetic energy electrons using the equations given above. CasaXPS offers these line-shapes in the form:

A(a,b,n)GL(p): Gaussian/Lorentzian product formula modified by an asymmetric form define above. The parameters a and b allow the asymptotic form of the asymmetric tail to change whilst also altering the shape of the asymmetry prior to attaining the asymptotic shape. Figure 4 shows one such class of profiles where the parameter a has been held fixed at 0.35 while b varies between 0.2 and 1. The third parameter n defines the width of a Gaussian used to convolute the basic shape of the profile.

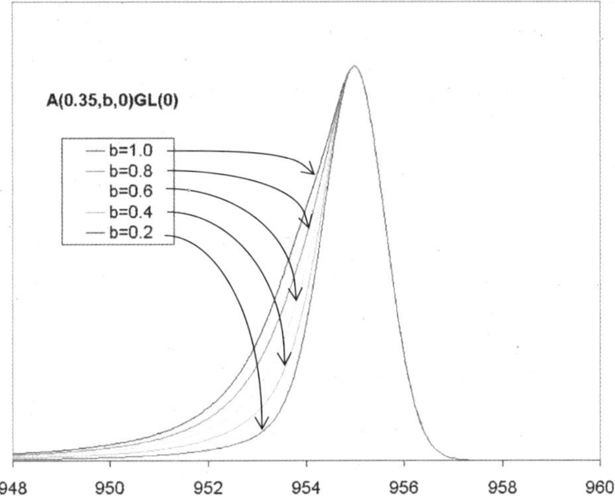

FIGURE 7•4 Asymmetric form due to Ulrik Gelius

The line-shapes defined by A(a,b,n)GL(p) and GL(p)K(b0,b1) are different from the Doniach Sunjic profile in that the Doniach-Sunjic profiles asymptotic limit is zero, while both of the *ad hoc* forms allow steps to be modelled within the line-shape itself. Another characteristic of these *ad hoc* line-shapes is that the peaks are defined in terms of intensity, position and FWHM taken from the underlying Voigt function. This is not unnatural in view of the fact that the area beneath all of these curves would be infinite without the introduction of some arbitrary cutoff limits.

DS(a,n) : The basic Doniach Sunjic profile is defined in terms of the asymmetry parameter a and a convolution width n. The profile defined above is numerically convoluted with a Gaussian whose width is determined from n to produce the final line-shape.

Although the Doniach Sunjic line-shape allows very good fits to experimental data, the infinite intensity defined by the curve makes this profile difficult to use under practical situations. Any scheme that limits the area through cutoff functions or ranges introduces inconsistencies into the peak parameters that make relationships such as stoichiometry invalid. The degree of asymmetry used to describe the data moves intensity modelled by the line-shape away from the primary peak position and towards the cutoff regions (Figure 7•5). For this reason it is essential to use the Doniach Sunjic line-shape in situations where the asymmetry parameters are close to the same value so that calculated areas can, in some sense, be compared.

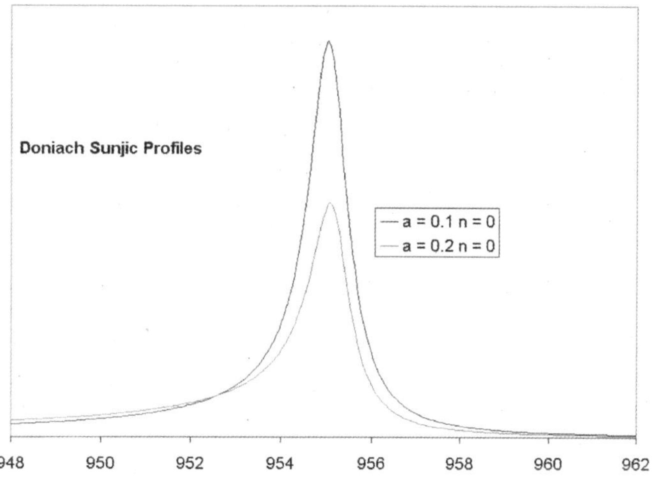

FIGURE 7•5 Doniach Sunjic profiles for asymmetry parameter equal to 0.1 and 0.2. Note how the asymptotic curves cross over illustrating how intensity is transferred awayfrom the peak maximum.

Modifications to the Doniach Sunjic function

The advantages of the Doniach Sunjic profile are lost when synthetic models are required for quantification purposes. It is necessary to introduce *ad hoc* cutoff behaviour before finite intensities can be reported and, once obtained, the peak areas can only be used in a quantification reports when relative sensitivity factors are know for the specific cutoff criterion adopted.

Empirical observations suggest that pure Doniach Sunjic profiles do not always yield good fits for electrons with higher kinetic energy (KE) when monochromatic X-ray sources are employed. For example, Al 2p doublets recorded using a monochromatic X-ray source appears less Lorentzian in nature (to the higher KE side of the peak) than the Doniach Sunjic shape will allow (Figure 7•6), however the same line measured using an Aluminium anode does exhibit the predicted (but somewhat broader) shape. The latter is thought to be due to the double influence of the Doniach Sunjic shape from the photoemission process as well as the X-ray line profile, while monochromatic X-rays have lost the Doniach Sunjic distribution as a consequence of the filtering process.

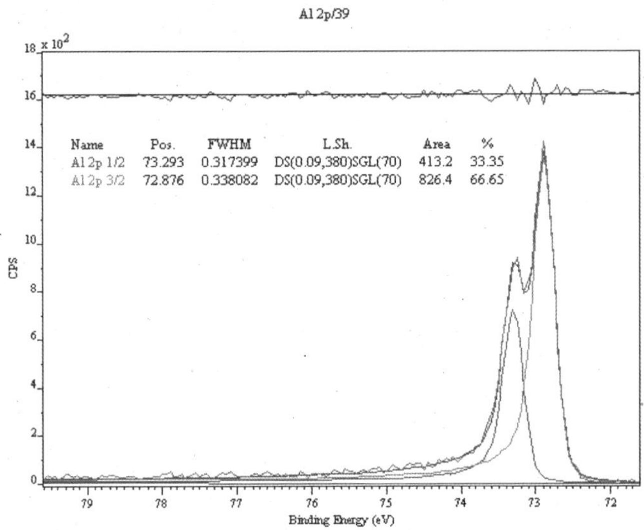

FIGURE 7•6 Al 2p Doublet from a monochromatic X-ray source.

In order to fit spectra such as that shown in Figure 7•6 the pure Doniach Sunjic shape has been modified. The lower KE asymmetry derives from the functional form for the Doniach Sunjic profile while the higher KE side of each component is a Voigt-type function. A numerical convolution is applied to the combined profiles to produce the line-shapes in Figure 7•6.

Given an approximation to the Doniach Sunjic profile of the form shown in Figure 7•6 then the possibility of characterizing the line-shape by the associated Voigt-type profile becomes feasible. The Voigt-type portion of the line-shape provides the position, width and intensity while an asymmetry index [1] measures the departure of the line-shape from a symmetric form. Such a regime fits well with the philosophy adopted by Shirley, where the background shape was used to reduce an asymmetric step to a form that could be characterized by a Voigt function. The asymmetry index is given by

$$a = 1 - \frac{fwhm_{left}}{fwhm_{right}}$$

The asymmetry index is reported on the Components Property Page for each line-shape.

H(a,n)GL(p) : Hybrid Doniach Sunjic/Gaussian-Lorentzian (product) line-shape.

H(a,n)SGL(p) : Hybrid Doniach Sunjic/Gaussian-Lorentzian (sum) line-shape.

The "H" form of the DS/GL hybrid line-shape is identical to the DS(a,n)GL(p) form except the position, FWHM and area parameters are all determined from the Voigt-type shape. Compare the area values reported in Figure 7•7 to those from Figure 7•6. The area taken from the Voigt-type portion is significantly less than the intensity taken from the Doniach Sunjic curve where a cutoff has been used. Another difference between the two forms is that the area reported by the "H" is much less sensitive to adjustments in the asymmetry parameter (numerical convolution does move shape information between the two halves of the line-shape).

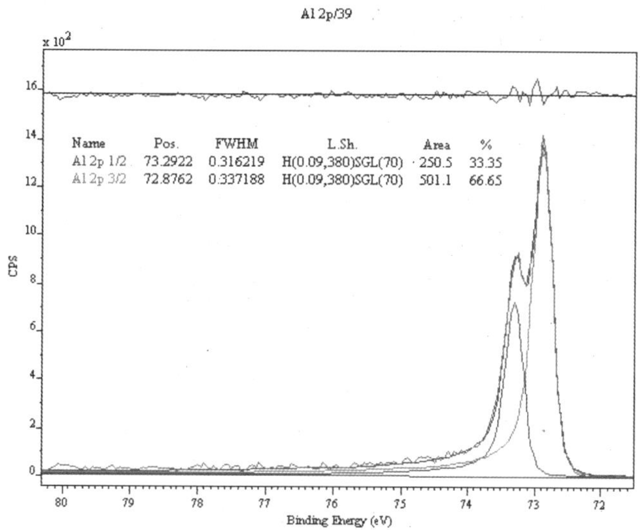

FIGURE 7•7 Hybrid DS/GL line-shape characterized by the Voigt-type portion.

A New Line-Shape

The hybrid form of a DS/GL line-shape is blended using a numerical convolution. The maximum for the two shapes diverge for larger asymmetry values, therefore characterizing the peak parameters by the Voigt-type portion is more difficult as the asymmetry becomes more pronounced. A new class of functions can be introduced to over come this problem, namely, a set of line-shapes that are a blend of the asymptotic behaviour from the DS line-shape with a Lorentzian.

The cosine term in the Doniach Sunjic profile rapidly approximates a step function as the asymmetry parameter increases, while for small values of this parameter the cosine term is responsible for the movements in the DS maximum and for a = 0 results in a pure Lorentzian form. The formula for the "F" profile combines these characteristics by blending the asymptotic behaviour of the DS profile with one of the Voigt-type functions using a linear mapping.

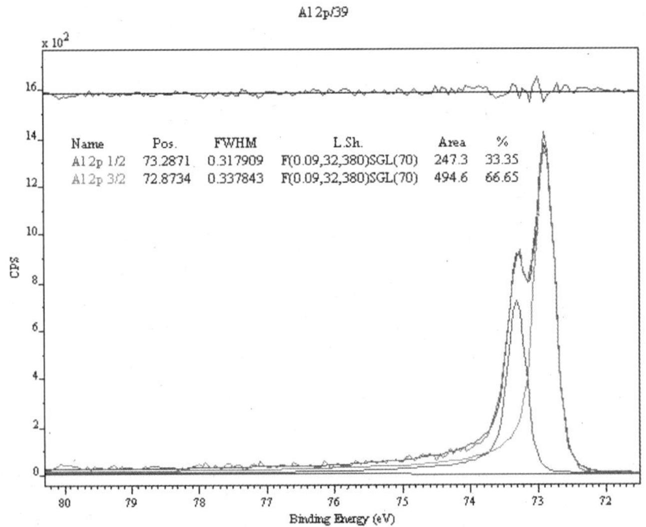

FIGURE 7•8 A new line-shape that blends the asymptotic behavior of a Doniach Sunjic profile with a Voigt-type function.

F(a,m,n)GL(p) : Asymptotic Doniach Sunjic/Gaussian-Lorentzian (product) linear mapping line-shape.

F(a,m,n)SGL(p) : Asymptotic Doniach Sunjic/Gaussian-Lorentzian (sum) linear mapping line-shape.

The linear mapping between the two functional forms is defined as a percentage through the m parameter. Figure 7•8 shows the same Al 2p data envelope as is seen in Figure 7•7, but the line-shapes are defined using the "F" functional form. In

Figure 7•8, the same asymmetry value has been used as the one in Figure 7•7 however the linear mapping parameter is set to 32% DS asymptotic form 68% Voigt-type function. The resulting profile is convoluted with a relatively wide Gaussian (characterized by the number 380) and accounts for the Lorentzian bias in the Voigt approximation.

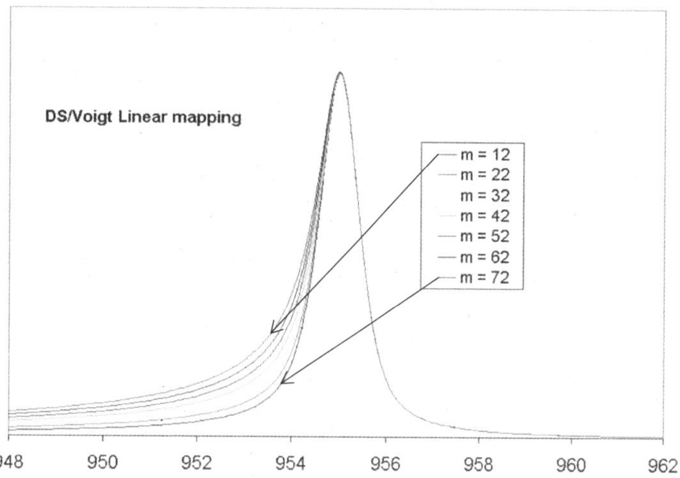

FIGURE 7•9 "F" function plotted over a range of values for the linear mapping parameter, a constant asymmetry parameter (0.09) and no numerical convolution applied.

Figure 7•9 shows how a pure Voigt-type function, SGL(70), is merged with the asymptotic form for a Doniach Sunjic profile (asymmetry parameter equal to 0.09) for a range of the linear mapping parameter. The characteristic shape of the Doniach Sunjic profile can be retained to some degree but is supplemented by a class of shapes that go beyond what is possible with a DS profile. The principal difference between the DS profile and the "F" function lies in the static position of the maximum for the class of profiles. The moving maximum is a feature of the DS profile and can be achieved to some extent by use of the numerical convolution option. Increasing the width of the Gaussian caused the line-shape to lean in towards the asymmetric side of the peak (Figure 7•8).

A further variation on a theme is also possible where the "F" function can be replaced by: $E(x, \alpha, m, F, E) = mVoigt(x)^{(1-\alpha)/2} + (1-m)Voigt(x)$

The "E" function merely allows the shape to the right to be a part of the asymmetric shape to the left.

Background Subtraction

The line-shapes described above are very dependant on the availability of background subtraction algorithms that complement their use. If a background is incorrectly removed from a spectrum then the merit of attempting to model the result with theoretically correct synthetic line-shape is less clear.

Tougaard and co-workers [6] have devoted much time to understanding the backgrounds that are present in XPS spectra. The transport of electrons through a material after excitation by X-rays can be described by equations [7] that involve a one-sided convolution of the recorded data with an energy-loss probability distribution. This loss function offers a prescription by which electrons leaving the sample with an initial energy may have their characteristic energy altered by the interactions with the surface. Each material has a characteristic loss function of varying complexity and Tougaard has written many papers describing methods for establishing both generally applicable approximations (so called, universal cross-sections) and specific forms for individual materials.

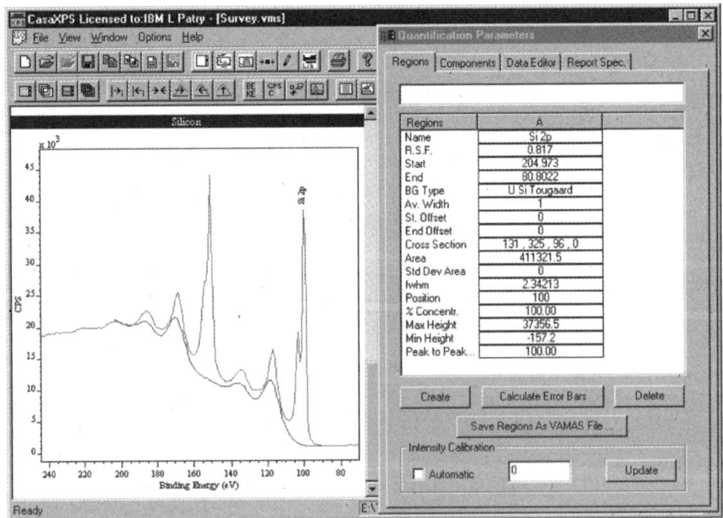

FIGURE 7•10 Three-parameter universal cross-section for Silicon.

The principal methods used by Tougaard for calculating the loss functions are theoretical dielectric response models and empirical estimates computed from spectra acquired using reflected energy-loss spectroscopy (REELS). Tougaard follows these calculations by fitting the results to a rational function, which models the resonance structure in the calculated loss-functions and allows the essential distribution to be described by a simple formula. These loss-functions are defined in terms of four parameters, three of which describe the shape of the rational function and one to allow for various band gaps in different materials. Tougaard refers to the distribution below as a "three-parameter universal cross-section" and has established values for a number of materials including Aluminium, Silicon, Silicon Dioxide and others [6]. The form of this universal cross-section is

$$P(T, B, C, D, T_0) = \begin{cases} \dfrac{BT}{(C - T^2) + DT^2} & T > T_0 \\ 0 & T \le T_0 \end{cases}$$

CasaXPS offers several ways of using the three-parameter universal cross-sections. The background type defined on the Regions Property Page of the Quantification Parameters dialog window (Figure 7•10) may be chosen from the following list:

U Si Tougaard (short form **U Si**): Three-parameter cross-section determined for Silicon.

U SiO2 Tougaard (short form **U SiO2**): Three-parameter cross-section determined for Silicon Dioxide.

U Al Tougaard (short form **U Al**): Three-parameter cross-section determined for Aluminium.

U Ge Tougaard (short form **U Ge**): Three-parameter cross-section determined for Germanium.

U Poly Tougaard (short form **U Poly**): Three-parameter cross-section determined for Polymers.

U 4 Tougaard (short form **U 4**): Custom three-parameter cross-section where the parameters are entered on the Regions Property Page (Figure 10) in the Cross Section text-field. The four numerical values entered in a comma separated list correspond to B,C, D and T0 in the energy-loss function above.

C 4 Tougaard (short form **C 4**): Alternative custom three-parameter cross-section where the parameters are entered on the Regions Property Page (Figure 7•10) in the Cross Section text-field. The four numerical values have the same meaning as those found in the U 4 Tougaard form, however T0 is used to shift the resonance structure rather than simple acting as a cutoff limit. It is therefore possible to retain the shape of the distribution but move the position of the maximum without changing the C and D parameters.

The custom Tougaard cross-sections offer a chance to change the background in response to the spectrum under investigation. Such an approach is not unprecedented [8]. "Direct, simultaneous determination of XPS background and inelastic differential cross sections using Tougaard's algorithm" by Dr. M. Jo describes a method that extracts loss-functions from the XPS data using optimization procedures applied to a cross-section defined in terms of Spline polynomials.
The rigid model offered by the three-parameter cross-section provides more guidance in the regions most influential in peak fitting, but will struggle to reproduce the asymptotic behaviour for the cross-section as well as the adjustments needed to describe the near peak region of the background. Nevertheless, a practical approach to the use of Tougaard backgrounds in peak fitting is to choose a background that works in conjunction with the synthetic peaks. The custom form for the cross-section allows the background to develop at the same time as the peak parameters and choice of line-shapes. Figure 7•6 through Figure 7•8 all use a custom three-parameter universal cross-section to describe the background under an Al 2p metal doublet; typically, researchers would choose Shirley background for such spectra.

Adjusting the three-parameter universal cross-section

The rational function for the three-parameter cross-section is defined in terms of three constants B, C and D. This approximation is the quotient of a linear polynomial and a quadratic in terms of T2, but it is the denominator that determines the position and strength of the resonance in the cross-section. The parameters C and D both determine the position, width and height of the resonance, but to a first order (small D) the position is influenced by the square root of C and the sharpness of the cross-section is determined from the size of D.
The distance of a complex pole from the real-line in the domain of a complex function is instrumental in determining the width of a resonance when the function is plotted with respect to a real variable. The D parameter principally moves the poles of the Tougaard cross-section away from the real-line, while C changes the projection of the pole onto the real-line. Small D corresponds to near singular behaviour in the cross-section and therefore results in shape structures located, to some degree, by C while the magnitude of the function can be moderated by adjusting the B parameter. It should be noted, however, that B adjusts the behaviour of the cross-section in the region close to zero energy-loss. This is precisely the region of interest for modelling a peak and so the values given to B and D can be influential in obtaining a reasonable synthetic model.

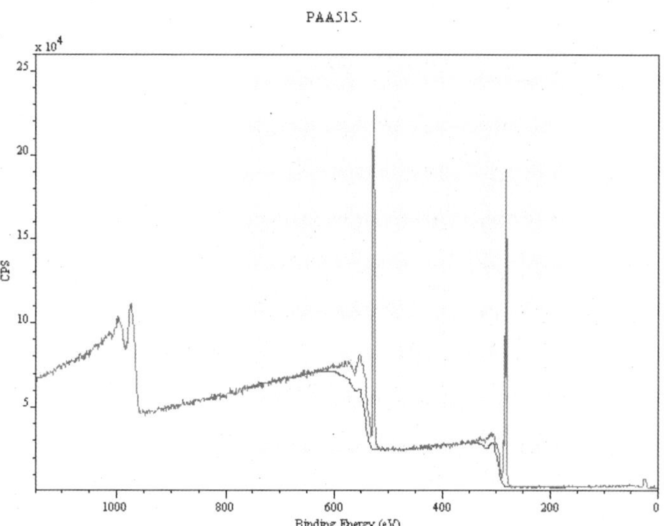

FIGURE 7•11 Wide scan spectrum from a PAA sample with a custom three-parameter cross-section applied.

The three-parameter universal cross-sections for specific materials may be used as a basis for determining a background. A wide scan spectrum allows the essential form for a background to be determined in the case of materials for which cross-sections are not available. The resulting background is easily transferred to any high-resolution spectra from the same sample (Figure 7•11 and Figure 7•12).

The meaning attributed to custom backgrounds should be viewed in the same way a synthetic model for a peak envelope is viewed. That is to say, only chemistry and knowledge of the sample allows a synthetic model to be constructed which has any meaning and this should also be the case for custom Tougaard backgrounds. The use of these backgrounds is entirely at the discretion of a researcher and bad backgrounds can be constructed in much the same way that synthetic models are open to abuse. However, a carefully constructed background can reveal features that may be overlooked or not accessible when a linear or Shirley form is applied. Figure 7•12 on page 164 shows an O 1s structure from PAA (poly(acrylic acid)) after a custom Tougaard background has been subtracted. The structures to the deeper binding energy side of the peak may be real or an artifact of the background algorithm, but a linear background applied to the same data would fail to ask any questions at all regarding these features.

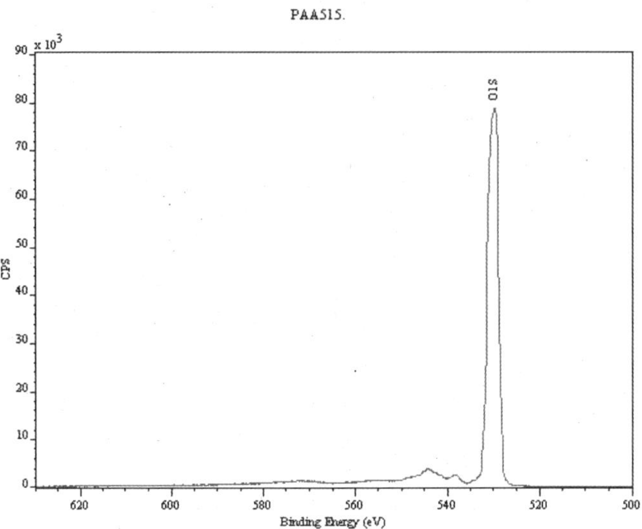

FIGURE 7•12 High-resolution spectrum from the same PAA sample shown in Figure 7•11. The custom Tougaard background has been subtracted from the data.

Simulating Spectra using Custom Tougaard Backgrounds

Given a set of synthetic peaks and a three-parameter universal cross-section it is possible to construct a spectrum with an associated background. The synthetic PAA spectrum in Figure 7•13 has been constructed from a set of Gaussian/Lorentzian line-shapes approximately positioned at the experimental values for the valance band, two C1s structures, two O AES peaks and one O 1s photoelectric peak. The widths and intensities are consistent with results for such a sample measured using a Scienta ESCA-300 at RUSTI, Daresbury Laboratory, UK, and the cross-section has been chosen similar to the one used in Figure 7•12. A sequence of iterations using these peaks and the three-parameter universal cross-section shown in Figure 7•14 converges to the synthetic spectrum in Figure 7•13.

This technique for generating theoretical spectra offers a way to understand the consequences of choosing a particular cross-section and if nothing else provides a basis from which real spectra can be viewed. Surface morphology and nano-structure causes most real spectra to exhibit background behaviour that is not characteristic of a material but the environment in which the element is found. Attempting to construct a spectrum based purely on a resonant cross-section model clearly shows the limitations of such approximations when applied to real samples.

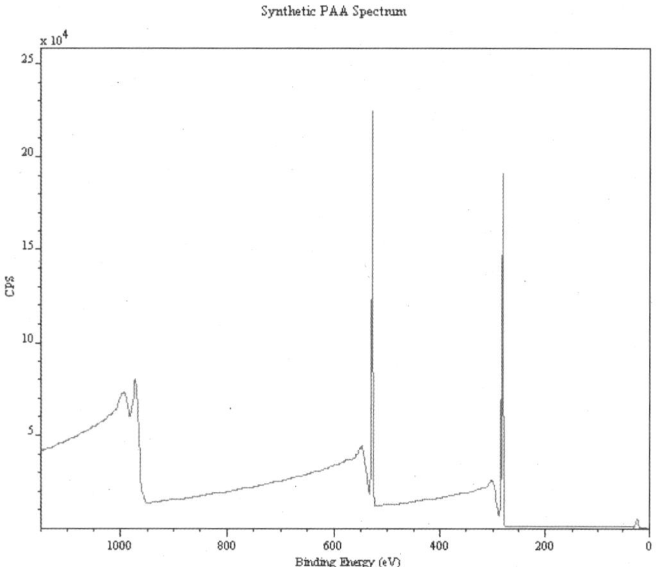

FIGURE 7•13 Synthetic PAA spectrum constructed from a set of Gaussian/Lorentzian line-shapes and a custom Tougaard cross-section.

Although synthetic modelling has little direct use for real samples, it does offer an analyst the opportunity to investigate algorithms used to probe real spectra. For example, numerous techniques have been proposed from Multivariate Statistics for reducing XPS spectra to meaningful chemical information [9] [10]. Procedures for generating simulated spectra allow data sets to be constructed with similar characteristics to the experimental data but with well-defined information. These algorithms can be applied to the simulated spectra in order to establish what information is extracted for a perfectly understood data set.

Everyday procedures such as curve fitting can be evaluated using simulated spectra and therefore the ability to produce spectra with known shapes can enhance understanding for students and researchers alike.

Spline Backgrounds

CasaXPS 2.1.8 offers backgrounds based upon the existing CasaXPS background types Shirley, Linear, Tougaard and (so called) None, but where the curve is decided by a set of cubic spline polynomials rather than the underlying functional forms. These new background types “Spline Linear” (abbreviation “sl”), Spline Shirley (“ss”), Spline Tougaard (“st”) and Spline None (“sn”) are flexible backgrounds and can be adjusted under mouse control or optimised as part of a synthetic peak model.

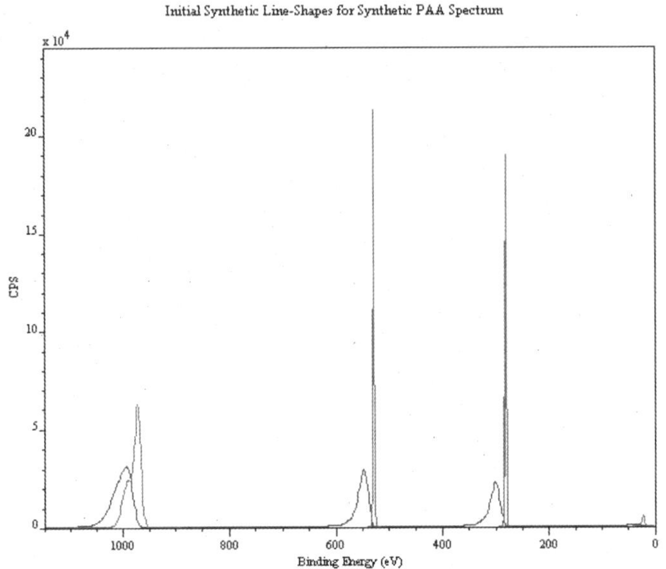

FIGURE 7•14 Initial synthetic components used to construct a PAA spectrum. The background curve (broader, energy loss peaks) represents the first iteration of the cross-section with the model data.

A spline is a mechanical device used by draughtsmen to draw aesthetically pleasing curves using pen and paper: the draughtsman fixes a set of points (knots) on the drawing, then bends a flexible strip of plastic or wood (the spline) around the knots and traces the shape onto the paper. Spline interpolation is the mathematical equivalent of this process and is achieved using a piece-wise cubic polynomial approximation to replace the mechanical spline. The smooth curve is achieved by requiring that adjacent cubic polynomials take on the value of the knots and that the first derivatives of these polynomials at the knots are equal. These conditions together with two end point conditions, such as assuming the second derivative at the two extreme knots is zero, allow a unique solution for the set of cubic polynomials for a given set of knots.

CasaXPS uses six knots to define five cubic polynomials cross a region, which determine the background shape for that region. The knots are evenly spaced between the start and end points of the region, while the intensity at a knot may be adjusted under mouse control. This is achieved on the Components Property Page, where selecting a point on the background and dragging the mouse to a new position reposi-

tions the nearest knot and therefore adjusts the shape of the spline background. If the knot is an internal knot then the spline is bent to a new shape, while if the knot corresponds to the end-points, the start and end offset region-parameters are adjusted resulting in the background gradient altering.
A new option on the Components Property Page allows the internal knots within a spline background to be optimised. Please note, the chi-square goodness-of-fit knows nothing about the physical world and therefore optimising a background, while improving the goodness-of-fit will probably have nothing to do with the true mechanisms unless the model optimised is perfectly defined and there is no noise in the data. Well-defined rigid models are the only models worth optimising. Backgrounds are the least-well understood shapes in XPS and as such optimisation is not recommended. The inclusion of the spline background type in CasaXPS is intended for use where all else fails: for example a peak positioned on a strong plasmon loss feature is not open to a Shirley background approximation, however a Spline Linear with a few judicious modifications may allow an analysis to proceed.
While optimisation of peaks and background is not recommended, optimising a background for a region without peaks does have a useful benefit. The spline background, for a well-chosen interval, will create a smooth approximation to the signal and so provides a means of estimating the noise in the data. The estimate for the noise will appear along with the residual trace in the form of the standard deviation in the normalised residual. If Poisson statistics are at work, then the value for the noise should be close to unity. With knowledge of this statistic for a given piece of data the target value for a peak-fit can be established.

References

[1] Doniach S. and Sunjic M., *J. Phys. 4C* **31**, 285 (1970)
[2] Shirley D. A., *Phys. Rev.*, **55**, 4709 (1972)
[3] Evans S., *Surf. Interface Anal.*, **17**, 85 (1991)
[4] Castle J. E. *et al*, *J. Electr. Spectr. Related Phenom*,**106**, 65 (2000)
[5] G. Wertheim, *J. Electron Spectrosc.* **6**, 239 (1975)
[6] Tougaard S., *Surf. Interface Anal.,* **25,** 137 (1997)
[7] Tougaard S., *Surf. Interface Anal.,* **11,** 453 (1988)
[8] Jo M., *Surface Science,* **320,** 191 (1994)
[9] Fiedor J.N., Proctor A., Houalla M. and Hercules D.M., *Surf. Interface Anal.*, **20,** 1 (1993)
[10] Do T., McIntyre N.S., Harshman R.A , Lundy M.E. and Splinter S.J., *Surf. nterface Anal.,* **27,** 618 (1999)

CHAPTER 8

Peak Fitting

Peak Fitting and Polymers

The availability of data from well-characterized samples such as those offered by Beamson and Briggs [1] owe much to the popularity of XPS as a tool for understanding the chemistry of polymers. A typical C 1s envelope [2] (Figure 8•1) includes structure that offers chemical information about a sample, but without some initial starting point it is difficult to construct an appropriate model for the data envelope.

The high-resolution spectrum in Figure 8•1 derives from poly (acrylic acid) PAA reacted with inorganic material (or partially reacted) in an acid-base reaction. If Gaussian/Lorentzian (GL) line-shapes are added in an arbitrary way, the curve fit yields little information about the sample other than to say that it deviates from the published data for PAA and therefore demonstrates the presence of additional chemistry at the surface. Figure 8•2 shows a synthetic model for this data envelope where three GL peaks have been added, then fitted using a Marquardt-Levenberg [3] optimization algorithm. No constraints have been applied and the result is a reasonable fit to the experimental data but the fitting parameters are not readily open to chemical interpretation. For example, the FWHM values are much bigger than would typically be expected from C 1s profiles for the given instrumental resolution. The three-peak model suggests further de-convolution is required before the sample can be fully understood.

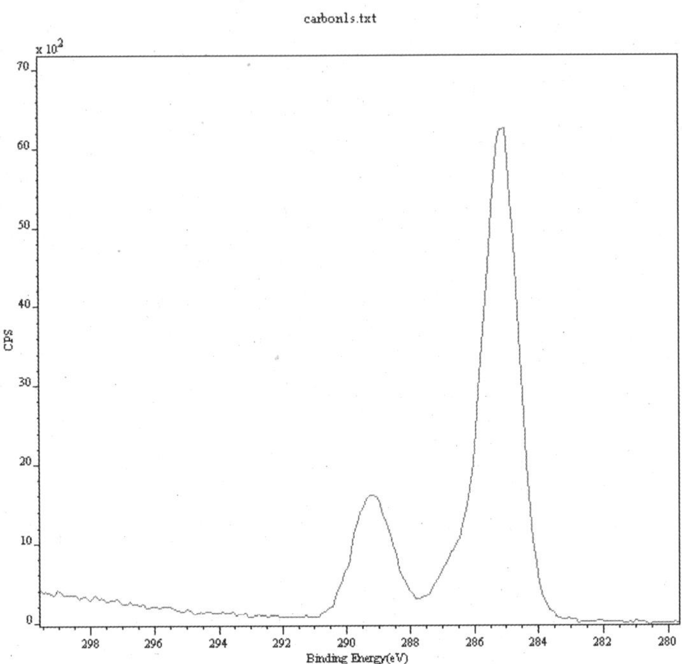

FIGURE 8•1 C 1s high-resolution spectrum taken from a polymer-based sample using a VG ESCALAB 220i at University College London, UK.

If it is assumed that the FWHM for a C 1s photoelectron peak is 1.1 eV (only a guess), then applying peaks with said constraint results in a peak fit shown in Figure 8•3. Two additional peaks appear in the model and further more three of the peaks look like they may have something to do with pure PAA. The PAA stoichiometry is still doubtful but the essential positions for a pure PAA envelope have more or less appeared. Further input is required to make sense of the new synthetic model.

To constrain a parameter so that it does not adjust during an optimization step, set the constraint interval to have the same value as the fixed value e.g., if it is required to set the FWHM value to 1.1, the constraint interval should be entered with the value "1.1,1.1". Actually, it is sufficient to set the parameter value outside the constraint range currently defined for the parameter

Spectra from Beamson and Briggs (available on CD in ISO 14976 format) offer the opportunity to examine more complex polymer data in the context of known synthetic models. Figure 8•4 on page 172 is an example of such data where a set of three line-shapes has been used to model the clean PAA C 1s data envelope. The important feature is that the stoichiometry and chemical shifts for the C 1s lines can be incorporated into the model and this information is then transferable to other polymer spectra. The peaks in Figure 8•4 are linked in area, but only the position of peak "C 1 b" and "C 1s a" are constrained by an offset.

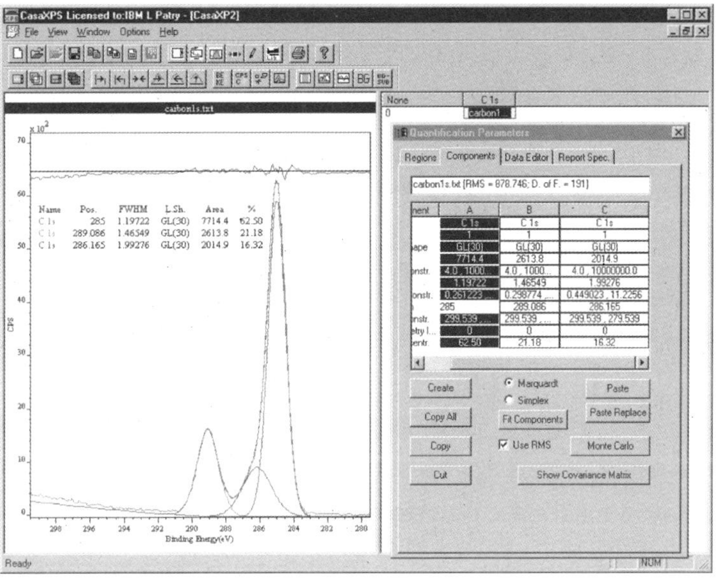

FIGURE 8•2 An initial fit to the C 1s spectra shown in Figure 8•1.

To link a component parameter the constraints must be adjusted as follows. Each synthetic component defined on the Quantification Parameter dialog, Components Property Page, appears as a column of parameters in the scrolled list shown in Figure 8•3. These columns are headed "A", "B", "C" and so on. To constrain the area of the component in column B to be half of the area of the component in column A, the area constraint in column B should be set to "A * 0.5". Similarly, to offset a component in column C by 0.2 from the component in column B, enter "B+0.2" in the position constraint field in column C.

The peak shapes from the Scienta ESCA-300 used to produce the Briggs and Beamson data may differ from a VG ESCALAB 220i (the source of the real data in Figure 8•1) or any other manufacturers instrument, but the essential structure should be suitable as a basis for the new model. Copying the pure PAA model into the data in Figure 8•3 leaves a residual that requires an adjustment for the two non-PAA peaks together with the introduction of a third peak.

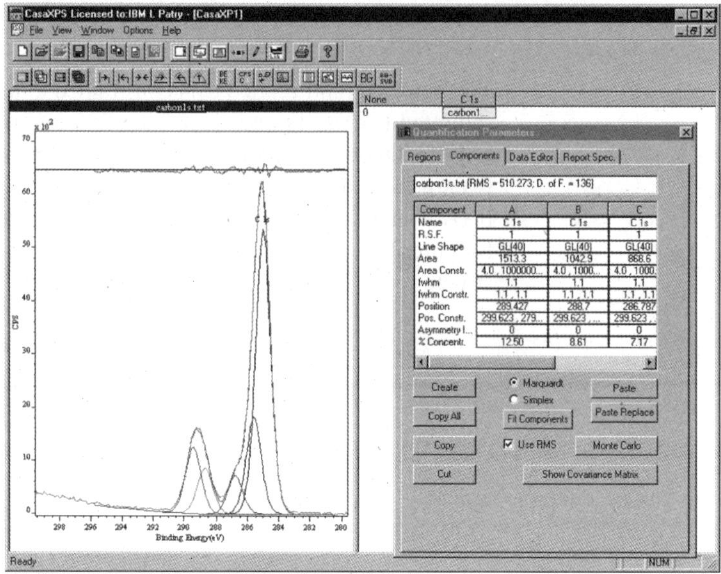

FIGURE 8•3 Same C 1s envelope as Figure 8•2 but the synthetic peaks are all constrained to have FWHM equal to 1.1 eV.

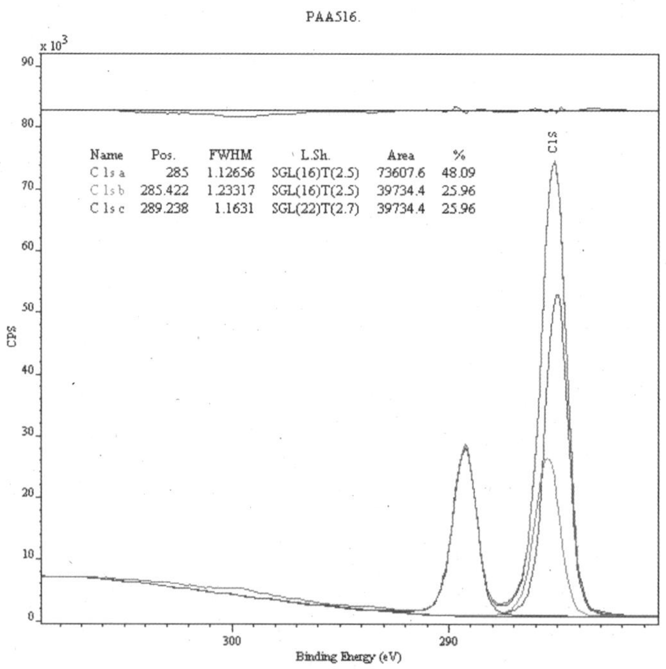

FIGURE 8•4 C 1s envelope from clean PAA acquired on a Scienta-300, RUSTI, Daresbury Laboratory UK.

The new peak in Figure 8•5 is constrained to be the same width and position as the saturated PAA C 1s peak located at 285 eV (BE). The area of this new peak is allowed to adjust at will and accounts for carbon with the same characteristics as the PAA peak at that position. The consequence of introducing the new peak is that the PAA synthetic model can adjust without breaking the stoichiometric relationships for pure PAA, while differences in the intensity of the saturated peak from the PAA structure are allowed for by this additional constrained component (Figure 8•5).

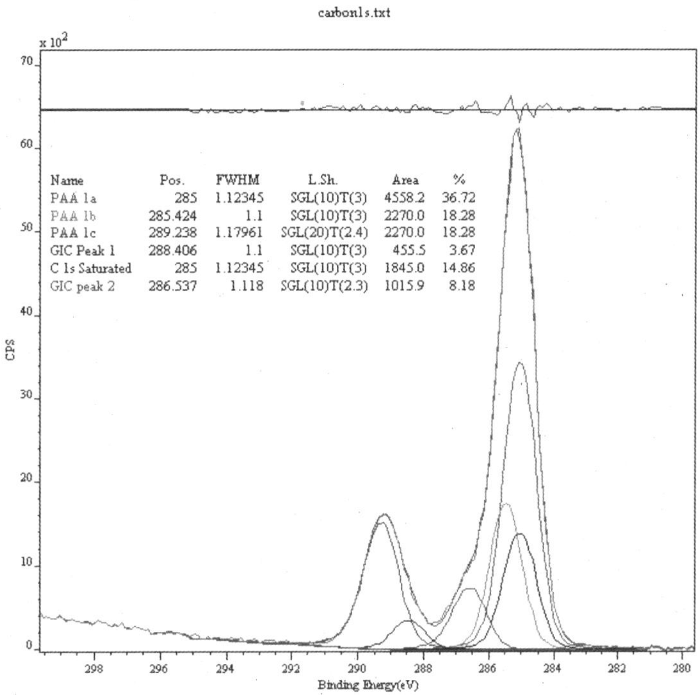

FIGURE 8•5 Final form for the synthetic model. The Glass Ionomer Cement (GIC) [2] C 1s envelope containing three peaks from PAA plus three additional peaks not seen in a clean PAA spectrum.

Synthetic models such as the one above can be tested using Monte Carlo simulation techniques [4] to assess the stability of the peak parameters with respect to noise in the data. Once a set of peaks and constraints has been developed, the Monte Carlo procedure simulates repeated identical experiments on the same sample, and for each simulation a new fit is determined for the peak parameters. The result of this procedure is a set of distributions for the individual parameters from which scatter plots may be constructed that highlight the way noise influences the data model. Figure 8•6 is a scatter plot for the normalized peak areas of the non-PAA peaks against the saturated C 1s peak from the pure PAA model. The peaks at 288.4 eV and 285 eV have areas that are anti-correlated with the pure PAA model (all the peak areas within the PAA model are constrained to one another,) but the synthetic component at 286.54 eV shows a degree of correlation with the intensity of the pure-PAA area.

At first glance the behavior of the GIC 2 peak at 286.54 eV is unexpected since one might think that two peaks next to one another should produce anti-correlated area distributions. The constraints have altered the concept of “next to” since the PAA sub-model spreads across the entire envelope and it becomes difficult to judge by eye what the influence of noise might be on the final result. This type of insight can only help to understand what constraints do to a fitting procedure as well as provide a rule-of-thumb estimate for error bars (multi-dimensional error distributions can seldom be described by a single number.)

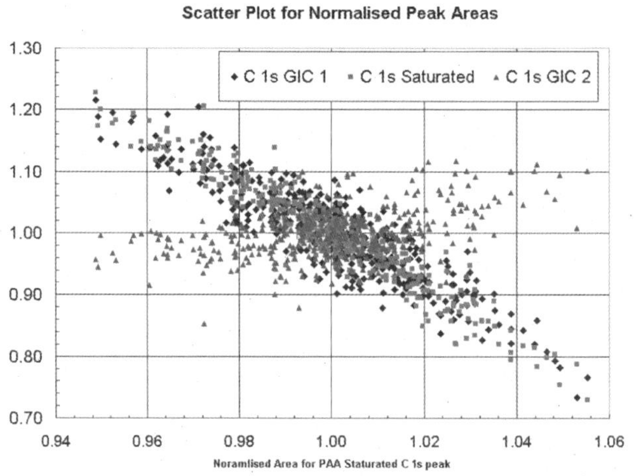

FIGURE 8•6 Monte Carlo simulation results for normalized peak areas. The three peaks associated with GIC are plotted against the saturated C 1s peak area from the pure PAA model.

The exact meaning for the model in Figure 8•5 is left to the experimentalist and may need changing in view of other input, however such a model is only possible when peak fitting routines offer mechanisms for fixing parameters with respect to one another. The role played by the pure PAA synthetic model is that of a foundation shape from which differences in the unknown polymer can be assessed. These additional peaks may still require further interpretation, but with the aid of chemical knowledge and supporting evidence a meaningful model can emerge from seemingly intractable data.

Multi-scan Peak Fitting

The follow example illustrates some of the issues associated with peak-fitting a unified model across a set of related high-resolution spectra. A summary of the steps involved is detailed below, as too is a description of how to copy and fit these models to other groups of spectra.

Spectra from a blend of Poly(methyl methacrylate) PMMA and Poly(styrene) PS was recorded on the SSI (SSX 100/206) with a monochromatic Al Kα X-ray line, a pass energy of 50 eV and an analyzed area of about 1.4 mm^2 (Figure 8•7).

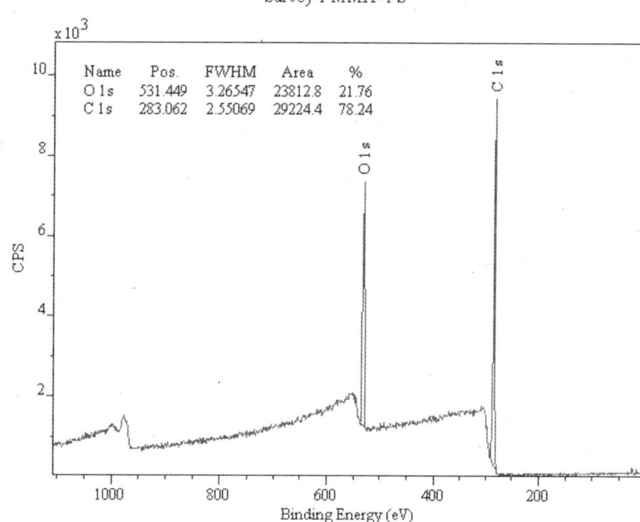

FIGURE 8•7 Blend of PMMA and PS recorded at CIFA [5].

Spectra from pure PMMA and PS [2] show that only PMMA has a contribution to the O 1s line seen in Figure 8•7. The PMMA C 1s and O 1s peaks can be decomposed in respectively four [C-(C,H), C-C=O, C-O and O-C=O) and two [O==, O-] components the areas of which follow the 2:1:1:1:1:1 ratio. The C 1s [C-(C,H] peak of PS appears at the same binding energy as the dominant C-(C,H) component in PMMA C 1s line.

The traditional method for analyzing the high-resolution spectra from such a material is to construct two separate peak models for the C 1s and O 1s spectra and then optimize the peak parameters without reference to the required relationship between C 1s and O 1s peaks components. The final peak fits are then judged by comparisons between the intensities for the two sets of peaks optimized separately. Alternatively, a PMMA peak model could be constructed for both the C 1s and O 1s lines as well as the PS C 1s line(s), which is then optimized as a unit. The outcome of such an optimization can differ from that of optimizing the peak fits separately.

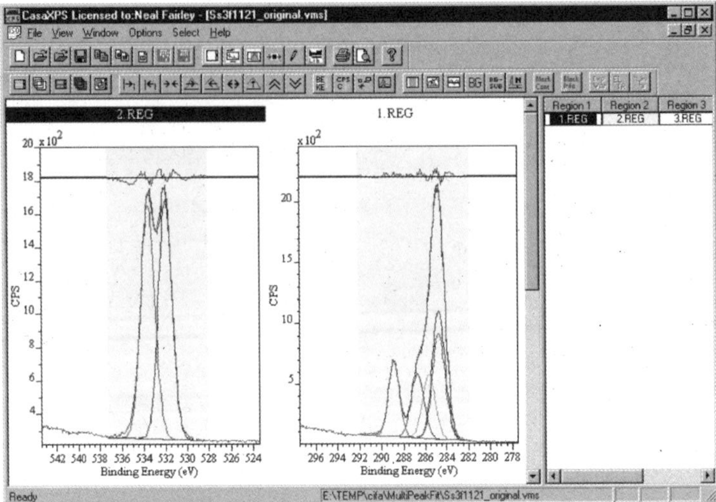

FIGURE 8•8 Separate peak-fits for the C 1s and O 1s high-resolution from the PMMA+PS blend.

Figure 8•8 and the corresponding quantification reported in Table 8•1 show the results of optimizing the C 1s and O 1s synthetic peaks based upon each to the separate spectral regions. The constraints shown in Table 8•1 do not include any information about the relationship between the PMMA C 1s and O 1s lines. Although the residuals displayed in Figure 8•8 are reasonable, the Atomic Concentrations in Table 8•1 deviate from the expected stoichiometry for the PMMA polymer.

TABLE 8•1.

Name	Energy	% Conc.	Pos Const	FWHM Const	Area Const	Const Id
C 1s-1	284.744	22.37	297.51 , 277.51	0.3433 , 8.582	E * 2	A
C 1s-2	284.744	18.38	A - 0	A * 1	4.0 , 10000000.0	B
C 1s-3	285.675	11.19	297.51, 277.51	A * 1	E * 1	C
C 1s-4	286.72	11.2	297.51 , 277.51	A * 1	E * 1	D
C 1s-5	288.872	11.21	297.51 , 277.51	0.2227 , 5.568	4.0 , 10000000.0	E
O 1s	532.031	12.82	543.51 , 523.51	0.5929 , 14.82	4.0 , 10000000.0	A
O 1s	533.628	12.84	543.51 , 523.51	0.2376 , 5.94	A * 1	B

Figure 8•9 on the other-hand has been fitted using additional constraints linking

FIGURE 8•9

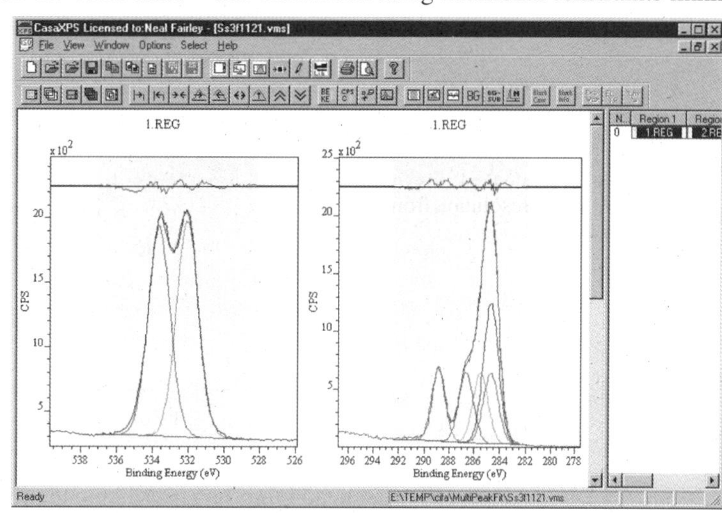

the areas for the PMMA C 1s lines to the O 1s lines shown in Table 8•2. The optimization is performed on the C 1s spectrum at the same time as the O 1s data and as a result the relative intensity for the saturated carbon peaks has altered between the models defined by Table 8•1 and Table 8•2. Without imposing the additional constraints between the C 1s and the O 1s PMMA lines in Table 8•2, it is difficult to visually determine how the C 1s envelope should be fitted in order to achieve the correct stoichiometry for the PMMA component. Obviously, individually optimized spectra will not have the necessary input to ensure the known relationships and therefore fitting the spectra as a unit has clear advantages.

TABLE 8•2.

Name	Energy	% Conc.	Pos Const	FWHM Const	Area Const	Const Id
C 1s-1	284.711	25.02	297.51 , 277.51	0.3433 , 8.582	E * 2	A
C 1s-2	284.711	12.43	A - 0	A * 1	4.0 , 10000000.0	B
C 1s-3	285.499	12.51	297.51, 277.51	A * 1	E * 1	C
C 1s-4	286.639	12.51	297.51 , 277.51	A * 1	E * 1	D
C 1s-5	288.866	12.51	297.51 , 277.51	0.2227 , 5.568	4.0 , 10000000.0	E
O 1s	532.032	12.51	543.51 , 523.51	0.5929 , 14.82	E*2.93	A
O 1s	533.623	12.51	543.51 , 523.51	0.2376 , 5.94	E*2.93	B

Performing a Multiple Region Optimization

Optimization across spectral regions is achieved by overlaying the spectra concerned in a single tile, creating all the components required as part of the model and pressing the "Fit Components" pushbutton on the Components property page. When more than one spectra appears in the active tile, performing a fit (by pressing the Fit Components button) causes CasaXPS to prompt the user with a dialog window asking whether the fit should be achieved using multiple spectra. If the Yes button is pressed on the dialog window, the optimization will use the components defined on the active spectrum in the active tile and then apply these components to each of the spectra displayed in the active tile. If the No button is pressed then only those components corresponding to the active spectrum will be optimized; the Marquardt optimization method will leave components with zero intensity inside the data region unchanged.

Constructing a Model Spanning Multiple Spectral Regions

CasaXPS allows synthetic components to be defined on a single spectrum where the location of a synthetic component does not have to belong to the energy range for the data in that spectrum. Thus both the C 1s and O 1s synthetic components can be defined on the same spectrum. If both spectra to be fitted are overlaid in a tile and the Components property page on the Quantification dialog window is visible, then creating components for both the C 1s and the O 1s will assign these components to the active spectrum in the active tile. When multiple spectra are overlaid, the active spectrum is the spectrum whose name appears in the top text-field of the Components property page. It is useful to note which spectrum is the active spectrum out of the set of spectra ultimately used in the optimization.

As a rule, when a new component is created CasaXPS attempts to find a reasonable set of parameters based upon the residual for the active spectrum. The active spectrum cannot be used to define peak parameters for components outside the range of the data and so an alternative strategy must be used for creating these out-of-range components. In order to create the basis components for the model, each of the individual spectral regions may be used to initially create the appropriate peaks. Once the peaks have been created on the individual spectra, the "Copy All" button on the Components property page allows these peaks to be moved from one spectrum to another. After assembling the entire set of peaks on the active spectrum, the appropriate constraints can be imposed.

Viewing the Components and their Residual

The normal mode of display shows each spectrum together with components and corresponding residual over the energy range for that spectrum. When peaks are defined on a single spectrum corresponding to other related spectra then an alternative display mode applies. An option on the Display property page on the Tile Parameters dialog window allows the components displayed to be switched to only those on the active spectrum. Once this switch is set, the components and residuals from the active spectrum of the active tile are plotted against each of the spectra displayed in the active tile. Figure 8•10 shows the Tile Parameter dialog window where the "Display Comps from First Spectrum" option is ticked. Note that without this display option ticked, the components outside the energy range of the active spectrum will not be visible, even though the overlaid spectra are displayed. If this option is not ticked, any residual plots will relate to the individual spectra and their specific components (if any) rather than the common set of components on the active spectrum.

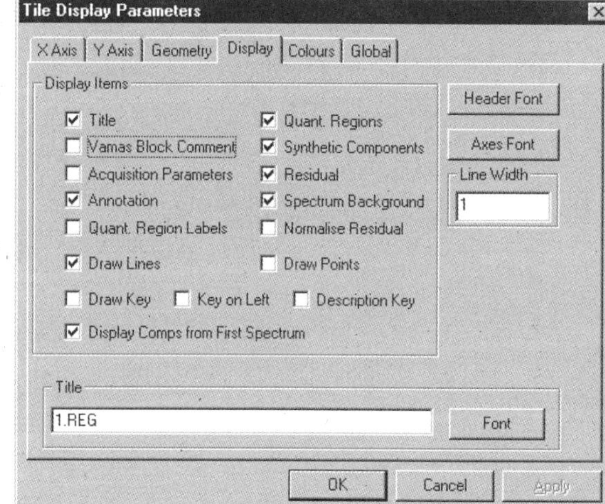

FIGURE 8•10 Display Comps from First Spectrum.

Issues associated with common peak models

- RSF values
- Transmission correction

Once a set of peaks for the C 1s and O 1s spectra are defined, the problem of relating intensities must be addressed. The raw spectra for the two regions differ with respect to the relative sensitivity for the elements, variation in the transmission efficiency of the instrument and mean-free-path adjustments. In the case of an SSI instrument the mean-free-path energy dependency is entered into the file created by CasaXPS as a second corresponding variable. The transmission variation is then accounted for within the RSF values. Generally, however, to compare data from different elements and/or acquired over different energy intervals requires the raw spectra to be corrected for transmission as well as adjustments made to any area constraints included in the combined model using RSF information.

Figure 8•11 shows the Components property page, where the peak in column A represents the C-(C,H) saturated hydrocarbon component from PMMA while column B is fixed with respect to column A in terms of position and FWHM. Column B represents the PS C-(C,H) line. Column C is a second PMMA C 1s line and each of the C 1s lines from PMMA are related in area to the C 1s line in Column E. All these C 1s peaks have identical RSF values and are measured over approximately the same energy range. This however is not true for the PMMA O 1s lines and although the energy-dependence variation can be removed by applying the Intensity Calib option, the RSF differences must be added to the model in the form of constraints. Note that in Table 8•2 the area constraints for the O 1s lines are set to the ratio of RSF value for the O 1s lines to the C 1s lines. The peak parameters for the area are relative to the active spectrum and so the O 1s areas must be scaled by the RSF values as well as the stoichiometry for the molecule. Stoichiometry dictates that the area for the O 1s lines should be the same and also equal to the area of the C 1s line in Column E (Table 8•2).

Summary

Steps required when peak-fitting across multiple high-resolution spectra:

- For each high-resolution spectrum involved in the peak-fit, define appropriate backgrounds.
- Overlay the spectra from each region used in the peak-fit.
- Select the display mode for the components and residual so that components from the active spectrum are displayed with respect to the spectra overlaid in the active tile. The Display property page (Figure 8•10) offers the “Display Comps from First Spectrum” option.

Add all the components to the active spectrum for each region displayed and apply constraints.

- Press the Fit Components pushbutton (Figure 8•11). A dialog window appears asking for conformation that a Multiple Peak-fit is required. Answer YES.

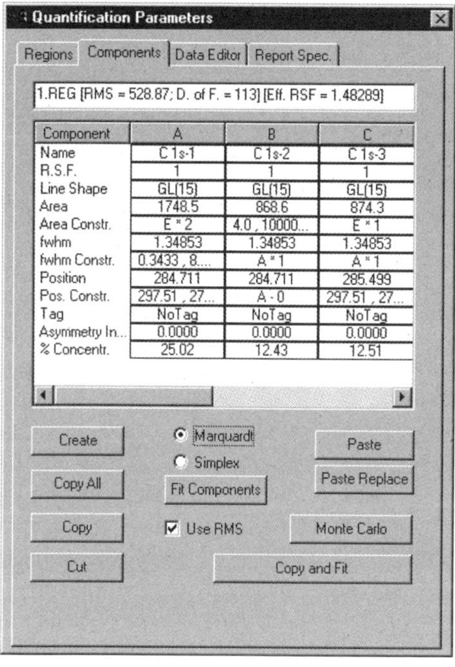

FIGURE 8•11 Components Property Page.

Automatic Fitting of Peaks Across Multiple Spectral Regions

Once a peak fit has been performed on a single tile containing multiple spectral regions, the set of peaks can be copied and auto-fitted to other overlaid spectra. This requires each set of spectra, for which an auto-fit is required, to appear in separate tiles in the scrolled list on the left-hand-side of the Experiment-frame. The scrolled list in the left-hand-side can be loaded with overlaid spectra by selecting the spectra using the right-hand-side browser view and then pressing the toolbar button. The result of pressing the toolbar button is to create a set of tiles (one per row selected in the right-hand-side) in which the selected spectra are overlaid on a row basis. Figure 8•12 on page 182 shows a depth profile where three rows in the right-hand-side are selected and subsequently displayed using the toolbar button. Spectra displayed in this format may be fitted to a synthetic model by pressing the Copy and Fit button on the Components property page. Note that only components will be copied by this operation, so background regions must be defined prior to the button being pressed.

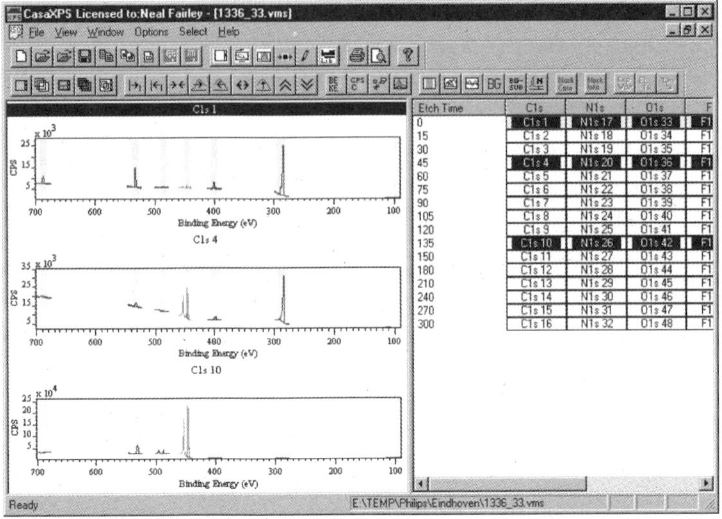

FIGURE 8•12 Each row selected in the right-hand-side of the experiment frame is displayed as overlaid spectra in the left-hand-side once the toolbar button is pressed.

Peak fits using multiple spectral regions: An example

Consider a dataset containing a number of spectral regions where the same set of regions are acquired repeatedly as part of a time-elapse experiment. The spectra recorded at a given time are subject to identical charging effects, however, each temporally dispersed set of spectra exhibits differing energy shifts due to sample charging effects. One means of charge compensating the data is to use the Al 2p peak shown in Figure 8•13 as a reference for the three peaks shown in the O 1s spectrum and shift each spectrum recorded at the same time by a common offset so that the Al 2p spectra are aligned throughout the time sequence. Alternatively, the positional relationship between the Al 2p and O 1s peaks can be maintained using a peak model that includes both the Al 2p and the O 1s peaks and then use the peak position constraint to guide the O 1s peak positions with respect to the A 2p peak position during the course of optimisation.

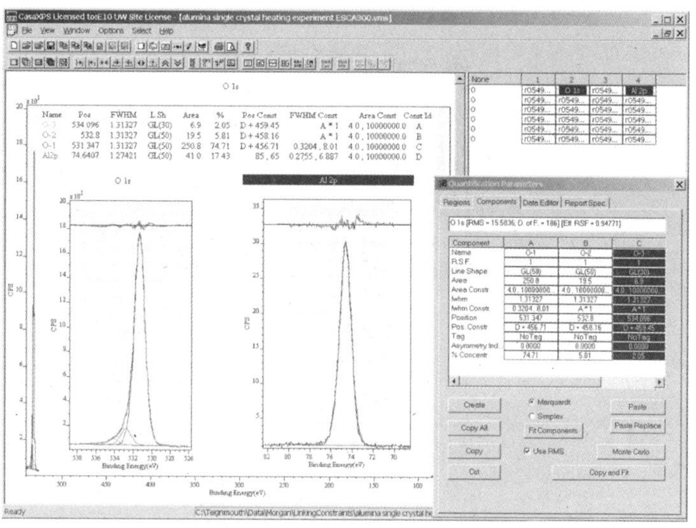

FIGURE 8•13 Synthetic peaks defined for the O 1s region, which includes both O 1s and Al 2p peak, displayed by overlaying the O 1s and Al 2p spectra in a single tile. Inset Tiles display the same data but zoomed into each of the respective spectral regions.

Constructing the Peak Model

When two or more spectral regions are displayed in a single tile, the active spectrum within that tile is the spectrum for which the block name appears in the top text-field on the Quantification Parameters dialog window (Figure 8•13). The combined peak model for the O 1s and the Al 2p spectra must be defined on the active spectrum, which for the situation shown in Figure 8•13 is the O 1s spectrum in each row. Once the four peaks, three O 1s and one Al 2p, are defined on the O 1s spectrum and the two spectral regions for the O 1s and Al 2p data are overlaid in a single display tile, pressing the Fit Components pushbutton causes a dialog window to warn that multiple sets of data will be used to optimise the fitting parameters (Figure 8•14). Press the Yes pushbutton to continue with the optimisation based upon both sets of data.

FIGURE 8•14 Dialog window that indicated more than one spectral region will be used to perform peak parameter optimisation.

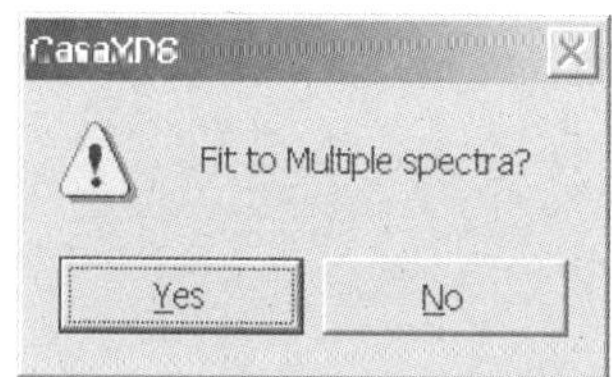

Viewing the Extended Peak Model on Overlaid Spectra

The synthetic peaks and corresponding residual displayed with the data is determined by a check box on the Tile Parameters dialog window shown in Figure 8•15 If the Display Comps from First Spectrum check box is ticked then the display shows only the components defined on the active spectrum. Note, a synthetic component is only displayed in energy intervals for which spectral data exists. Selecting the Display Comps from First Spectrum check box causes the overlaid spectral regions to be applied to the components defined on the active spectrum and therefore the components and residual for the extended data become visible in the overlay tile.

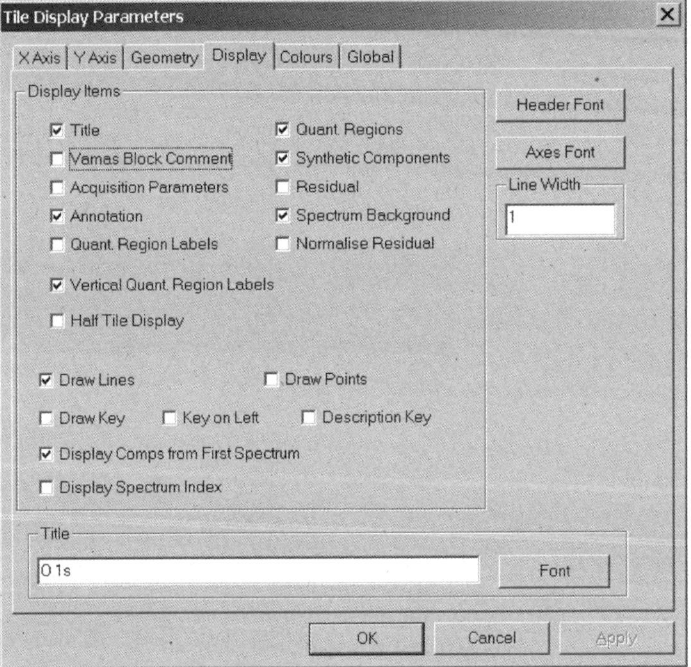

FIGURE 8•15 Display Comps from First Spectrum check box.

Copying Peak Models

The auto-fit option uses the display tile to determine when the peak model should derive from the active spectrum. The Copy and Fit pushbutton on the Components property page of the Quantification Parameters dialog window will copy the peak model from the active spectrum in the active tile and attempt to fit this model to all spectra loaded into the current list of display tiles. Note that the background regions must be defined before the Copy and Fit pushbutton is invoked since the copy action only concerns the synthetic peaks. To transfer and fit the peaks in the current example:

1. Propagate any regions required to define the background for the data involved in the peak-fit. In this case both the O 1s and the Al 2p regions should be transferred to the other members of the experiment using the propagate mechanism.
2. Select the two columns of O 1s and Al 2p spectra as shown in Figure 8•16.
3. Press the toolbar display button. A scrolled-list of display tiles is created where each tile contains the spectra overlaid from each row of selected tiles in the right-hand browser view.

4. Now press the Copy and Fit pushbutton on the Components property page of the Quantification Parameters dialog window. A dialog window offers information about how many tiles will be affected by the operation and provides a chance to abort the action. If the Yes pushbutton is chosen on the popup dialog window then the peak-fit from the active tile will be copied to each active spectra in the scrolled-list of display tiles and a fit to the data is performed.

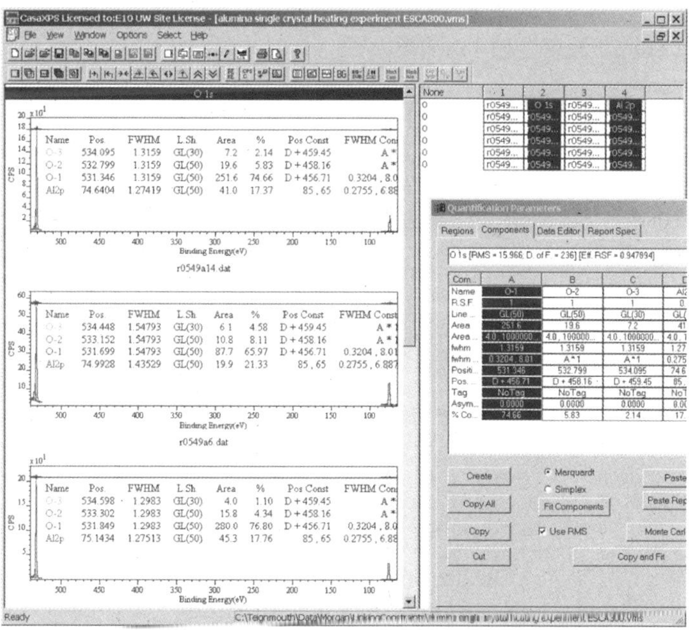

FIGURE 8•16 List of display tiles showing each row of selected O 1s/Al 2p spectra display overlaid in preparation for the Copy and Fit operation.

Curve-Fitting using Models Defined from Data

Figure 8•17 shows a set of C 1s spectra where the problem is to assess the relative contributions from the spectra annotated with C 1s 2 (HSE), C 1s 3 (DBS) and C 1s 4 (Fap ERT) to the spectrum FavE black. Relationships of this nature can be investigated using line-shapes defined from the spectra, where these line-shapes act as synthetic components in a peak model for the FavE black data.

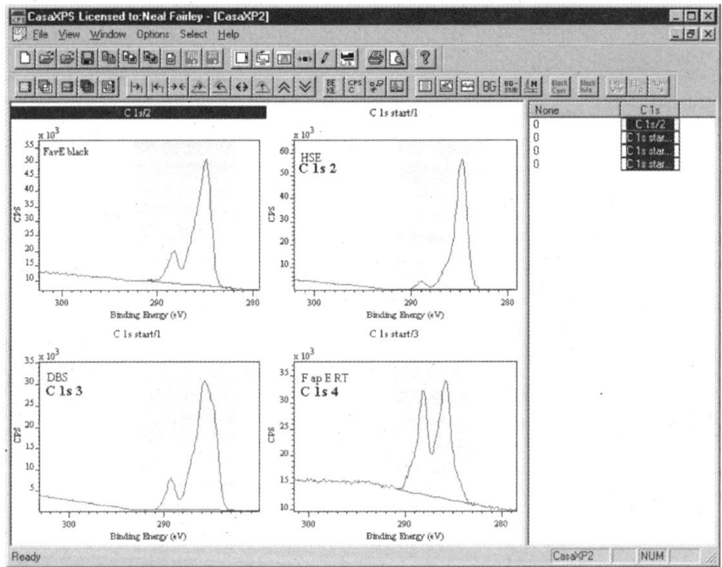

FIGURE 8•17 Initial set of C 1s envelopes.

Defining line-shapes from the data

Each spectrum from which a line-shape is to be defined must have a background removed from the data, and therefore a quantification region should be created for each of the basis spectra (Quantification Parameters Dialog Window, Regions Property Page). Once a region exists for each target spectrum the data should be both background subtracted and scaled using the CPS Scale option on the Calculator Property Page of the Processing Dialog Window. A scale factor of unity will result in a processed spectrum that is appropriate for comparison purposes.

The processed data should then be preserved to new ISO blocks within the same file. Select the background subtracted spectra in the right-hand-side of the Experiment Frame, then right-clicking the mouse over the right-hand-side of the Experiment

Frame and ticking the Data Only checkbox before pressing the OK button on the Append Blocks dialog window.

Figure 8•18 shows the Experiment Frame after the processed blocks have been appended to the current file. Note that the transition fields within the data blocks have been edited to include an M character. Altering the transition field causes the processed block to appear in a new column. The Block Id fields were also edited to unique names, e.g. C 1s 2, C 1s 3 and C 1s 4. These names will be used to identify the data blocks as the source for the line-shapes used to model the FavE black spectrum

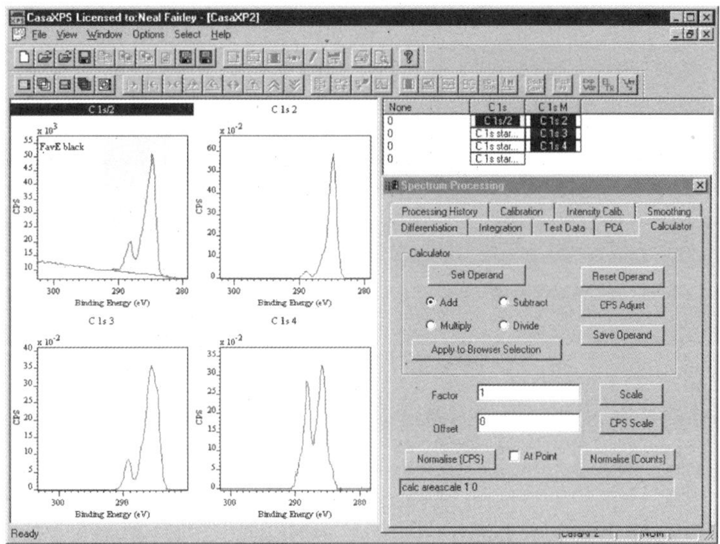

FIGURE 8•18 Line-shapes from the basis spectra C 1s 2, C 1s 3 and C 1s 4 created from the original spectra using the Calculator Property Page and then appending the processed data to the same file. Note that the new blocks have been copied using the Data Only flag set. This ensures the line-shapes are permanent and cannot revert back to the unprocessed data.

Once the data have been prepared as described above, all that remains is to enter the Block Id strings into the Line Shape edit-fields on the Components Property Page (Figure 8•19).

NB Only the Simplex Method provides consistent optimization results when Data is used to define line-shapes. The Marquardt method relies on derivatives that are not well defined for complex peak structures and noise.

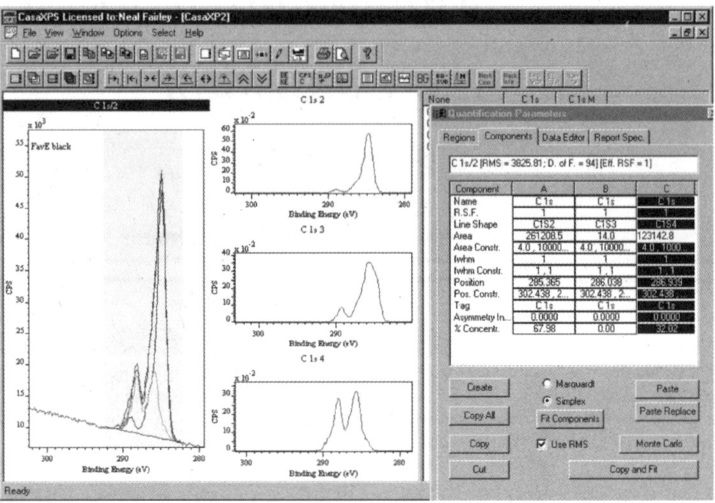

FIGURE 8•19 Peak fit using the three line-shapes defined from data.

References

([1] - [5] only referenced from the text)

[1] Beamson G. and Briggs D., "The XPS of Polymers Database" Surface Spectra Ltd (2000)

[2] Jones F.H., Hadley P.C., Hutton B.M., Eccles A.J., Steele T.A., Billington R.W. and Pearson G.J., "Fluoride uptake by glass ionomer cements: a surface analysis approach" , Submitted to "*Biomaterials*" (2003).

[3] Press W.H. et al, "Numerical Recipes in C", Cambridge University Press (1988)

[4] Cumpson P. J. and Seah M. P., "Random Uncertainties in AES and XPS", *Surface and Interface Analysis*, **18** 361 (1992)

[5] Michel Genet (private communication)

[6] Tougaard, S."Practical Algorithm for Background Subtraction", *Surface Science*, **216**, 343-360 (1989).

[7] Carley, A.F. and Morgan, P.H., "Computational Methods in the Chemical Sciences", 1989, Ellis Horwood Limited

[8] Conny, J. M. and Powell, C. "Standard test data for estimating peak-parameter errors in XPS" *Surface and Interface Analysis*, **26**, 939-956 (1998)

CHAPTER 9

Manipulating Spectra - the Calculator

CasaXPS provides a "spectrum Calculator" which enables the normal arithmetic operations (addition, subtraction, multiplication and division) to take place between spectra or spectral regions.The Calculator is a means of managing and extracting information used to identify spectral features and then to take that information on to quantify regions and components. Calculator operations may be performed in a "one-to-one" or "one-to-many" fashion.

The Calculator User Interface

The Calculator section of the Calculator property page seen in Figure 9•1 on page 190 represents those options associated with the definition and manipulation of the temporary block information referred to as the **operand**.

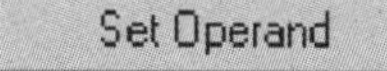

The operand is set from the active spectrum (first data block selected and displayed in the active tile). On pressing the Set Operand button, the current temporary block maintained by the Calculator property page is replaced by the data block for the active spectrum in the active tile. A trace from the operand is displayed in the active tile whenever the Calculator property page is active.

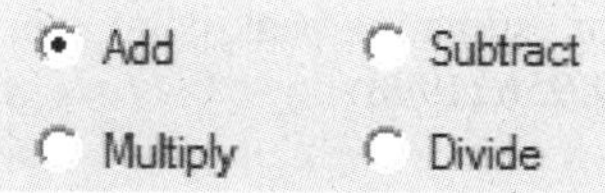

The operand is combined as a right operand with the current browser selection using the operator selected from these radio buttons. The operator must be chosen before the Apply to Browser Selection button is pressed.

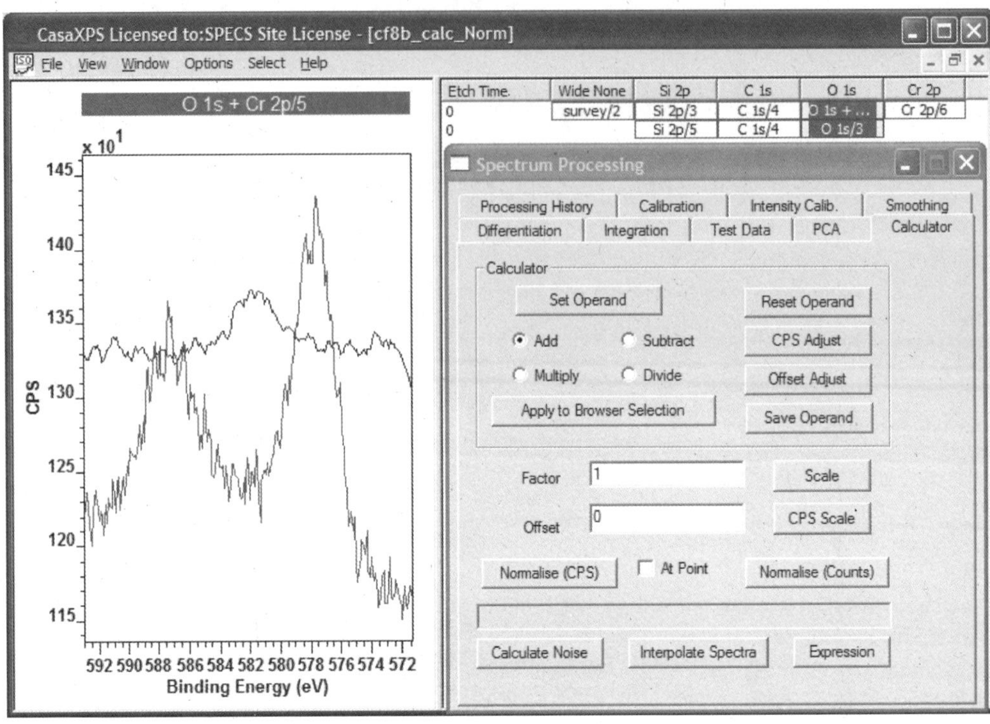

FIGURE 9•1 Calculator Property Page

Apply to Browser Selection

The data from the current operand is used as the right-operand for the chosen operation and is applied to each left-operand corresponding to the data blocks selected in the Experiment Browser. The result of the operation is stored in the process data for the selected block. Resetting the processing operations using the Processing History property (page 201) will restore the raw data, thus undoing the action of the calculator.

Reset Operand

The data held within the operand may be adjusted relative to the spectrum displayed in the active tile. These adjustments can take the form of button actions or mouse controlled offsets, shifts and scaling. When adjusting the operand using the mouse, the Shift-Key must be held down. If the mouse is left-clicked at a point in the active tile whilst holding down the Shift-Key, the operand is offset in order to position the operand trace through the point at which the mouse click was performed. To scale and shift the operand the mouse should be dragged whilst at the same time the Shift-Key is held down. The vertical size of the resulting box determines the scaling applied to the operand data and the horizontal dimension produces a shift in the energy position. Pointing to a data channel on the operand with the Shift-Key held down and

then dragging the mouse to the comparable position on the active spectrum will move the operand over the top of the active spectrum.
The action of the Reset Operand is to scale the data in the operand to match the ordinate range for the active spectrum in the active tile. That is to say, the operand data is forced to be visible within the active tile and therefore in a state where the mouse actions can be used to position the operand relative to the active spectrum.

CPS Adjust

Pressing the CPS Adjust button alters the dwell-time and number of scans within the operand to match the data block corresponding active spectrum.

Offset Adjust

Adjusts the data in the operand to ensure all the ordinate values are non-positive. For further discussion of the application of the offset adjust button, see page 199.

Save Operand

Once an operand has been adjusted appropriately, the block maintained by the Calculator property page can be saved to an Experiment Frame using the Save Operand button.

The remainder of the options on the Calculator property page act directly on the active spectrum or indicated data blocks. For some operations, the result of the operation is a new Experiment Frame containing the block generated by the action of a button: for most the processed data in the active spectrum is altered by the operation.

Scale

The ordinates within the active spectrum are scaled and offset using the values entered into the Factor and Offset text-fields on the Calculator property page. A quantification region must be defined on the spectrum. The background subtracted data is scaled by the value entered in the Factor text-field and added to the background plus the value entered in the Offset text-field.

CPS Scale

Analogous to the Scale option, however the data is normalise to the area of the first region defined on the spectrum before scaling is performed. The background is removed from the data and the offset value added to the normalised scaled ordinates.

Normalise (CPS)

Normalisation based on a data channel within a spectrum is performed using this button. The action of the button is modified by the “At Point” tick box. If the tick-box is deselected, the normalisation is performed using the data channel with maximum intensity, however when ticked, the normalisation is performed using a datum indicated by a mouse action. To define the desired position at which the normalization is to be performed, hold the Shift Key down and left-click on the spectrum. The feedback text on the Calculator property page will indicate the value for the energy determined by the Shift-Left-Click action. Each datum in the spectrum is divided by the intensity at the selected bin, where the intensity is measured in counts per second.

Normalise (Counts)

Normalisation based on the maximum intensity in a spectrum, where the intensity is measured in counts.

Calculate Noise

The Calculate Noise option offers a means of visualising the noise in a spectrum. The procedure repeatedly applies a linear least squares fit to a linear polynomial over localised sections of the spectra with the view to minimising the information content at a local level. The difference between the spectrum and the smoothed curve replaces the processed data in the block. The procedure is most effective away from peaks.

Interpolate Spectra

Two spectra overlaid in the active tile are used to generate a sequence of data blocks, where the new blocks contain data derived from linear interpolation between the two original spectra. Linear combinations of the two spectra are generated where the percentage step size for the interpolating parameter is entered in the Factor text-field on the Calculator property page. The original spectra are assign experimental variables of zero and one hundred, so for example, if the Factor text-field is set to "1", then the number of interpolated spectra will be one hundred and one. A new Experiment Frame is created containing the interpolated spectra.

Expression

An arbitrary arithmetic expression including the data blocks from a given Experiment Frame is used to create a new block within a new Experiment Frame. Each data block within an ISO file is assigned an index number starting with zero. The individual blocks are referenced, when included in an expression, using the block index in the format vb0, vb1, vb2, etc. and combined using constants and brackets as seen in the Block Expression dialog window shown in Figure 9•2. Once an expression is entered into the text-field and the OK button pressed, a new block is created in a new Experiment Frame; the block comment contains an entry indicating the name of the original ISO file and the expression used to create the data within the new block.

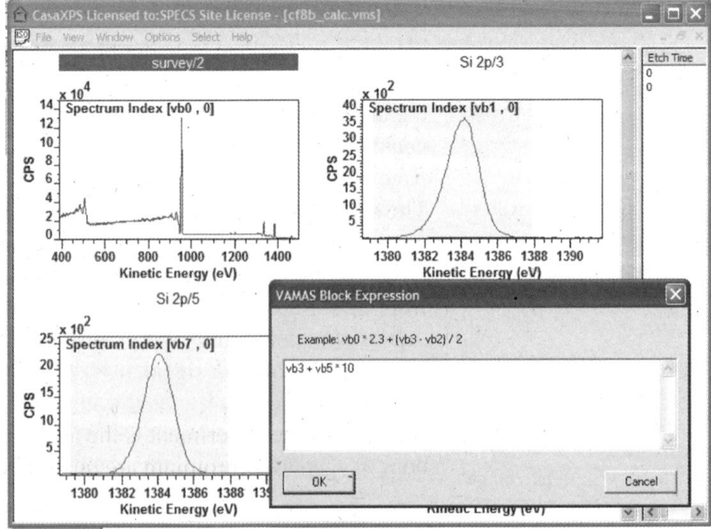

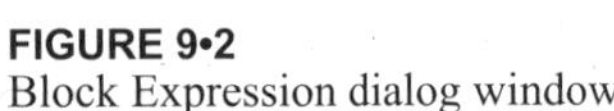

FIGURE 9•2
Block Expression dialog window

Displaying the VAMAS block index:

Figure 9•2 shows four tiles (one hidden by the Block Expression window) each displaying a single spectrum. At the top of each bounding box a string displays information about the source for the data shown in the tile, namely the Block Index and the index for the Corresponding Variable. The index for the Corresponding Variable on display is typically index zero, while the vb# (Block Index) depends on which block is drawn in the display tile. The option for displaying the spectrum index information is located on the Display property page of the Tile Display Parameters dialog window (see page 72).

Background Subtraction: an example

The first step in any quantification is to define backgrounds for spectral features. The most commonly used backgrounds are linear, Shirley or Tougaard, all of which owe their popularity to ease-of-use rather than the ability of these shapes adequately to describe the background beneath a photoelectric peak. Nevertheless, despite their failings, the desire of most analysts is to apply one of these algorithms to all peaks used for quantification in a consistent manner and gain virtue by producing repeatable results. Further, the choice of background is often coupled with a particular set of sensitivity factors, so there is a strong desire to reduce all data to an appropriate state in which the preferred background is easily deployed.

The data set chosen to illustrate the Calculator capability comes from a surface modification experiment where an "after" spectrum contains a small Cr 2p doublet located on a tail (energy loss structure) from a large O 1s peak. In the absence of the O 1s peak, most analysts might choose to use a Shirley background as a means of defining the intensity for the Cr 2p doublet; but the energy loss structure from the oxygen results in the chromium doublet being "tilted" with respect to its normal appearance in the absence of oxygen. Furthermore, the O 1s energy loss structure contains minor peaks lying within the limits of the chromium doublet region (Figure 9•3 on page 194) and these loss peaks are of significant size compared to the Cr 2p doublet. The spectra shown in Figure 9•3 were acquired from modified and unmodified surfaces, where the Cr 2p doublet appears in the spectrum after modification of the material. The hope is therefore to remove the contribution from the O 1s loss structure from the Cr 2p peaks in the modified material using the unmodified O 1s loss structure as a model, before applying a standard Shirley background to define the intensity of the chromium signal.

The goal of the experiment is the analysis of a set of samples for which silicon, carbon, oxygen and chromium are quantified using narrow scan spectra.

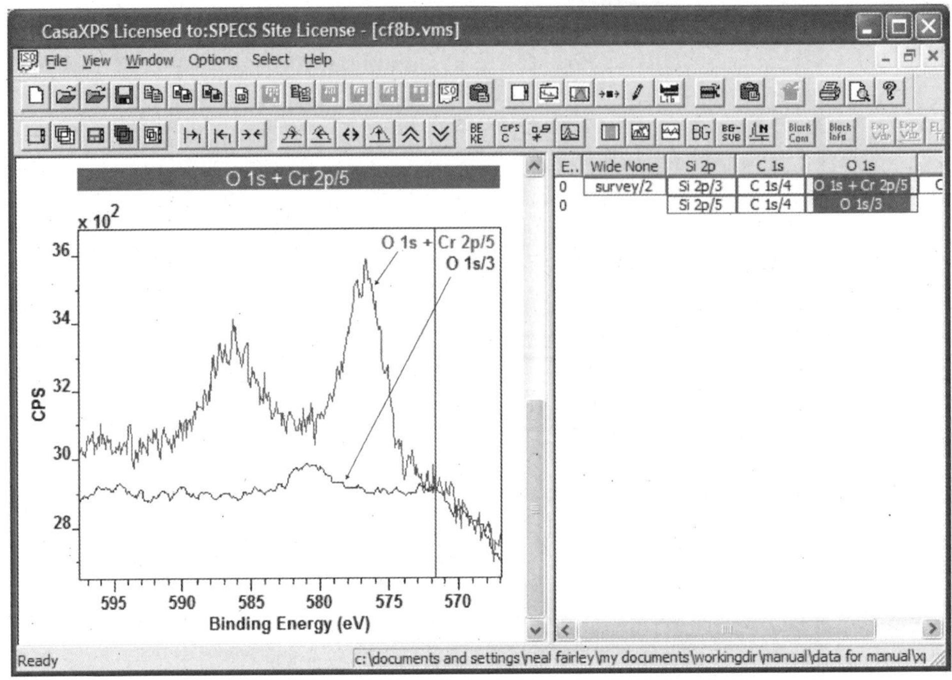

FIGURE 9•3 Comparison of the Cr 2p energy interval for modified and unmodified materials

The O 1s peaks shown in Figure 9•4 illustrate the nature of the Calculator's task more clearly. The lower trace is the data taken from the unmodified material and, although the count rates are similar, the data from the modified material differs potentially in magnitude and relative position of the spectral features. Before attempting to remove the distortions from the Cr 2p doublet due to the O 1s loss structure by subtracting the unmodified data from the modified data, the two sets of data must be calibrated in energy and scaled in intensity in such a way as to retain the essential quantification information.

The Experiment Frame in Figure 9•3 displays a zoomed section of the O 1s traces displayed in Figure 9•4, the two traces in Figure 9•3 being visualised using a display option which normalises the displayed data at a point indicated by the vertical line. The spectra as displayed in Figure 9•3 look reasonably well matched, although the choice for the normalisation point required an element of trial and error before settling on the position seen there. While visually acceptable, the same procedure for scaling the data is far less acceptable when the difference of the scaled data is intended for use in quantification against other intensities. When quantification is involved, the method for scaling and shifting the data must be as systematic as possible.

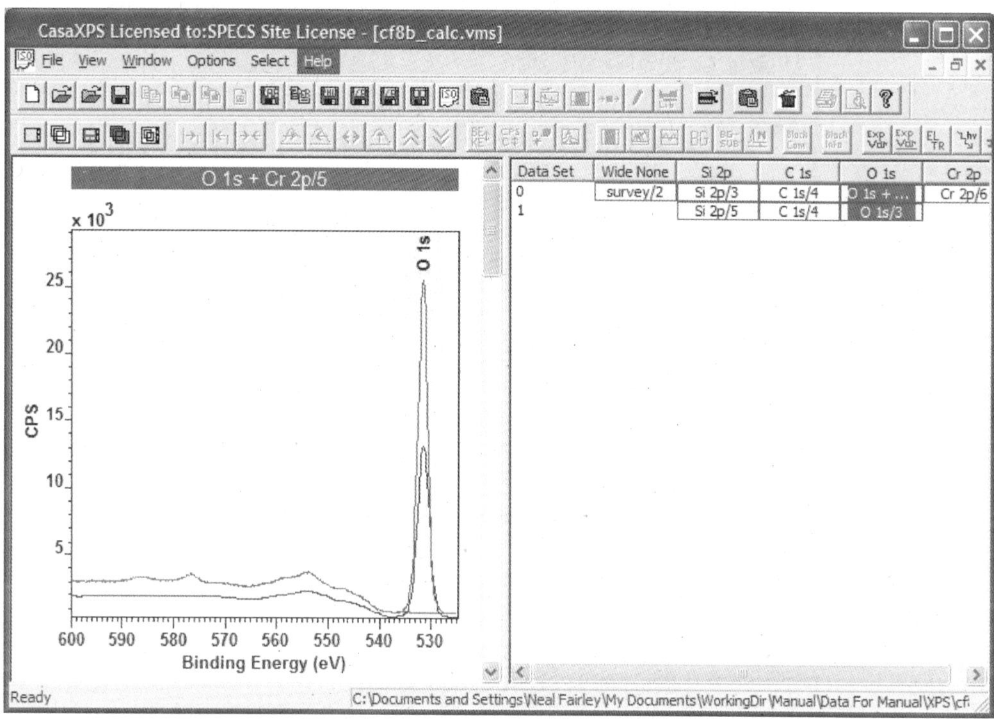

FIGURE 9•4 Experiment Frame showing the data blocks from the modified (top row) and unmodified (bottom row) samples

Normalising the Data

The procedure for normalising a data-set in preparation for use with the Calculator is performed using the functions from the Calibration property page of the Spectrum Processing dialog window (Figure 9•5 on page 196: see also Figure 6•12 on page 114). The Calibration property page in Figure 9•5 is defined ready for the data set shown in Figure 9•4, where the Si 2p spectra in each data set are about to be used to compute the necessary adjustments to both intensity and energy. The idea is that the two samples are essential equivalent in terms of the Si 2p spectra and that any differences in position and intensity of these two independently acquired Si 2p spectra can be accounted for within the acquisition process. That is to say, if each spectrum within a data set is energy calibrated and intensity scaled using a normalisation constant determined from the Si 2p peak, then provided the same procedure is applied to both data sets, the background characteristics on the Cr 2p data can be accommodated by subtracting the two sets of scaled spectra.

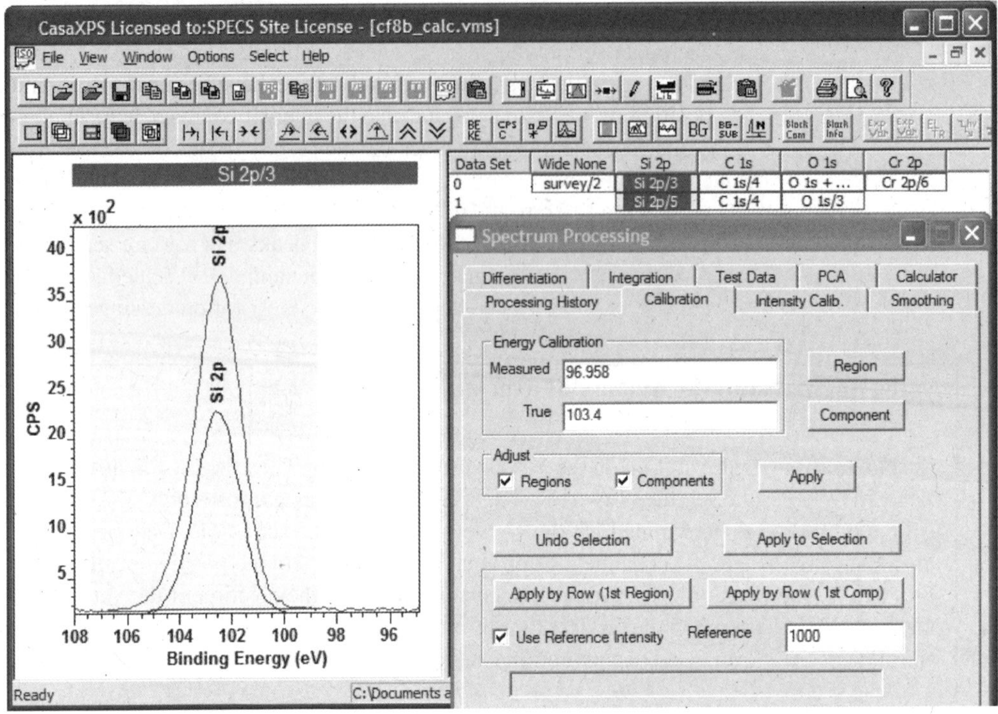

FIGURE 9•5 Calibration Property Page where the parameters are set-up ready to scale and energy calibrate a data set.

Furthermore, making these same adjustments to each spectrum in the data based on the counts per second intensity from the Si 2p data will leave the quantification table invariant with respect to the transformation.

The steps to perform these transformations are as follows:

1. Define a quantification region on each of the Si 2p spectra.
2. Enter the position for the Si 2p peak in the True text-field on the Calibration property page.
3. Since regions are defined on the Si 2p spectra as part of this procedure, the tick-boxes in the section labelled Adjust for both regions and components should be ticked.
4. Ensure the tick-box labelled Use Reference Intensity is also ticked.
5. Ensure only the Si 2p blocks are selected in the right-hand-side of the Experiment Frame.
6. Finally, press the button labelled Apply By Row (1st Region)

The consequence of these steps is the creation of a new Experiment Frame in which each data set now appears as energy calibrated and intensity adjusted spectra. The Si 2p region defined on the original data set have been used to compute the measured position of the Si 2p peak and also the intensity of the peak in counts per sec electron Volts. These computed values for the two Si 2p spectra are used to shift each of the spectra on a row by row basis and the intensity of each spectrum is scaled so that the area of the Si 2p peaks in the new Experiment Frame is equal to the arbitrary value entered in the Reference text-field. All peaks in a row are scaled by this same value resulting in rows of data from different samples all scaled so the intensities are directly comparable. The data set is now ready for processing by the Calculator.

Note: Normalising the display of overlaid spectra: 

The **Normalise Display** option on the **Y Axis** property page of **Tile Display Parameters** dialog window (page 70) enables the scaling of two or more spectra for display purposes only; the spectra are displayed using the same axes determined for the active spectrum in the active tile. Data are scaled using the intensities from a single data channel within each spectrum, where the position of the data channel is determined by the location of a vertical line cursor. To adjust the position of the vertical cursor the Shift Key must be held down while the left-mouse-button is click at a point on the active tile.
The state of the Normalise Display option can be toggled using the toolbar button.

The new Experiment Frame generated by the calibration step is assigned the same name as the original ISO file, but with the string _Norm concatenated with the base name of the file (see the title bar for the CasaXPS main window in Figure 9•6 on page 198). The data in the normalised file is scaled in intensity measured in CPSeV, therefore the dwell-time and number of scans fields in the original file are no longer meaningful and so are set to unity in the normalised data blocks. A further point worth noting is that the area field reported for the Si 2p spectra on the Regions property page of the Quantification Parameters dialog window is equal to the requested scale value, which means that transmission and escape depth corrections are also accounted for during the intensity scaling procedure. The transmission and escape depth parameter are transferred to the new file of scaled spectra so quantification can proceed in an identical fashion to the equivalent computation performed on the original file.

The intensity calibration might equally well be performed based on a synthetic component rather than a quantification region. If the intensity for the normalisation is determined from a synthetic component, it may be a surprise to observe the area parameter from the synthetic component, following the normalisation, is not necessarily equal to the reference intensity, which was the case for normalisation based on regions.

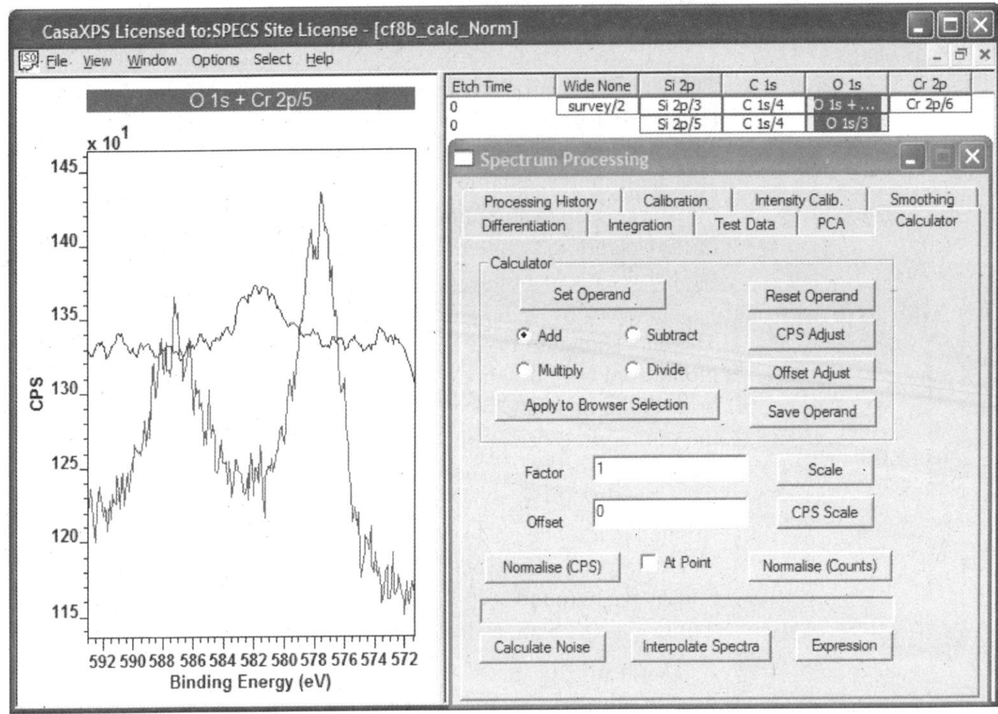

FIGURE 9•6 Experiment window with Calculator Property Page

Indeed, if transmission and/or escape depth corrections are included in the calculation, the area parameter for the synthetic component used to normalise the spectra will not equal the reference value. This is because the area value seen on the Components property page of the Quantification Parameters dialog window is a parameter for a peak model and is therefore an input value. The values used in a quantification report are output parameters and according to the rules used in CasaXPS, only output values are adjusted for transmission and escape depth, therefore the area reported for a component in a quantification report will agree with the reference intensity used to normalise the spectra. In the case of the quantification region, the area is computed from the data and therefore an output; hence the transmission and escape depth corrections are already included area as determined from the region.

Subtracting the Spectra

The Calculator property page module in the Spectrum Processing dialog window (Figure 9•6) offers a range of options for performing arithmetic on spectra. The calculator is structured so that a spectrum is loading into a temporary location associated with the Calculator property page and from this temporary location the data may be applied to any number of spectra indicated via the Browser, where the result of the operation is passed into the processed data for each of the spectra included in the Browser selection.

The data shown in Figure 9•6 are displayed using a zoomed section of the O 1s regions from both the modified and unmodified materials following the energy calibration and intensity normalisation procedure described above. After normalisation, the unmodified background lies above the Cr 2p doublet peaks, although data are scaled appropriately. The calculator requires the data from the unmodified sample to be loaded into the temporary location and the trace in the temporary location positioned below the Cr 2p doublet by an offset operation. Once prepared, the temporary data can be subtracted from the Cr 2p data to yield a spectrum for which a Shirley background is applicable. Using the normalised Experiment Frame, the steps within CasaXPS are as follows:

1. Display the O 1s spectrum from the unmodified sample in the active tile and press the Set Operand button on the Calculator property page. The trace associated with the temporary location on the Calculator property page is displayed over the top of the data in the active tile and will continue to be displayed for as long as the Calculator property page is the top page on the Spectrum Processing dialog window.
2. Double-click the Cr 2p narrow scan spectrum block from the modified material. The Cr 2p spectrum will display in the active tile and, provided the calculator page is active, the calculator operand (temporary location data) will appear also in the active tile.
3. Press the Offset Adjust button (Figure 9•6). The action associated with the Offset Adjust button causes a negative offset to be applied to each data channel of the temporary calculator operand. For XPS spectra, this ensures each data from the calculator operand is below the target spectrum; in this case the Cr 2p data.
4. Select the Cr 2p VAMAS block in the right-hand-side of the Experiment Frame. Only those spectra selected in the browser will be operated upon by the calculator. Further more, if the same operation is required for a set of Cr 2p spectra taken from different samples, say, then selecting all the Cr 2p spectra at the same time permits the action to be applied on bulk.
5. Select the type of arithmetic operation required. In this example the radio button for subtraction should be chosen.

6. Press the button labelled Apply to Browser Selection. The current processed data in the Cr 2p spectrum is replaced by the section of the O 1s spectrum subtracted from the Cr 2p data.

The exercise is completed by defining appropriate quantification regions for the four narrow scan spectra (C 1s, Si 2p, O 1s and the computed Cr 2p data) and the creating a quantification table either using the Quantification property page on the Annotation dialog window or the Report Spec property page of the Quantification Parameters dialog window.

The result of the quantification based on the normalised data and the use of the calculator is superior to a quantification performed on the original data. Apart from eliminating the general trend of the O 1s loss background, a first order correction for the resonant loss feature clearly visible in Figure 9•6 is also possible. For small Cr 2p peaks, the background due to scattering of the O 1s transition would certainly result in an over estimate for the intensity of the Cr 2p signal if the procedure described above were not followed.

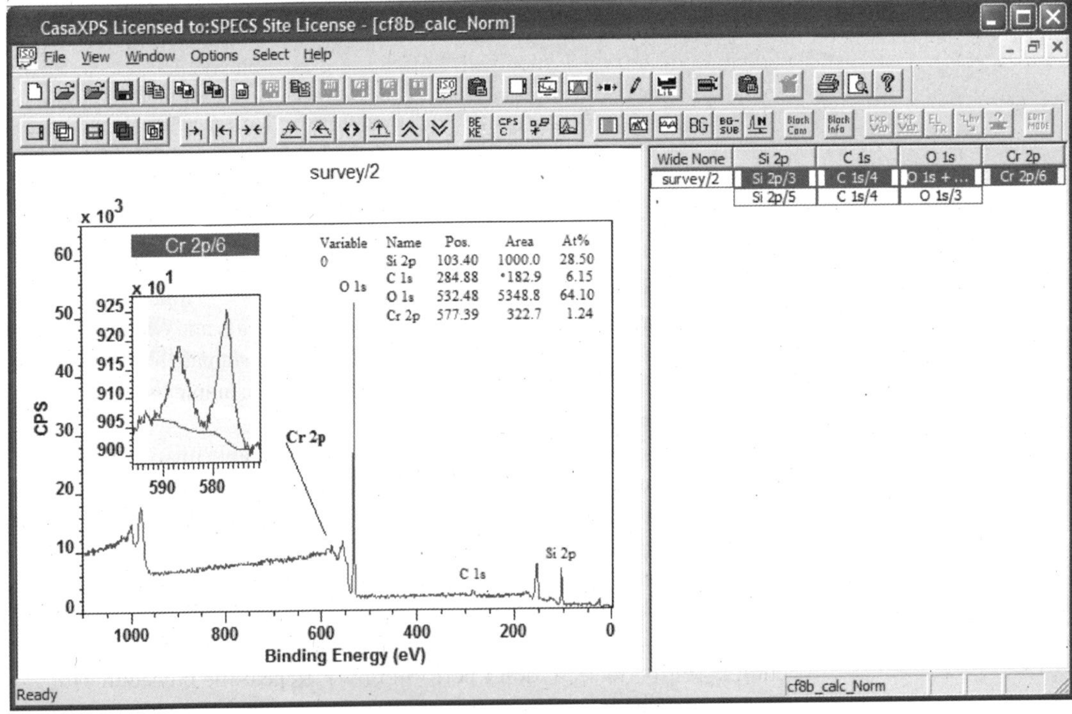

FIGURE 9•7 Final quantification

A Note of Caution

Any manipulation of the data resulting in intensity adjustments invalidates the use of Monte Carlo error analysis. Uncertainties determined from quantification regions or synthetic peaks require the data to obey Poisson statistics; any scaling of the data will remove the information needed to analyse the errors and therefore error bars can not be computed for data modified by the calculator.

Processing History: a general note

Each spectrum file data-block can be processed using any of the available options. The "Processing History" property page provides the means of viewing the operations that have been applied to the data. Further, the processing may be reversed entirely or selectively. The "Reset" button removes all the processing operations and restores the data to that originally supplied in the ISO 14976 file. Alternatively, by selecting a set of lines within the scrolled list, then pressing the "Apply Selection" button, the indicated set of processing operations are performed on the original data. (Selection of items follows the normal Microsoft conventions - left mouse button for a single item, left click then shift left click for an inclusive list, or control click for separated items). Thus the previous processing history is replaced by only those actions selected prior to pressing the button.

Processing for a block of data can be globally applied to a set of data blocks. Select these blocks within the Browser, then right click on the corresponding display view to reveal a dialog window for propagating operations such as curve fitting and processing. Choose the required actions then press the "Apply" button. A second window appears showing the progress of the propagation and offers a "Stop" button. If the "Stop" button is pressed, the propagation will terminate following the completion of the current action. Note that some actions, for example curve fitting, may take a significant period of time to complete.

Not all actions for processing the data may be applicable to all the selected blocks. To prevent, for example, the energy calibration for the current block from being propagated to other blocks, the history item can be flagged so that it is not propagated along with other appropriate actions.

To exclude a processing operation from a propagate action, select the item in the history list, then press the button labelled "Propagate Flag". The entry in the list for the selected processing option will change to indicate that the flag for propagating the action is set to "false" ("don't perform this"). Repeat the procedure (toggle) to remove the exclusion.

Charge Correction

For both conducting and non-conducting samples not electrically connected to the instrument, the electrical potential at the surface during analysis floats to a stable state different from the potential of the instrument itself, one result of which can be spectra with peak positions offset in energy from the expected values. The most common cause of shifts in peak positions is an analysis of insulating materials, where, assuming the adoption of some reasonably adequate charge compensation mechanism, XPS spectra are possible without deformation of the peak-shapes, but for which a calibration of the binding energy scale in the form of a uniform energy adjustment is required. CasaXPS corrects for energy shifts using the Calibration property page on the Spectrum Processing dialog window. There are several methods for energy calibrating a file of spectra and the following section describes the steps involved.

The information required to shift the peaks to a new position is simply two values representing the measured and the true peak positions. Although there are several strategies for energy calibration, in essence all represent different means for determining these two values.

Energy Calibration Based on Locating a Synthetic Peak

A data envelope is the sum of all the individual peaks beneath the measured curve and while a maximum intensity is identifiable, the position of the maximum does not necessarily directly relate to a given transition. The relative intensities of the underlying transitions may cause the position of the maximum to vary whilst the location of these constituent peaks are fixed and determined by physics. The desire is therefore to identify the peaks contributing to the data envelope and then energy-calibrate the spectrum so as to position one of these synthetic peaks at a known energy. Consider the example shown in Figure 9•8, where a carbon 1s spectrum taken from a nylon sample illustrates the case in point. Four synthetic components are used to model the C 1s data and these peaks must be fitted to the envelope before the position of the lowest binding energy component is used to charge reference the sample. A sequence of CasaXPS steps leading to the energy calibration of the C 1s spectrum is as follows:

1. Create a quantification region defining the background to the C 1s peak.
2. Create a peak model for the data.
3. Select the synthetic component on the Component property page and press the Component push-button on the Calibration property page (Figure 9•8). The values for the Measured and True text-fields are updated from the position of the

selected component and, if the selected component name matches as entry in the Element Library, the True text-field is updated from the library.

4. Make any adjustments to the True text-field necessary to define the position of the selected component.
5. Tick the boxes labelled Regions and Components in the Adjust section on the Calibration property page.
6. Press the Apply button on the Calibration property page.

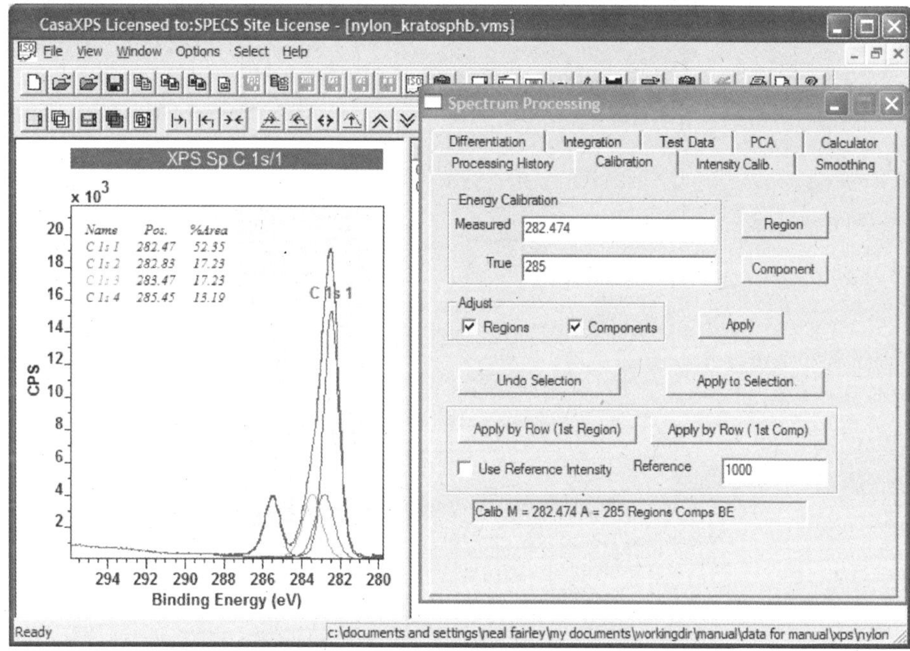

FIGURE 9•8

For the example in Figure 9•8, a detailed description of the above steps follows.

Creating the Background

Regions and Components are created on the Quantification Parameters dialog window. The Regions property page is partially displayed over the CasaXPS main window in Figure 9•9, where the parameters for a single quantification region are listed under the heading A. To create a region, the button on the Regions property page labelled Create is pressed. Initially, the limits for the quantification region, and therefore the energy interval over which the background is defined, are determined from the current zoom state for the displayed spectrum.

After creation, the limits specified via the Start and End parameters in column A

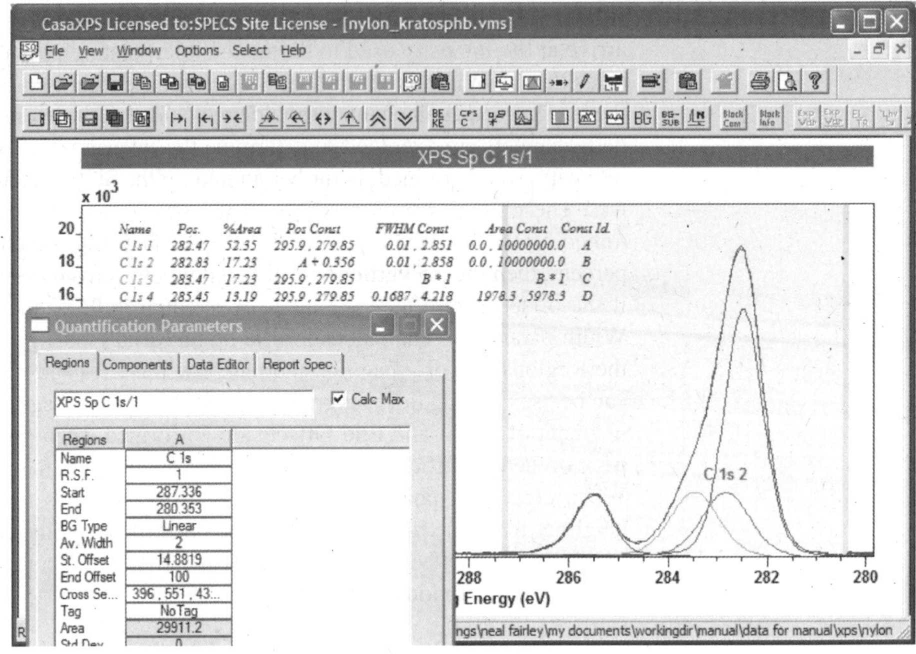

FIGURE 9•9

may be entered manually or adjusted using the mouse. Each field on the Regions property page table can be selected using the cursor and the left mouse button. On selection, the parameter field becomes a text-field and a new value can be entered; the value is only accepted when the Enter key is pressed on the keyboard. Alternatively, provided the Region property page is the top-most property page on the Quantification Parameters dialog window, the cursor when dragged using the left-mouse button allows the limits for the quantification region to be adjusted by eye. Two grey vertical lines mark the extremities of the quantification region and the appearance of these grey limits indicates the Start and End parameters are susceptible to alteration via the mouse. The background type is set to Linear, meaning a linear polynomial is used to interpolate the background defined by two energy/intensity coordinates at the start and end limits of the quantification region. The start and end energies for the quantification region are chosen so that the interpolating polynomial best describes the background below the peaks. The intensity at which the background is defined at an interval limit is, in this example, modified by two quantification region parameters: the Av Width parameter and the Start/End Offset parameter. The Av Width parameter specifies the number of data channels either side of the start and end channels over which the intensities should be averaged to arrive at the intensity used to define the interpolating polynomial. Figure 9•9 indicates the Av Width is set to two, which specifies that two intensities either side of the background limit channel should be included in the average. That is, in total five data channels are averaged to determine the intensity at a limit (Figure 9•10). The intensity, so determined, is further altered by the St. Offset and End Offset parameters. These offset parameters represent a percentage offset for the intensity determined by the Av Width parameter. If the offset were specified to be one hundred percent, then the background would go down to zero-intensity at the limit, whereas if the offset is zero then the intensity is equal to the value determined by the Av Width parameter. The values for the St. and End Offset parameters are entered on the Regions property page or alternatively, the left-hand-mouse button and cursor can be used to drag the background up and down at the region limits. The mouse adjustment of the St and End Offsets are performed when the Components property page of the Quantification Parameters dialog window is the top-most property page. When creating a peak model, it is often advantageous to tweak the background height at the same time as the peak parameters; hence the mouse action for the St Offset and End Offset values is associated with the Components property page rather than the more obvious Regions property page.

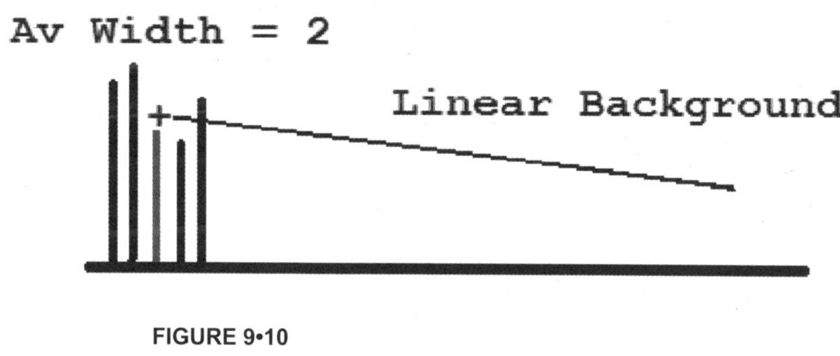

FIGURE 9•10

Creating the Peak Model

Creating a synthetic peak model for an unknown material is challenging, at best. For this reason a well known material, namely nylon, is chosen as the example data. Given the C 1s spectrum in Figure 9•9, the first response is to reach for The XPS of Polymers Database edited by Beamson and Briggs. The data envelope and least-squares optimization using synthetic line shapes offers numerous possible peak models none of which are likely to represent the true chemistry of the surface. Faced with an unresolved set of peaks clearly responsible for the structure to lower binding energy of the two resolved shapes in Figure 9•9, simply adding peaks until the best residual is obtained would result in the inclusion of anywhere between three and six synthetic peaks. Thus, by using the experience and chemical knowledge of Beamson and Briggs, a four component model describing the C 1s spectrum of nylon is selected. Once again, knowing the number of peaks again is not sufficient to create an accurate description of the data simply based upon line-shapes and unconstrained optimised peak parameters. The challenge is therefore to select line-shapes and constrain the optimisation parameters to ensure the peak model conforms to a hypothetical stoichiometry. The data in Figure 9•9 were collected from a Kratos Axis 165, while the Beamson and Briggs data derives from a Scienta ESCA 300. Line-shapes from two different instruments are subject to variation due to the response of the analyser and also the efficiency of the charge compensation mechanism. As a consequence, the line-shapes used in the Beamson and Briggs database are not necessarily applicable to the Axis 165 data, thus leading to a degree of self-expression when choosing the Gaussian/Lorentzian and asymmetry mixtures, but nevertheless, the guidance offered by the Beamson and Briggs model is invaluable. The peak model shown in Figure 9•9 is constructed using the Components property page (Figure 9•11) on the Quantification Parameters dialog window. The four components are created by pressing the Create button followed by adjusting the newly created peak to a plausible position and size. The mechanism for modifying the parameters of a synthetic

FIGURE 9•11

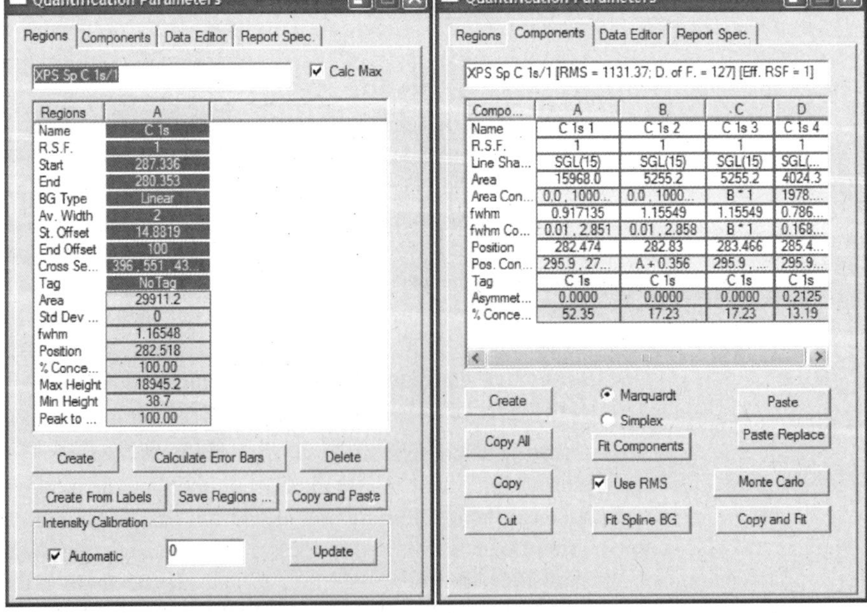

peak involve mouse actions, where the left mouse button allows the position and height of a peak to be adjusted; dragging the position of the peak maximum causes these two parameters to update. The FWHM of a peak is also adjustable under mouse control; however the action of dragging the peak from about the half height on the side of a peak updates the FWHM and the height. The height adjusts at the same time as the FWHM to maintain the area of the peak. To change the FWHM alone, the Shift keyboard key must be held down during the drag movement. As new peaks are added to the model, the size and position of each newly created peak is determined from the greatest variation in the residual between the data and the current synthetic envelope. All the component parameters are offered on the Components property page and can be manually entered using the same type of text-field available on the Regions property page.

Without constraining the component parameters, when optimised, these parameters take on values determined by mathematics. In an ideal world, the line-shapes and background would be exactly those required to describe the physics responsible for the spectral shapes, also the data would be devoid on noise. Within this ideal world, the optimisation step would stand a chance of returning meaningful values for the parameters; however, simply the presence of noise in the spectra would be enough to cause problems to unconstrained optimisation, let alone the absence of the true background and line-shapes from the calculation. When analysing real spectra, the influence of noise and the lack of the exact functional forms for the line-shapes forces the use of parametric constraints. The Component property page tabulates the components (Figure 9•11) where each set of component parameters are displayed in a column beneath a header labelled with an alphabetic character. Constraints between peak parameters are defined in terms of these labels heading the columns. Figure 9•9 on page 204 includes an annotation table in which the constraint relationships are listed. For example, the highly correlated peaks in columns A and B on the Components property page (Figure 9•11 on page 206) require a position constraint in column B of the form "A + offset" to ensure the expected chemical shifts are realised in the model. Similarly, the FWHM between columns B and C are constrained to be identical; the area relationship between B and C is also defined by the Area Constraint field in column C using the string "B * 1". After adding these constraints to the four component peak model, optimisation using the Marquardt algorithm returns a set of component parameters which are both optimal when describing the data and also plausible when describing the chemistry.

Energy Calibration

The peak model in Figure 9•9, once optimised, is intended to represent the relative positions for the chemically shifted C 1s peaks. The remaining step is then to apply an energy shift to the data to align one of the component peaks with a known line position, so that the model is positioned in an absolute sense. Again following Beamson and Briggs, the position for the component labelled C 1s 1 is assumed to be 285eV binding energy. Using the Calibration property page, the objective is therefore to enter into the Measured text field (Figure 9•8 on page 203) the value for the un-calibrated C 1s 1 peak position and the value of 285 in the True text-field. Whilst these values could be entered manually, a link between the selected component on the Component property and the Measured text-field on the Calibration property page allows the value to be automatically transferred. There is a further interaction between the True text-field and the element library also involving the selected component on the Components property page. Transferring the position of the component labelled C 1s 1 is as simple as pressing the header button labelled A, causing the column below the header to turn blue, then pressing the Component button on the Calibration property page. If the name field in the selected component matches an entry in the current element library, then the energy for the matched entry is entered into the True text-field. If no match is found then the True text-fields is updated from the value for the component position. In either event, the Measured text-field is updated from the position field of the selected component on the Component property page. If the value entered into the True text-field is not the intended position for the selected component, the correct value should be entered.

Although the two values for the Measured and True text-fields are sufficient information for calibrating the spectrum, the specification is not complete until the decision is made regarding the action to be applied to the currently existing region and component parameters. Tick boxes within the section headed Adjust optionally allow regions and/or components to be shifted at the same time as the data. For the current example, both tick boxes should be ticked, the result of which is the region and components are also move at the same time as the spectrum is calibrated.

Applying the shift determined from the two input parameters can be applied to the spectrum in the active tile via the Apply button or, if more than one spectrum requires an identical shift, the calibration can be applied using the Apply to Selection button (Figure 9•8). The latter button refers to the selection of blocks in the right-hand-side of the experiment frame.

Undo and Retrospective Calibration Adjustments

The scenario described in the previous section, in which the position of a synthetic peak determines the energy calibration for a set of spectra, is sometimes performed following an initial investigation of the material using a survey spectrum. For example, an elemental quantification table may suffice for some data and so the usual steps of calibrating the spectra based on the position of the C 1s 1 line taken from the low resolution survey data, followed by creating quantification regions is enough to characterise the material. However, following the initial investigation, it may become apparent that chemical state information is required and therefore one or more high resolution spectra are modelled using synthetic components. Once a peak model is involved, the precision of the component assignments becomes of more importance than the position of the C 1s peak in the survey spectrum, and as such the initial calibration must be replaced by a new energy shift determined from one of the components in a similar fashion to the previous section. In CasaXPS, although it is possible to apply more than one calibration command to a spectrum, the actual energy shift is determined from the last in a sequence of such commands. Data already shifted, however will be located with respect to the current calibration and so it is necessary to undo the current calibration before computing the new energy shift. There are several issues relating to this process and these will be discussed in the following example.

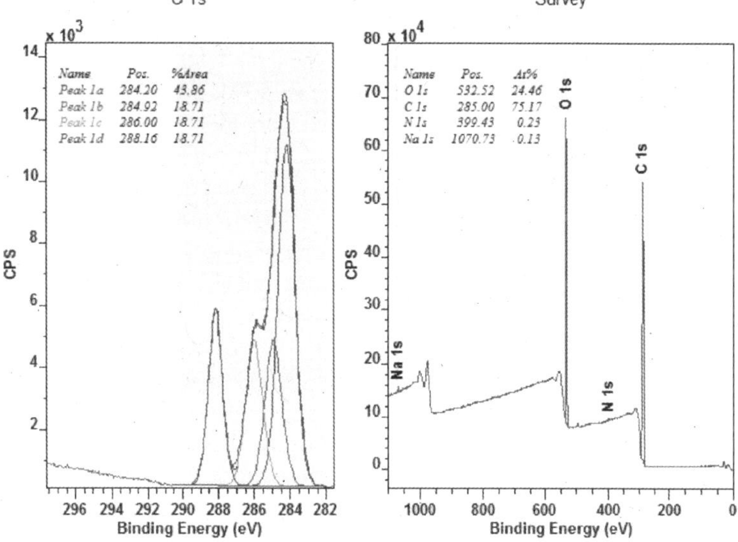

FIGURE 9•12
High Resolution C 1s region and survey spectra, both spectra energy-calibrated using the C 1s peak from the survey spectrum.

Consider the spectra in Figure 9•12. The survey spectrum is initially quantified by assigning quantification regions to each of the peaks representative of the elemental intensities, and then using the C 1s region on the survey spectrum, an energy calibration is performed to ensure the peak position for the C 1s maximum is located at binding energy 285.00eV. The data from the high resolution spectrum is calibrated using the same difference between the measured and the true energies determined from the survey C 1s region. At the time of calibration, the high resolution C 1s data has no regions or synthetic components defined. If at a later time, regions and components are added to the high resolution C 1s spectrum, the situation will exist where the spectrum is shifted in energy, however the region and components are relative to the calibrated data. If the data is re-calibrated to position one of the model peaks, simply undoing the calibration step from the Processing History property page will leave the region and components at the positions determined from the previously calibrated spectrum, while returning the spectrum itself to the un-calibrated locations. It is therefore necessary to manipulate the data with a little more care when make such adjustments; these more subtle adjustments are performed on the Calibration property page. The problem with simply using the Processing History property page to remove the calibration is that the processing history has no record of the regions and components created after the initial calibration step. To remedy the missing information the Calibration property page includes a button for undoing the calibration on a selection of spectrum blocks where the intention to move or not move the region and/or components is indicated, prior to pressing the Undo Selection push-button, by the tick boxes in the section headed Adjust labelled Regions and Components (Figure 9•13).

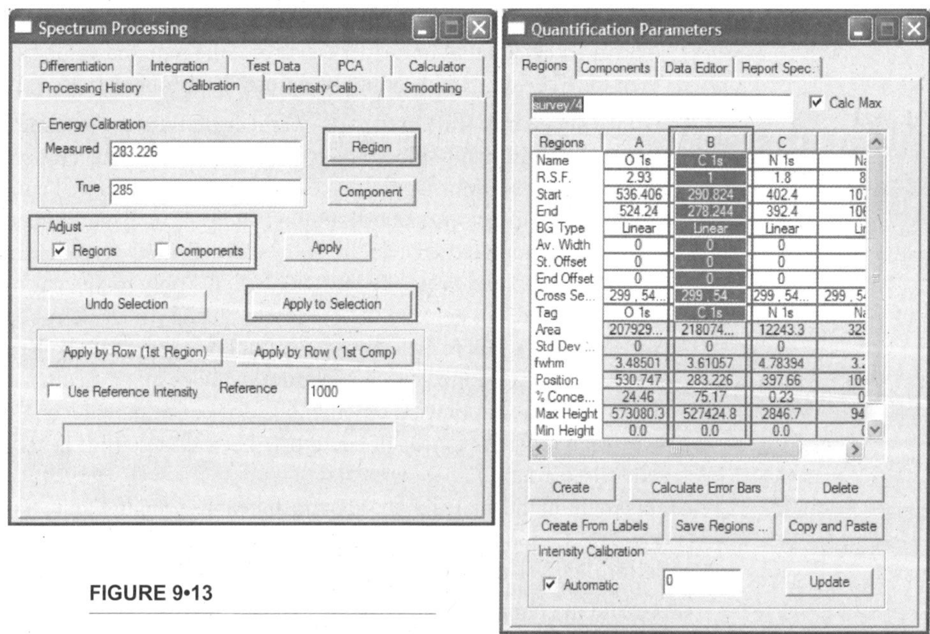

FIGURE 9•13

The steps, performed on these data during the initial calibration based on the survey spectrum and then subsequent re-calibration based on a model peak, are as follows:

1. Create quantification regions on the survey spectrum. Since for this particular spectrum the photoelectric lines are easily identified using the Find Peaks button on either the Element Table or the Periodic Table property pages of the Ele-

ment library, the Create Regions button on these same pages can be used to initially create the quantification regions. The location of these initially created regions must be checked and re-aligned, if necessary, using the Regions property page on the Quantification Parameters dialog window. Display the Quantification Parameters dialog window and press the Reset toolbar button for initializing the zoom-list. Then press the Zoom-Out toolbar button to cycle through a zoomed display of these newly created regions making any adjustments as necessary.

2. Define the Measured and the True binding energies for the C 1s peak in the survey spectrum. Using both the Regions property page on the Quantification dialog window and the Calibration property page on the Spectrum Processing dialog window, select the C 1s region from the Regions property page as illustrated in Figure 9•13 and press the Region button indicated on the Calibration property page (also Figure 9•13). The value for the Measured text-field is determined from the selected Region on the Regions property page, while the True value is loaded from the element library entry for which the Name field from the selected region produces a match. If the element library energy is not the required value for the data in question, then the True text-field should be adjusted appropriately.
3. Apply the energy shift to the spectra measured under the same sample charge state. Using the right-hand-side of the Experiment frame, select those blocks for which the calibration is appropriate; in this case both the survey and the high resolution spectra are acquired under the same charging conditions and therefore all should be included in the calibration step. Once the blocks are selected, tick the Regions tick-box in the Adjust section and press the button labelled Apply to Selection on the Calibration property page. An entry in the processing history is made for each block selected at the time the Apply to Selection push-button is pressed. The Regions tick-box must be ticked because the survey spectrum includes quantification regions and without instructing the calibration command to do so, the start and end energies for the quantification regions would not be adjusted at the same time as the energy scale for the underlying data.

At this point in the analysis, the spectra appear as seen in Figure 9•12 on page 209, however the peak model displayed in Figure 9•12 is yet to be created and so only the energy scale for the C 1s data is affected by the calibration command.

4. Create the region and four synthetic components (as seen in Figure 9•12) displayed in the C 1s tile. The peak model uses a linear background and four Gaussian Lorentzian synthetic components. The model is guided towards the peak positions tabulated in Figure 9•12 using constraints for the area of each of the three less intense lines, where the constraint simply forces all three to have identical area. One further constraint forces the FWHM of the peaks labelled Peak 1b and Peak 1c to take on a common value. Without constraints, the four peaks would optimise to a mathematical minimum almost certainly unrelated to the true line positions; the primary rea

sons for the difference between the mathematical optimum and the chemical optimum are the uncertainty in the true line-shapes coupled with the presence of noise in the data. The constraints for the components in Figure 9•12 are entered into the constraint fields on the Components property page, where the specific constraint for the FWHM of the two peaks Peak 1b and Peak 1c is entered in the FWHM constraint field in column headed C and is defined using the character heading B, namely B*1. The constraint should read: the FWHM of C is equal to the FWHM of B multiplied by one.

Having created the peak model for the C 1s spectrum in Figure 9•12, it is apparent the components within the model should be the means by which the energy calibration is determined. A common assignment for the carbon components is to set the position for the peak labelled Peak 1a to binding energy 285.00 and therefore the calibration based on the survey must be redone.

5. Undo the current calibration for all spectra from which no regions or components where created after the initial calibration step. Assuming none of the spectra involved are processed using other spectrum processing options, select all those spectra for which no regions and/or components have been defined since the survey scan was used to calibrate the data set and overlay the spectra in the active tile. Press the Reset All button on the Processing History property page. Note, the calibration command recorded in the Processing History provides the information to determine whether the regions and/or the components were present and moved when the calibration was performed. If a region or component was defined on the data when the calibration was performed, the calibration command will include the strings "Regions" or "Comps", for example "Calib M = 283.226 A = 285 Regions Comps BE" appears in the Processing History list for a spectrum for which regions and components were present when the command was applied and the command included the instruction to move, in energy, the quantification items at the same time as the spectral bins. Provided the correspondence between the regions and components on the data matches the existence of these key words in the calibration command, the calibration can be undone using the processing history. If a miss-match exists between these values, then the Calibration property page must be used to selectively undo the calibration command.
6. Undo the calibration for those spectra for which regions and components have been added subsequent to the initial calibration step. For the current example, the C 1s region and components were created following the calibration of the spectral data and so removing the calibration for the C 1s data requires the use of the Undo Selection button on the Calibration property page (Figure 9•13 on page 210). The Undo Selection button acts on those blocks selected in the right-hand-side of the Experiment Frame and the positions of quantification items are adjusted based upon the state of the Regions and Comps tick-boxes in the Adjust section. Regardless of the order in which the regions and components were created in relation to the initial calibration, the Undo Selection button will either move the regions and components at the same time as the spectral bins or leave these quantification items alone based on the settings of the tick boxes in the Adjust section (Figure 9•13).

7. Apply the new calibration based on the lowest binding energy C 1s synthetic component from the high resolution peak model. The procedure for defining the new calibration is identical to the example using nylon in the previous section. Figure 9•14 displays the C 1s high resolution spectrum as an inset tile within the tile displaying the survey spectrum, where both spectra are calibrated using the shift determined from the synthetic peak labelled Peak 1a.

FIGURE 9•14

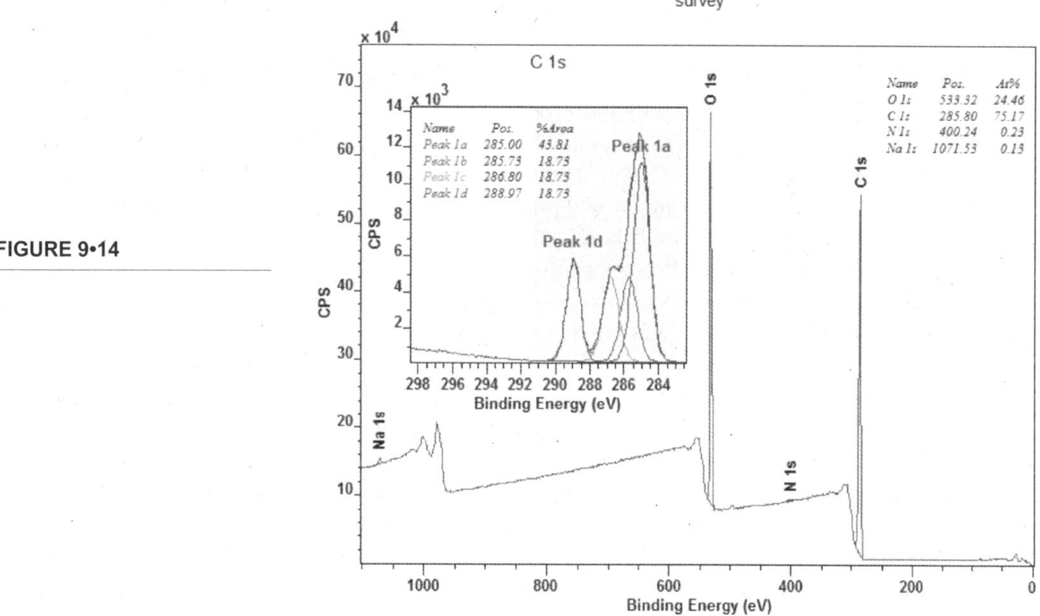

Locating a Peak Maximum without Quantification Regions

So far, the method for defining the measured peak position has relied on the use of a quantification region or a synthetic component. There are situations where this method is limiting, particularly for data where significant charge compensation is required and where the batch processing option is applicable. The problem with using quantification regions to identify a peak position is that the range over which a peak may appear in the un-calibrated data is possibly large compared to the interval over which the intensity for a peak is computed. It is therefore necessary to separate the function of identifying a peak maximum for use with the calibration command from the interval of the spectrum defined by a quantification region. The primary requirement for calibrating a spectrum before quantification regions are defined comes from the batch processing feature. Here, a template file containing all the processing and quantification items is used to define the steps required to process a raw file of equivalent data. The successful positioning of the quantification items is dependent on the spectra first being energy corrected, without which the limits to the quantification regions might miss the intended peak.

An alternative means of computing the measured peak maximum within a range of data channels is possible using a range of energies in the form of a pair of comma separated values entered in the Measured text-field on the Calibration property page. A True value must be assign as before, however when a comma separated pair of energy values are entered in the Measured text-field, the set of data bins defined by the energy interval are searched for the bin containing the maximum counts, then this bin plus adjacent bins are used to compute the energy corresponding to the maximum of a quadratic polynomial approximating the data near the maximum intensity in a least-squares sense. When the Apply button is pressed, the spectrum is shifted to locate the maximum determined from the quadratic at the value specified in the True text-field. The intensities used in the calculation are not background subtracted, therefore the calibration may differ from the peak position subsequently computed from a quantification region defined using a standard background type. A similar situation would exist if the spectrum were calibrated computing the measured value from a quantification region with background type linear and then subsequently changing the background type to Shirley; the effective peak maximum would move as a consequence of the different background approximations. A further feature of this method for determining the energy shift is that if the Apply to Selection push-button is used instead of the Apply button, then each spectrum within the selection is calibrated using a command string including a reference to the block from which the energy shift was determined. The significance of the reference to the originating block lies in the use of this form of calibration when applied via the batch processing mechanism. The reference to the block involves the absolute index of the block within the file and, while applicable to batch processing, is not so useful when propagating processing commands via the Browser Actions dialog window. In fact, for the most part, calibration is not the type of command for which propagation is appropriate since either the charge compensation is applicable to all spectra within a file, in which case the Apply to Selection button is used, or the charge state varies throughout an experiment, for example a depth profile, and so no one energy shift is easily applied to all spectra. The latter case is dealt with using the techniques described in the section on energy calibration and intensity normalisation.

Reference

[1] Michel Genet (private communication)

CHAPTER 10

Batch Processing

Batch processing is aimed at situations where a set of essentially equivalent samples require identical data reduction steps. The objective is to provide a simple sequence of user interactions resulting in a report as well as optionally printing of the spectra in a predefined format. CasaXPS 2.2.52 offers the following actions:

- Automatic data conversion to the ISO 14976 file format
- Data reduction steps bases upon a previously processed file
- Printing after formatting the data using a .tff file
- Generation of a report based upon a report specification file.

Once the necessary configuration files are defined, the whole procedure amounts to selecting the template file and the raw data directory, then waiting for the results.

Overview

The configuration files used as part of the batch processing feature appear in a directory called CasaXPS.BAT in the same directory as the executable file CasaXPS.exe (other CasaXPS directories such as CasaXPS.DEF and CasaXPS.HLP may also appear in the same directory). For each set of samples for which batch processing is required there will be at least one file in the CasaXPS.BAT directory, the so called template file. A template file is merely a .vms file containing the same number of data blocks in the same order as will be present in the data files acquired from each of the samples being analysed. Any processing, annotation, regions and components required for each sample should be applied to the data in the template file. When batch processing is invoked, these operations from the template file are applied to each file in a target data directory.

Accompanying the template file, are (optionally) two files with the same base-name but with file extensions .tff and .rpt. The .tff file must contain information generated from the template file relating to the tile display format required when printing the data. If printing is not required, then the CasaXPS.BAT directory should not contain a corresponding .tff file to the template file. Similarly, the .rpt file is an ASCII file specifying the format for a row of data in the final quantification report. If no .rpt file corresponding to the template file exists in the CasaXPS.BAT directory, then a default report is generated once batch processing is complete. Defining the display state used for printing the spectra is described in "Case Study 5: Displaying Spectra & Saving Display Formats" on page 38.

The objective for the .rpt file is to allow all the quantification items derived from the regions and components to be made available in a report, where these quantities can be combined using any arithmetic combination, including functions such as SQRT, LOG (natural log i.e. base e), EXP, SIN and COS. With the aid of these mathematical functions, intensities may be used to compute values for, say, overlayer thicknesses as well as error bars for such calculations. The emphasis is therefore on exporting the data for use elsewhere, for example in a spreadsheet program where the ultimate report may includes conditional display colours, more involved computation and/or graphical display, all previously defined on one sheet using data taken from a data entry sheet accepting the format defined by the .rpt file. The following describes defining the .rpt files.

Setting-up the .rpt File

The .rpt file is an ASCII file best prepared in a simple text editor such as Notepad. Key words and defined strings are used to annotate the report and extract the quantification details from the processed data files.

TABLE 10•1. Syntax for entries in the .rpt file

newline	Inserts a new line in the header section of the report
header;<text-string>	Defines a text string for use as a header to the report. Each header so defined appears in a separate cell when pasted into a spreadsheet.
text;<text-string>	Defines a text string that will appear embedded in the report and is repeated for each row.
blank	Inserts a blank cell in the ultimate spreadsheet.

TABLE 10•1. Syntax for entries in the .rpt file

vf;FILE	If defined, the filename for the original file for the sample entered in a row.
vf;DESC	If defined, the file description for the sample entered in a row.
vf;DATE	The date when the data in the row was acquired.
vf;BATCH	If defined, the batch identification string for the row of data
vf;header;dir	Enter the data directory name into the header section of the report.
vb<index>;<keyword>	Extracts information from a VAMAS block defined using the VAMAS block index (an integer) and a keyword (see following table). The VAMAS block indices run from 0 to the number of blocks in the file minus one.
vb<index>;<keyword>;r<index>	Extracts information from a Region in a VAMAS block defined using the VAMAS block index (an integer) and a keyword (see following table) followed by the letter r (for region) and the region index within the VAMAS block. The VAMAS block indices run from 0 to the number of blocks in the file minus one. The region index runs from 0 to the number of regions in a VAMAS block minus one. The region order is defined by the columns in the Regions property page.
vb<index>;<keyword>;c<index>	Extracts information from a Component in a VAMAS block defined using the VAMAS block index (an integer) and a keyword (see following table) followed by the letter c (for Component) and the component index within the VAMAS block. The VAMAS block indices run from 0 to the number of blocks in the file minus one. The component index runs from 0 to the number of components in a VAMAS block minus one. The component order is defined by the columns in the Components property page.

TABLE 10•2. Definition of keywords used in Table 10•1.

FIT	Goodness of fit metric (ChiSQ or RMS)
EXP_VAR	Experimental Variable assigned to the spectrum block
BLOCKID	Block identification string
SAMPLEID	Sample identification string for block
INTENSITY	Intensity measured in CPSeV (area under peak)
POSITION	Binding Energy of the peak maximum
FWHM	Full Width at Half Maxium
STDEV	Error in measured intensity for peak area. (Regions only)
MAX_HEIGHT	Intensity at maximum counts (depends on technique)
MIN_HEIGHT	Intensity at minimum counts (depends on technique)
RSF	Relative Sensitivity Factor used to compute Intensity
PEAK_TO_PEAK	Difference in maximum and minimum intensity within a Region
MASS	Mass value in Daltons (amu) assigned to a Region or Component.

The syntax definitions shown in Table 10•1 and Table 10•2 are illustrated in Figure 10•1, where an example .rpt file demonstrates how the individual quantification items are referenced and provides examples of arithmetic expressions involving intensities and standard deviations. Note the use of the SQRT function, which allows the uncertainty in the ratio calculation to be determined.

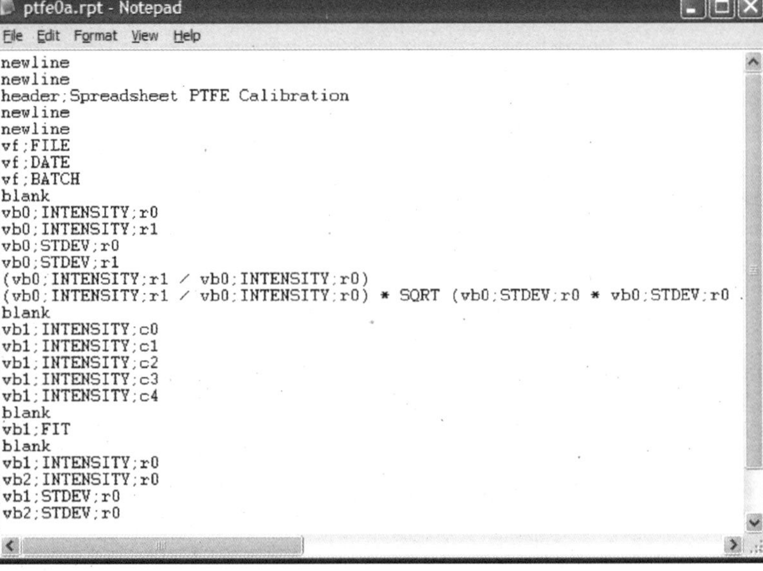


```
newline
newline
header;Spreadsheet PTFE Calibration
newline
newline
vf;FILE
vf;DATE
vf;BATCH
blank
vb0;INTENSITY;r0
vb0;INTENSITY;r1
vb0;STDEV;r0
vb0;STDEV;r1
(vb0;INTENSITY;r1 / vb0;INTENSITY;r0)
(vb0;INTENSITY;r1 / vb0;INTENSITY;r0) * SQRT (vb0;STDEV;r0 * vb0;STDEV;r0 .
blank
vb1;INTENSITY;c0
vb1;INTENSITY;c1
vb1;INTENSITY;c2
vb1;INTENSITY;c3
vb1;INTENSITY;c4
blank
vb1;FIT
blank
vb1;INTENSITY;r0
vb2;INTENSITY;r0
vb1;STDEV;r0
vb2;STDEV;r0
```

FIGURE 10•1 An example of an .rpt file.

Energy Calibration for Batch Processing

Calibration of spectra is typically different between samples and therefore a mechanism for instructing the calibration to proceed based upon the location of a peak from within a new file is required. To this end, a range of energies may be specified for the measured calibration energy, shown in Figure 10•2. Given a range of energies for the measured value, the exact energy used in the calibration for each sample is determined from the corresponding peak for which the range calibration is applied in the template file. If the Apply to Selection option in used, then the other spectra so calibrated will include a reference to the spectrum used to initiate the energy calibration.

The value determined for the Measured field in Figure 10•2 is computed from the data within the specified range. Note the energies specified using the range syntax refers to the original bins assigned at acquisition time. That is, previous calibrations do not alter the data bins used by the new calibration mechanism. Furthermore, the value so determined is obtained by interpolating the data via a quadratic polynomial, where a linear least square criterion is used to determine the polynomial.

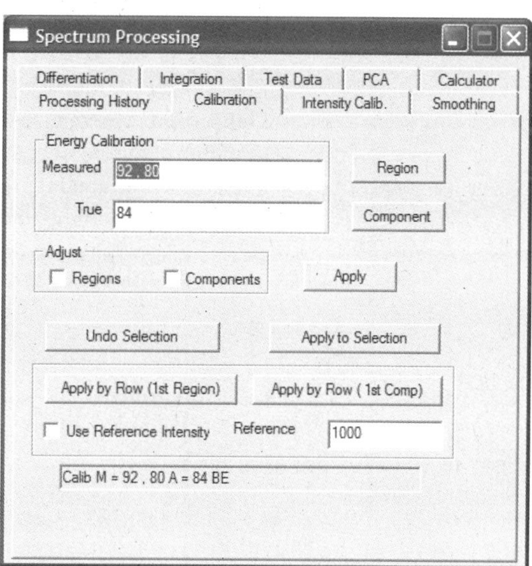

FIGURE 10•2 Energy Calibration Property Page showing the new range calibration syntax.

Batch Processing, non-printing, with the Default Report File

In this example, a set of samples representing a sequence of self assembled monolayers on gold, are analysed using identical acquisition and processing procedures. While the propagate option in CasaXPS is well suited to such tasks, the processing of these data can be achieved with minimal user interactions via the batch processing mechanism. The principal virtue of the batch mechanism lies in the ability of an experienced scientist to set-up the prescription for analysing the samples, after which a person with less XPS experience or even limited experience of using CasaXPS can generate a report in three easy steps.

Defining the Batch Processing

It is important that each sample is analysed with the same sequence of acquisition steps. That is, the resulting files must contain identical acquisition regions in terms of number and block order within the files. A representative spectrum file is prepared with all the required processing and saved in the directory CasaXPS.BAT located in the same directory as the executable file CasaXPS.exe. Since the default report will be used for these samples the report specification file is not required, the absence of which in the CasaXPS.BAT directory is the flag to use the default report. Also, the similar absence of a .tff file with the same base-name as the .vms file prevents any printing of spectra during the batch processing.
The template file shown in Figure 10•3 consists of four blocks, where the Au 4f region is used to energy calibration all four regions in the file. The new range calibration procedure described above is employed to calibrate the data.

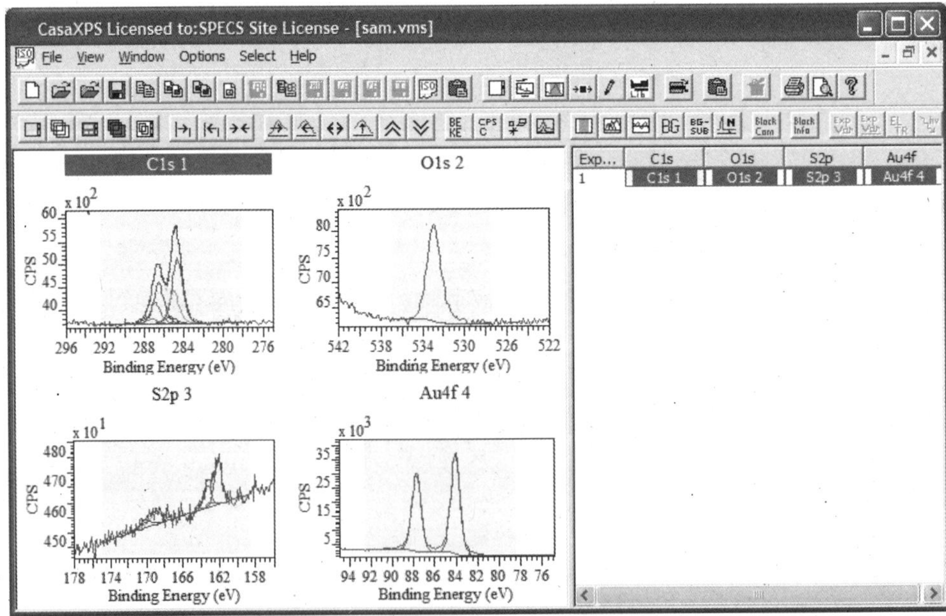

FIGURE 10•3 The template file sam.vms is stored in the CasaXPS.BAT directory. Data supplied by Philips Research, The Netherlands.

Applying the Batch Processing

After acquisition, the data files should be collected into a separate directory. In this example the data has been acquired using a PHI Quantum and therefore a set of .spe binary files generated. These .spe files must be stored in a directory with no other .spe files other than those measured from the intended sample set.
Three simple steps are then required to process a set of file.

1. From the File Menu, select the Batch Processing menu item shown in Figure 10•4

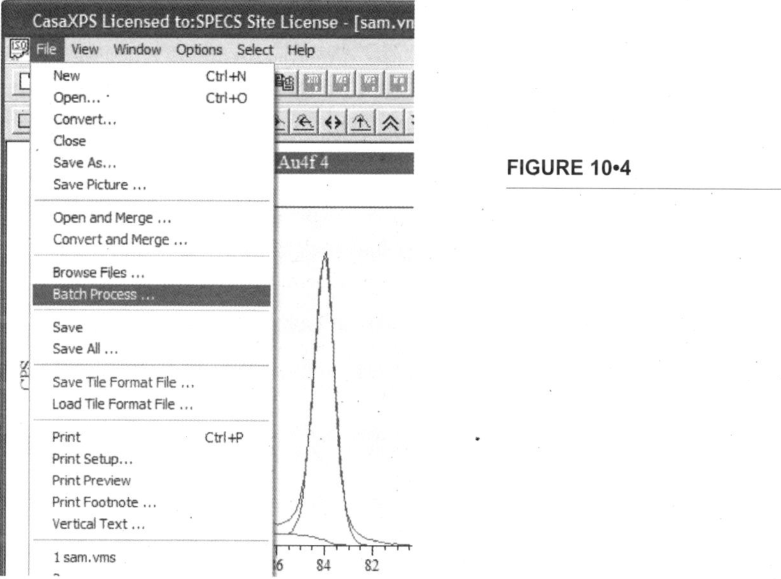

FIGURE 10•4

2. Select the appropriate template file from the list offered in Figure 10•5 and press the Apply Template button.

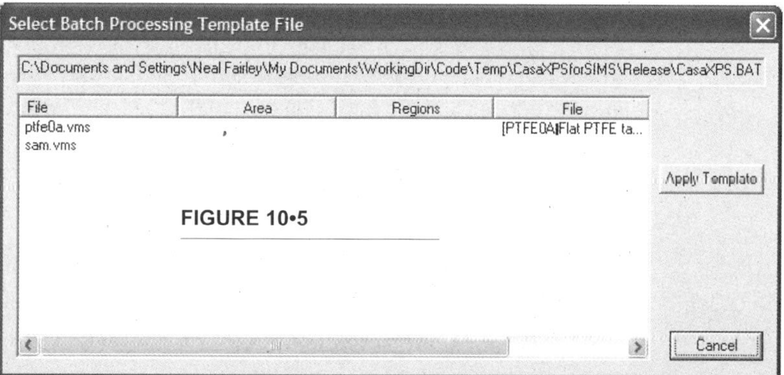

FIGURE 10•5

3. Select one of the .spe files within the File Dialog window and press Open.

It is now a case of waiting for the results to appear in a clipboard selection dialog window (Figure 10•6). The default report is a TAB spaced file which can be copied via the clipboard into a spreadsheet program or alternatively the report can be saved to file.

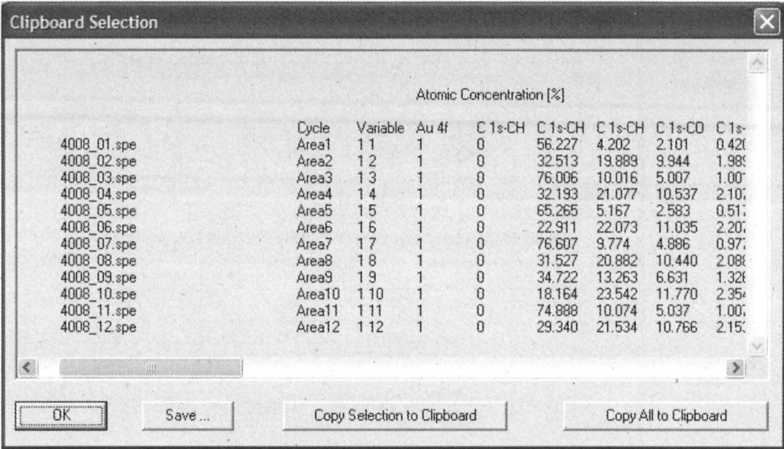

	Cycle	Variable	Au 4f	C 1s-CH	C 1s-CH	C 1s-CH	C 1s-CO	C 1s-
			Atomic Concentration [%]					
4008_01.spe	Area1	1 1	1	0	56.227	4.202	2.101	0.42
4008_02.spe	Area2	1 2	1	0	32.513	19.889	9.944	1.98
4008_03.spe	Area3	1 3	1	0	76.006	10.016	5.007	1.00
4008_04.spe	Area4	1 4	1	0	32.193	21.077	10.537	2.10
4008_05.spe	Area5	1 5	1	0	65.265	5.167	2.583	0.51
4008_06.spe	Area6	1 6	1	0	22.911	22.073	11.035	2.20
4008_07.spe	Area7	1 7	1	0	76.607	9.774	4.886	0.97
4008_08.spe	Area8	1 8	1	0	31.527	20.882	10.440	2.08
4008_09.spe	Area9	1 9	1	0	34.722	13.263	6.631	1.32
4008_10.spe	Area10	1 10	1	0	18.164	23.542	11.770	2.35
4008_11.spe	Area11	1 11	1	0	74.888	10.074	5.037	1.00
4008_12.spe	Area12	1 12	1	0	29.340	21.534	10.766	2.15

FIGURE 10•6 Default report after batch processing completes.

Batch Processing with Printing and .rpt File

In this example, a sequence of measurements is made across the surface of a silicon wafer where the ratio of metallic silicon to silicon oxide is required. For each sample the atomic concentration determined from regions on the survey spectrum are combined using formula in a .rpt file; also the intensities for the silicon states, as measured using a peak model, are processed from the data and the ratio computed as part of the report. A template .vms file must be prepared as shown in Figure 10•7 a .tff file for the display state is also prepare as well as a .rpt file defining the report format (Figure 10•8). All three files in the CasaXPS.BAT directory have the same basename and the inclusion of the .tff and .rpt files causes the printing of the specta and the generation of the report seen in Figure 10•9, respectively.

The steps for batch processing with printing and a custom report are identical to the steps described above. The only difference is that the additional .tff and .rpt files are required.

A batch of twenty sets of spectra was acquired on an Axis Nova in the Kratos Applications Lab. where each experiment was performed at a different location on the silicon sample along a line scan. The silicon wafer had previously been etched at specific locations. Such analyses can be set up by an experienced XPS analyst and left to a technician for batch processing the results.

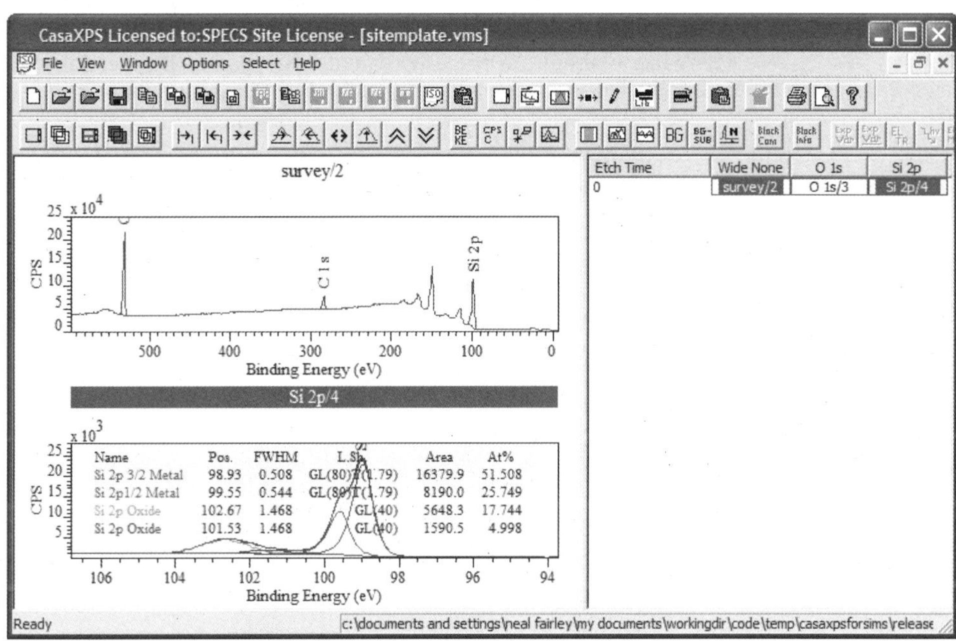

FIGURE 10•7 Template file SiTemplate.vms for processing and annotation (saved in CasaXPS.BAT).

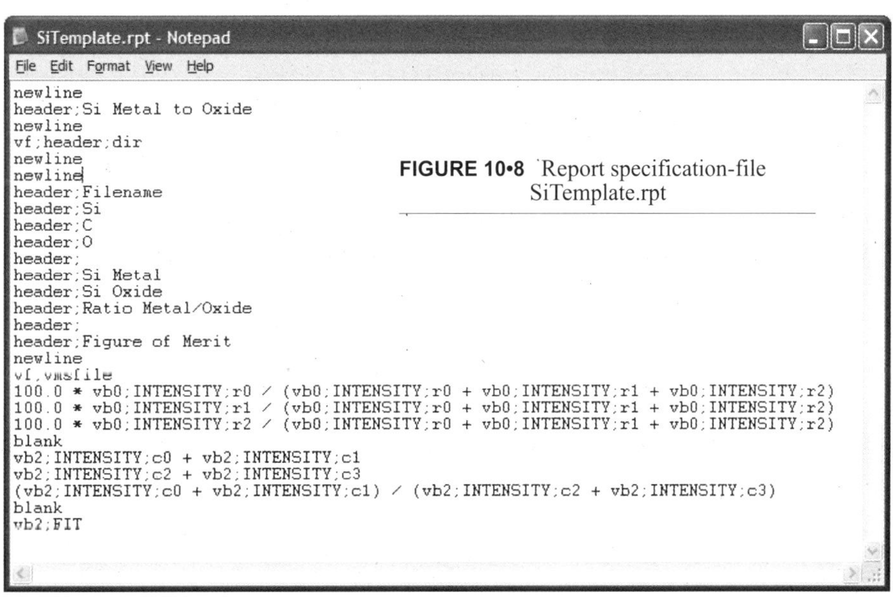

SiTemplate.rpt - Notepad

```
newline
header;Si Metal to Oxide
newline
vf;header;dir
newline
newline
header;Filename
header;Si
header;C
header;O
header;
header;Si Metal
header;Si Oxide
header;Ratio Metal/Oxide
header;
header;Figure of Merit
newline
vf;vmsfile
100.0 * vb0;INTENSITY;r0 / (vb0;INTENSITY;r0 + vb0;INTENSITY;r1 + vb0;INTENSITY;r2)
100.0 * vb0;INTENSITY;r1 / (vb0;INTENSITY;r0 + vb0;INTENSITY;r1 + vb0;INTENSITY;r2)
100.0 * vb0;INTENSITY;r2 / (vb0;INTENSITY;r0 + vb0;INTENSITY;r1 + vb0;INTENSITY;r2)
blank
vb2;INTENSITY;c0 + vb2;INTENSITY;c1
vb2;INTENSITY;c2 + vb2;INTENSITY;c3
(vb2;INTENSITY;c0 + vb2;INTENSITY;c1) / (vb2;INTENSITY;c2 + vb2;INTENSITY;c3)
blank
vb2;FIT
```

FIGURE 10•8 Report specification-file SiTemplate.rpt

Microsoft Excel - Book2

	A	B	C	D	E	F	G	H	I	J	K
1											
2	Si Metal to Oxide										
3	C:\Documents and Settings\Neal Fairley\My Documents\WorkingDir\Data\XPS\Kratos Data\Si Batch Data										
4											
5	Filename	Si	C	O		Si Metal	Si Oxide	Ratio Meta		Figure of Merit	
6	nf_01.vms	57.8448	15.7596	26.3956		11434.4	3523.13	3.24554		1296.24	
7	nf_02.vms	59.1201	13.7818	27.0982		10533.6	4754.52	2.2155		1359.63	
8	nf_03.vms	56.8444	14.6371	28.5185		11666.1	4042.14	2.88611		1631.05	
9	nf_04.vms	57.5816	14.6011	27.8173		11553.6	3685.52	3.13487		1656.81	
10	nf_05.vms	56.6516	17.3038	26.0447		11313.7	3637.65	3.11017		1305.46	
11	nf_06.vms	58.7471	15.3094	25.9436		11411.5	3578.43	3.18896		1417.44	
12	nf_07.vms	58.6452	14.5013	26.8534		11626.1	3683.27	3.15646		1374.59	
13	nf_08.vms	58.4367	14.9462	26.6171		11660.1	3649.27	3.19519		1471.68	
14	nf_09.vms	58.9075	13.1665	27.926		11792.4	3861.11	3.05416		1571.47	
15	nf_11.vms	58.8807	12.3978	28.7215		11770.3	3699.17	3.18186		1802.5	
16	nf_12.vms	57.8767	14.0194	28.104		11873	3495.01	3.39713		1638.72	
17	nf_13.vms	58.2597	13.2264	28.5139		12027.1	3544.72	3.39295		1631.69	
18	nf_14.vms	58.5069	13.5201	27.973		12272.1	3510.68	3.49566		1803.33	
19	nf_15.vms	58.918	12.9125	28.1695		12419.5	3521.34	3.52691		1680.21	
20	nf_16.vms	59.2892	12.8658	27.845		12436.8	3457.05	3.59752		1665.89	
21	nf_17.vms	59.4654	12.9808	27.5538		12457.6	3305.1	3.76922		1719.29	
22	nf_18.vms	59.7558	13.0756	27.1686		12549.2	3299.32	3.80357		1454.89	
23	nf_19.vms	60.7235	11.8125	27.4641		12507.7	3244.77	3.85474		1597.06	
24	nf_20.vms	60.9555	12.0231	27.0214		12559.5	3161.53	3.97261		1470.88	
25											

Sheet1 / Sheet2 / Sheet3

Ready

FIGURE 10•9 Report generated by the file SiTemplate.rpt after pasting into Excel

Section 3 - Mathematical Methods and Image Processing

CHAPTER 11

Monte Carlo Methods - Uncertainties in Peak Fits

Optimization routines are completely deterministic, in that, for the same set of parameters, used in the same functional forms, applied to the same set of data, with the same initial and final conditions will result in the same set of parameters on termination. Vary any of the above conditions and the result from the optimization routine will change in some respect. One method for assessing the uncertainty in the parameters for a peak model is to vary these optimization conditions by repeating an experiment on (what are at least hoped to be) identical samples. Then for each set of data apply the same optimization routine to the same synthetic model and so determine a distribution for the set of parameters used to quantify a sample. Such a procedure will vary almost every aspect of the measurement process and so result in a distribution for the parameters that truly represent the nature of the experiment.

The basis for such an approach as described above lies in the assumption that there exists a set of parameters (only known to nature) that does not depend on any optimization routine nor any other errors introduced into the measurement process, and these true values will lie inside the region of the n-dimensional parameter space defined by the set of outcomes to this sequence of experiments. Obviously, if the synthetic model does not describe a set of parameters in tune with nature, the results may be in more doubt than the measured distribution might suggest. However, given that all is well then the task is to offer a means of understanding the uncertainties in the peak parameters within the context of these parameter distributions.

Peak identification in XPS spectra represents a challenge since synthetic models more often than not involve overlapping line-shapes (see Figure 11•1 on page 227), the consequence of which is correlated optimization parameters. That is to say, if a single data envelope results from two overlapping peaks and if one of these underlying peaks is reduced in intensity then in order to describe the same data envelope the other must increase in intensity. Adjustments to peak parameters of this nature are inherent to any optimization procedure and the choice between the possible combinations of these peak intensities is made based upon a chi-square or root-mean-square metric [1]. The problem is therefore to identify the point at which these goodness-of-fit metrics fail to produce results that can be believed, and provide some means of illustrating the degree of uncertainty.

Given a distribution for each of the parameters, the best way to describe the nature of the uncertainties is to offer an error matrix as well as a tabulated form for the distributions. The error matrix provides numerical values from which the degree of correlation can be assessed while scatter plots taken from some subset of these distributions allows visual inspection for the same information. Ideally a single number for each parameter would provide the simplest means of stating the uncertainties, but as the old adage goes "To every complex problem there is a simple solution that is wrong", and so it is for correlated parameters. The unfortunate fact is that if the peaks weren't correlated then synthetic models would be unnecessary.

Monte Carlo Data Sets

Repeating an experiment many times is not always practical and for most XPS spectra peak models are developed based upon a single acquisition sequence. Estimates for uncertainties in the peak parameters must be made using assumptions about the noise typical of XPS spectra. The essence of Monte Carlo methods for estimating uncertainties is to take a data set, remove the noise from the data, then repeatedly add noise to the synthesized data to generate a set of simulated experimental results and apply the set of operations to that artificial data. The results represent a distribution for each of the parameters from which an error matrix and tabulated parameter distributions can be extracted.

Error Matrix

The result of a Monte Carlo simulation for a peak fit takes the form of m optimization parameters from each of the n simulated data sets. An error matrix derived from these parameter distributions is defined to be an m x m matrix as follows:

$$e_{ij} = \frac{1}{n-m} \sum_{k=1}^{n} (x_{sik} - \hat{x}_i)(x_{sjk} - \hat{x}_i)$$

Where the x_{sik} are parameter values calculated for each of the simulation steps and each distribution is centred with respect to the mean rather than the initial parameter value.

The standard error in each parameter is given by $\sqrt{e_{ii}}$

and the correlation between parameters *i* and *j* is given by $\dfrac{e_{ij}}{\sqrt{e_{ii}e_{jj}}}$

An alternative method for estimating uncertainties in the peak parameters is to quote the inverse of the Hessian matrix used in the Marquardt Levenberg optimization routine. So the question is, why bother with Monte Carlo simulation when the very information desired is offered by the optimization procedure itself? The answer comes in two forms. Firstly, the error matrix derived from Monte Carlo is not limited to the problem of just fitting the peaks. The simulation procedure can also take account of the background as well as the peaks, and so determine a global error matrix for the calculation rather than one that focuses on the stability of the peak parameters alone. Secondly, the Hessian matrix has a tendency to be near singular when the Marquardt algorithm terminates. Indeed the optimization routine follow a sequence of steps that switch between a direct solution of the Hessian matrix and a solution of a regularized form of the matrix. The regularization step occurs when the direct Hessian inversion fails to produce an improvement in the fitting parameters and a common reason for this to occur is that the Hessian is ill conditioned. Under these circumstances the covariance matrix derived from the Hessian should not be trusted.

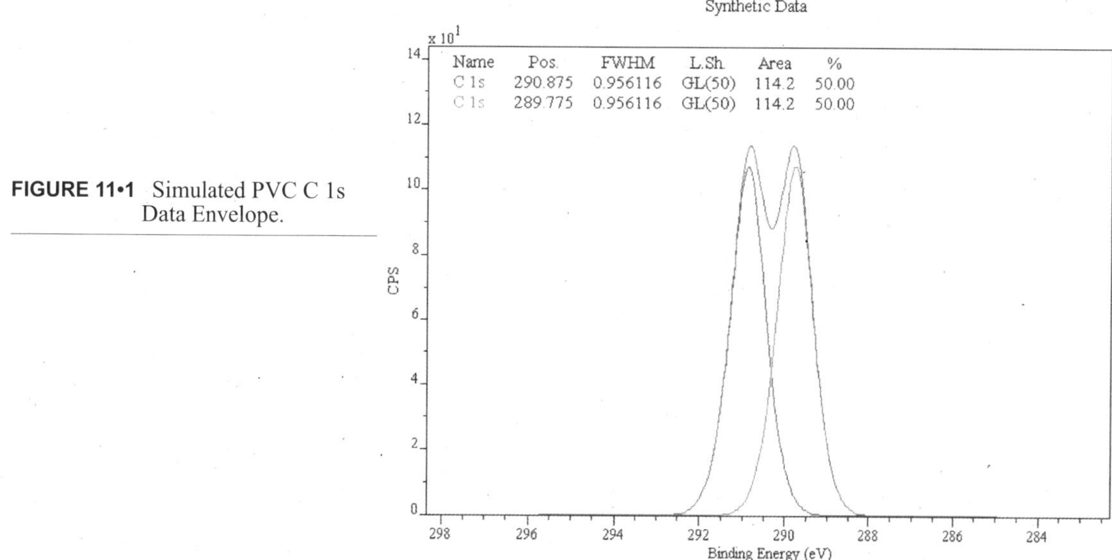

FIGURE 11•1 Simulated PVC C 1s Data Envelope.

Monte Carlo: A Simple Example

The best way to introduce the ideas behind Monte Carlo methods and how the results can be interpreted is to look at a simple example.

Consider the data envelope in Figure 11•1 on page 227. The spectrum is a synthetic envelope created from two GL (50) line-shapes without any background and where the peaks are separated by an energy gap consistent with C 1s lines in a PVC spectrum. In the absence of noise and experimental error the optimization routine always returns the peak parameters tabulated on the spectrum in Figure 11•1. The next step in the simulation is to introduce noise onto the data envelope that is consistent with noise found in experimental XPS spectra, i.e. variations about the channel intensity of magnitude related to the square root of the counts. Figure 11•2 shows the data envelope from Figure 11•1 after noise has been added and the peak parameters refitted.

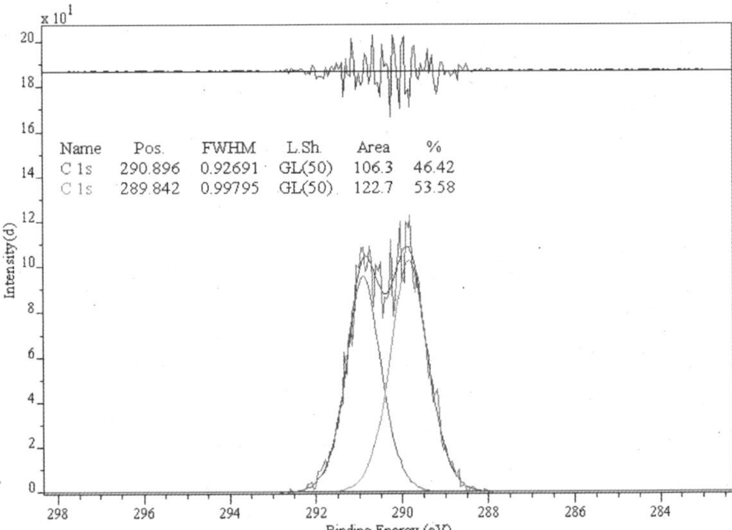

FIGURE 11•2 Simulated C 1s data where noise has been introduced to the system.

This procedure yields the first set of simulation results. If repeated many times, the output is six distributions, one for each of the peak parameters involved and these can be used to assess the influence of experimental noise on these quantities. Note that this differs in some respects from adopting a purely experimentally determined parameter distribution. The initial starting point for the peak parameters will not be identical for an experimental data set since the experimental data may be subject to sample charging, and any errors in the measurement procedure that can not simply be described by the random nature of the counting system will be omitted. A Monte Carlo simulated data set only tests the stability of a model with respect random noise and therefore may neglect other factors present in an experimental data set.

TABLE 11•1. Error Matrix for the C 1s spectrum in Figure 11•1.

	1:Area	1:Pos'n	1:FWHM	2:Area	2:Pos'n	2:FWHM
1:Area	39.230	0.122	0.196	-31.512	0.128	-0.186
1:Pos'n	0.122	0.001	0.001	-0.120	0.000	-0.001
1:FWHM	0.196	0.001	0.002	-0.185	0.001	-0.001
2:Area	-31.512	-0.120	-0.185	38.330	-0.125	0.202
2:Pos'n	0.128	0.000	0.001	-0.125	0.001	-0.001
2:FWHM	-0.186	-0.001	-0.001	0.202	-0.001	0.002

Table 11•1 is the error matrix that results from a Monte Carlo simulation for the spectrum and synthetic model in Figure 11•1 on page 227. If viewed as uncorrelated error distributions (which is certainly not true) then one standard deviation in each of the parameters is given by the square root of the diagonal elements in the matrix. It may be seen that the areas of the two peaks are anti-correlated, where the quantity $-31.512 / \sqrt{(39.230 \times 38.330)}$ characterizes the degree of interaction between the two parameters. The minus sign indicates the parameters are anti-correlated. A scatter plot constructed from the two distributions for the peak areas is shown in Figure 11•3 on page 230, where the anti-correlation is obvious from the bias in the scatter along a direction 135° to the positive x axis. An alternative scatter plot (Figure 11•4) for the area and FWHM parameters taken from peak C 1s 1 (Figure 11•1) shows that these two parameters exhibit positive correlation.

The uncertainty for correlated parameters can be estimated from plots such as those shown in Figure 11•3 and Figure 11•4. By projecting horizontal or vertical lines onto the parameter axes in such a way that the appropriate number of points lie between the projection lines for a given confidence limit, the uncertainty for a parameter can be assessed in the context of others. Again, the uncertainty for a parameter should be viewed in the context of all other correlated distributions, and yet the ellipsoid in 6-dimensional parameter space is difficult to quantify; the procedure based on scatter plots should be seen as merely a step in the right direction rather than arriving at the ultimate destination. In the case of the peak area shown in both Figure 11•3 and Figure 11•4, the estimate taken from the diagonal element of the error matrix would seem to be reasonable. Both scatter plots show that about 65% of the points lie inside projection lines positioned at about 0.95 and 1.05. This interval represents about ±5% of 114.2CPSeV (see Figure 11•1) and is not too different from the uncertainty taken from the error matrix ±6.2CSPeV.

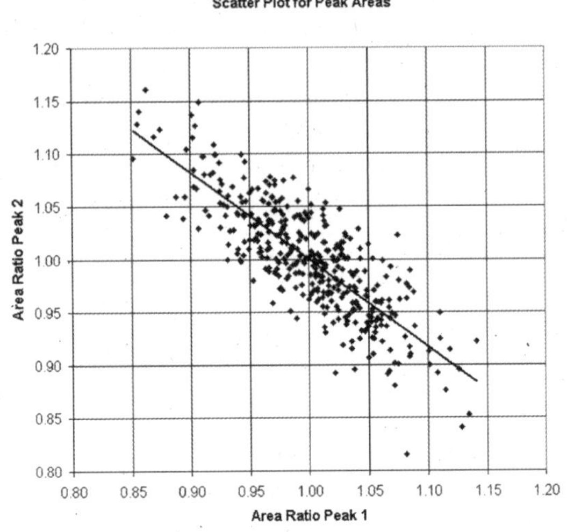

FIGURE 11•3 Scatter plot showing the anti-correlation between the peak area parameter distributions.

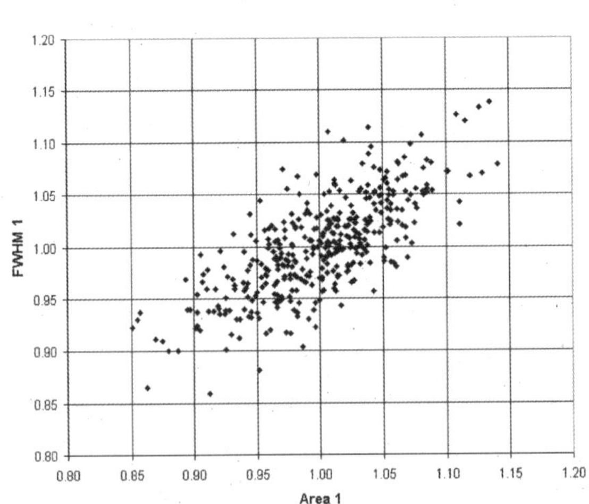

FIGURE 11•4 Scatter plot between Area and FWHM parameters for Peak C 1s 1 in Figure 11•3

All the parameter distributions are reported relative to the initial parameters used to define the data envelope. As a result the relative parameter distributions may not be symmetrical about the initial value and asymmetric confidence intervals are possible. Note that the error matrix is calculated from the distributions centred on the mean parameter value for the distribution, not the initial values.

One of the real advantages of using Monte Carlo error analysis is that it highlights when a quantification parameter is poorly determined by the combination of model and optimization procedure. It also allows the influence of constraints within a model to be evaluated. Adding information about chemical shifts, relative peaks widths and/or peak areas can alter the manner in which noise adjusts the parameters from their initial value. Monte Carlo derived scatter plots can often help to understand how rigid models based upon chemical knowledge can reduce the range of outcomes for a given set of parameters.

Quantification Issues

Quantification of AES/XPS spectra is routinely performed using combinations of intensities from integration regions and synthetic components. The purpose of this section is to point out some of the less obvious consequences that result from the specification of the background associated with the recorded peaks.

Monte Carlo End Point Determination

Consider an all too common situation where a pair of unresolved peaks is the subject of interest and the total intensity is the value that is calculated from the data.

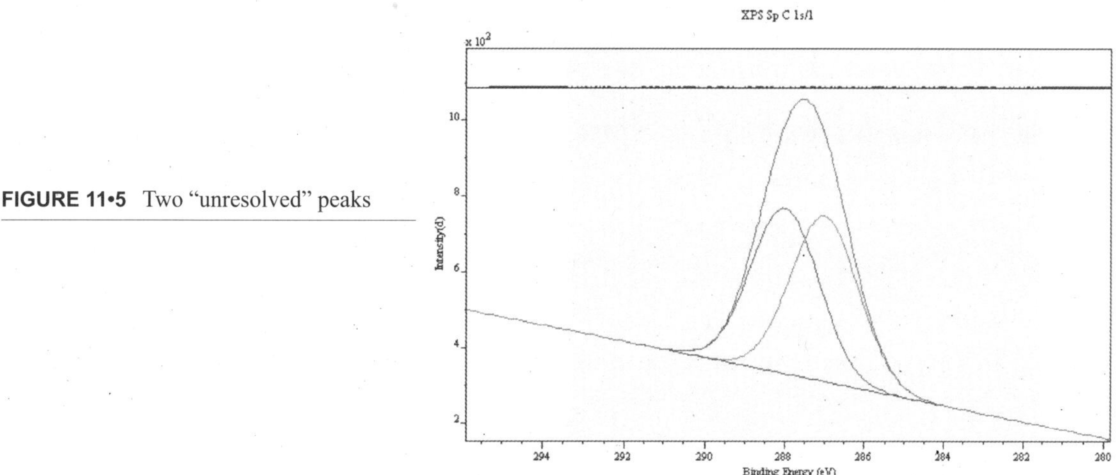

FIGURE 11•5 Two "unresolved" peaks

The data in this example (Figure 11•5) are purely synthetic; two Gaussian peaks of equal intensity and equal FWHM of 2eV are offset by 1eV on a linear background. This provides an envelope open to analysis by Monte Carlo simulation that will provide visual insight into the errors associated with noise in the data.

Integration Region Limits

The purpose of this analysis is to illustrate the consequences of choosing inappropriate end points for the integration region. Note that the peak-fitting procedure is not at issue since only the raw (unfitted) data are used to calculate the intensities for the combined peaks.

Figure 11•4 shows a scatter plot where each point is a normalized area plotted against simulation index. These areas are calculated from the original data envelope after the introduction of pseudo random data and the computed area is referenced to the initial area before noise was superimposed on the spectrum.

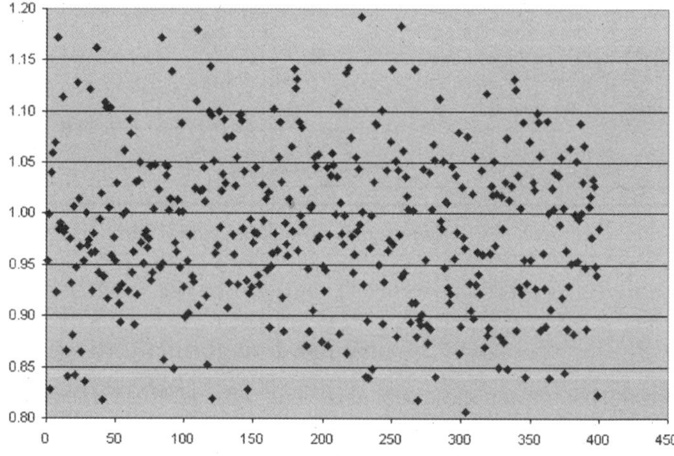

FIGURE 11•6 Peak areas from one end point

Each calculation includes determining a new linear background from the information in the simulated noise. In Figure 11•6 the background is computed using only one point at each end of the integration region while Figure 11•7 on page 233 is the equivalent experiment but in this case an average of 21 points defines the background limits.

It is clear from the two scatter plots that the intensity calculation using just one data point at each end of the integration region is much more sensitive to noise. The introduction of more data channels when determining the background improves the situation.

Not all peaks are isolated and so under these situations it falls to the acquisition phase to improve the statistics for the data channels that determine the background. An acquisition system that accurately determines the points used to define the back-

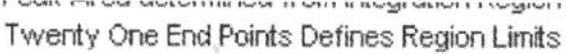

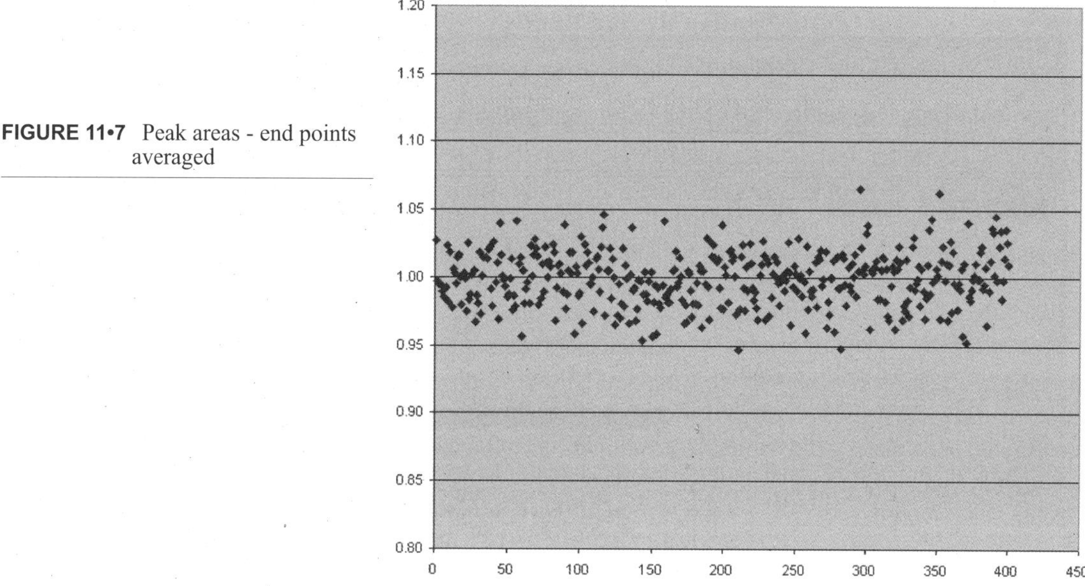

FIGURE 11•7 Peak areas - end points averaged

ground will yield intensities with significantly more precision than those that treat the end point as equal to any other in the acquisition region

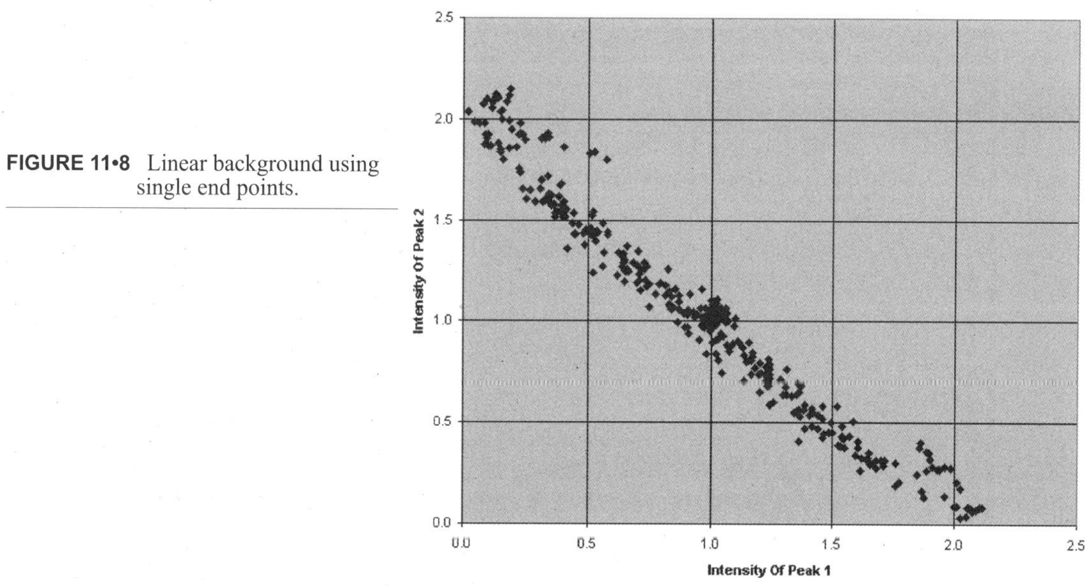

FIGURE 11•8 Linear background using single end points.

Intensities Determined by Peak-Fitting

Two peaks of equal width and area but separated by 1eV represent an intractable problem for a peak-fitting algorithm if the intensity of the individual peaks is required and no further information is available that can be used to provide constraints for the peak fitting parameters.

Figure 11•8 shows a Monte Carlo simulation for the peaks shown in Figure 11•5. Each Monte Carlo simulation step takes the envelope for the two peaks plus the background then superimposes a normal deviate of mean zero and standard deviation one scaled by the square root of the counts in each data channel. The envelope constructed in this way simulates a sequence of experiments performed on the same sample, so that the only variation in the data is the random noise inherent in the acquisition process. For each Monte Carlo envelope the peak parameters are refitted. The points used in the scatter plot are determined from the intensities for the two peaks after automatically fitting the peak parameters (Area, FWHM and Position) to the new data envelope; these intensities are normalized with respect to the initial values for the peak areas. It is apparent from Figure 11•8 that the uncertainty in the individual peak intensities may be in error by as much as 100% although the total area is determined much more accurately (Figure 11•9). The scatter in the computed intensities in Figure 11•9 should be compared with that for the intensities shown in Figure 11•6 on page 232.

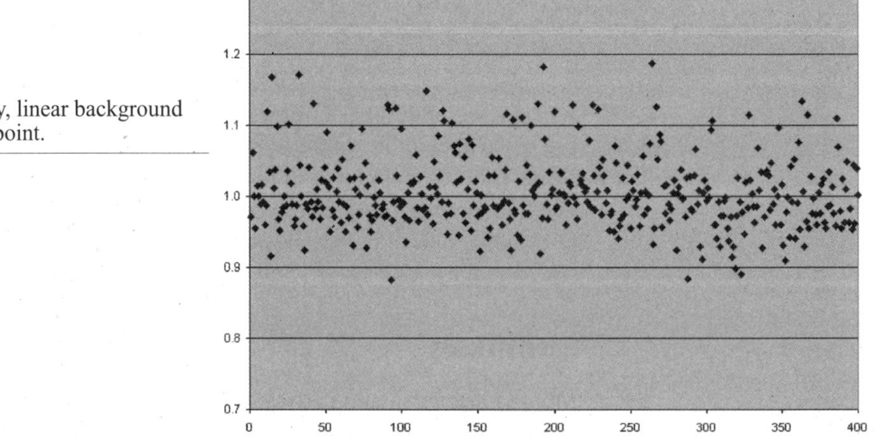

FIGURE 11•9 Total intensity, linear background from one end point.

In both cases the background used for the Monte Carlo simulations shown in Figure 11•8 and Figure 11•9 are determined using one end point for the integration region. The asymmetry in the distribution derives from the errors introduced by the background calculation that occurs prior to performing the peak fitting procedure. A comparison of the distributions suggests that the intensity determined form the peak fitting procedure will be more accurate than the integration region approach although there is a tendency to over estimate the total area when the sum of two fitted peak intensities is used and the background calculation is sensitive to noise. Again, the use of more acquisition channels to define the integration region end points improves the precision in the results for intensities calculated from the sum of the fitted peak areas (Figure 11•10). In addition, when the uncertainty in the background is reduced, asymmetry in the distribution of area ratios is improved dramatically - compare Figure 11•9 and Figure 11•10.

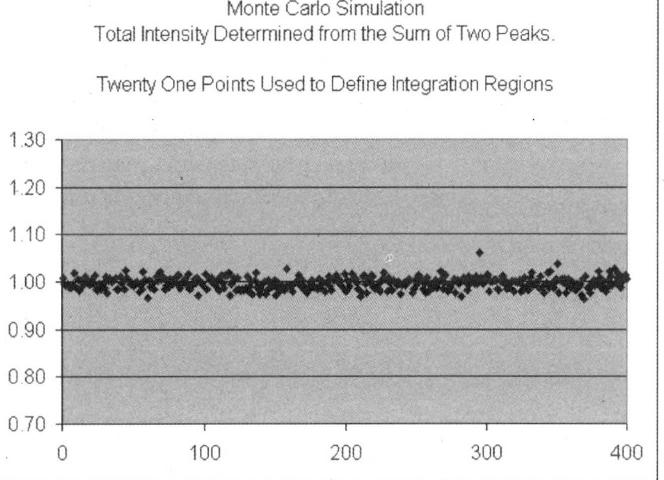

FIGURE 11•10 Linear background with twenty-one end points.

Introducing additional information into the background calculation does not improve the uncertainty in the intensities for the individual peaks determined by a peak fitting procedure, but it does remove some of the correlations in the peak intensities due to the background. These correlations are visible in Figure 11•8 at the extremes of the distribution but are not present when the same experiment is performed but with an improved background specification (Figure 11•11 on page 236).

Summary

Monte Carlo simulation provides a valuable tool for understanding the uncertainties associated with data reduction for XPS/AES spectra. The results presented here are available in other forms from other statistical data packages, but the visual feedback offered by CasaXPS Monte Carlo simulations provides an insight for analysts who feel less comfortable with a mathematical description of the same concepts.

Practical routine use of Casa Monte Carlo simulation procedures should guide the analyst towards a better understanding of the errors involved in extracting information from XPS/AES spectra, and so (hopefully) lead towards more securely founded analysis.

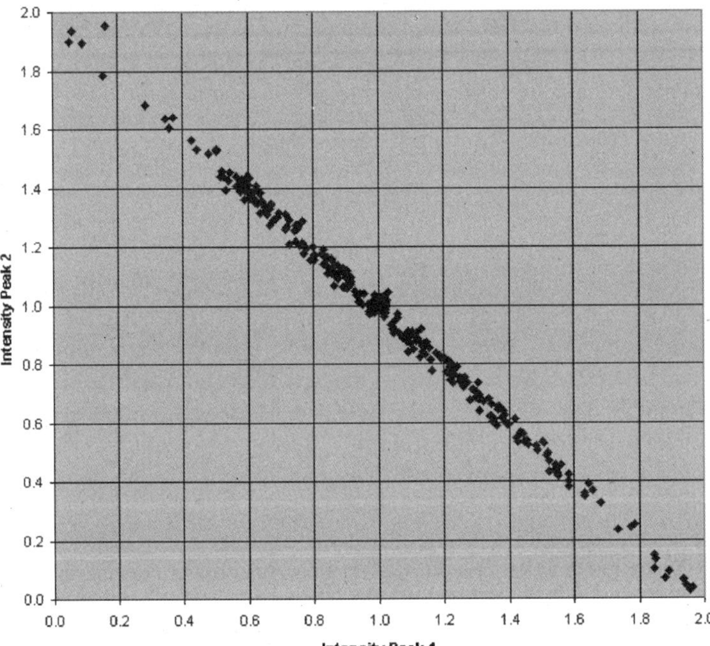

FIGURE 11•11 Individual intensities - Linear background with twenty-one end points.

References

[1] see "8) What is the difference between the Chi Square and RMS Figure-of-Merit?" on page 353
[2] K. Harrison and L.B. Hazell, *Surf. Interface Anal.* **18**, 368 (1992)
[3] P.J. Cumpson and M.P.Seah, "Random Uncertainties in AES and XPS: Peak Energies, Areas and Quantification." NPL Report DMM(A) 26',May 1991; *Surf. Interface Anal.* **18**, 345 (1992) and 18, 361 (1992)
[4] S. Evans, *Surf. Interface Anal.* **18**, 323 (1992).

CHAPTER 12

Principal Component Analysis

Modern XPS instruments are capable of generating copious quantities of data. When faced with such overwhelming information, the desire is strong to reduce the data to a small number of characteristic values from which *true* knowledge of a sample can be claimed. The success of such an approach relies heavily on the stability of the data reduction steps with respect to small but irrelevant changes in the data yet also sufficiently sensitive to reveal subtle but important differences held within. For XPS spectra, the traditional tools for reducing large number of data channels to meaningful characteristic values are peak integration and non-linear least squares optimization using synthetic components. Both tools, to a greater or lesser degree, rely on the experience of the analyst performing the experiment and while quantification based on regions is general stable with respect to changes in the data, optimized synthetic components require more care, but offers chemical state information often lost within relatively coarse integration regions. Given extensive data sets coupled with these relatively manually-intensive data reduction options, a movement towards a statistical approach to characterising such data is a natural development. Principal Component Analysis (PCA) and related techniques are mathematical procedures for manipulating a data set into a form containing a reduced number of characteristic quantities. The usefulness of these techniques, as with all statistical analyses, depends on an analyst's ability to interpret the results in the context of the sample. Interpreting the quantities generated by PCA is strongly coupled to an understanding of the underlying theory and an appreciation of the influence of irrelevant data fluctuations on the results.

This chapter is concerned with explaining the theory behind PCA and attempting to place this theory in the context of XPS data. The discussion will begin with spectral data, however all statements are equally applicable to image data and ultimately, image analysis via PCA combined with the traditional XPS data reduction tools is the real success of this theory when applied through CasaXPS.

Data and Vectors

An XPS spectrum is a set of acquisition bins, where energy separated bins accumulate counts for a given period of time. The usual interpretation of these data are discrete approximations to a continuous function plotted using counts per second against binding energy. However, an alternative view is to interpret the counts in each energy bin as a coordinate value for one dimension in a multidimensional vector space. Therefore each spectrum may be seen as a vector in an m-dimensional space, where m is the number of energy bins. In isolation, the vector interpretation of the data is far less natural than the continuous function perspective, however if a spectrum is merely one of many such spectra, for example a spectrum included as part of a depth profile experiment, then the vector description becomes more advantageous for the following reason. While an experiment may contain numerous spectra, not all spectra are entirely independent of one another. From a practical perspective, an analysis of a depth profile using a synthetic peak model consisting of a small number of Gaussian/Lorentzian line-shapes is only possible if these line-shapes, when combined appropriately, offer an adequate description for each spectrum in the profile sequence. The use of synthetic models for this purpose is evidence that the set of spectra from a depth profile, say, must include much redundancy. It is precisely the objective of multivariate statistics to isolate common trends within a large sample of data and provide quantitative values characteristic of these trends. These statistical techniques are essentially based on numerical algorithms for manipulating sets of co-ordinates and therefore the vector interpretation becomes more appropriate for large sets of spectra.

Co-ordinate Systems, Basis Vectors and PCA

The co-ordinate system represented in Figure 12•1 is an extension of the two and three dimensional Cartesian co-ordinate system. Indeed the Cartesian co-ordinate system is natural in the sense that the unit vectors in the direction of the mutually-orthogonal co-ordinate axes are easily written down in terms of the digits one and zero. While natural from a mathematical viewpoint, the objective of mathematics is to provide a general description, often at the expense of the specific and while the Cartesian system works well to simplify general mathematical descriptions, the specifics of a data set can be hidden by the use of Cartesian co-ordinates. That is to say, the Cartesian co-ordinate axes are positioned without regard to the data being represented. The fact is that any m-dimensional space described using the Cartesian unit

FIGURE 12•1

$$\begin{bmatrix} S_{1i} \\ S_{2i} \\ S_{3i} \\ \cdot \\ \cdot \\ \cdot \\ S_{mi} \end{bmatrix} = S_{1i}\begin{bmatrix} 1 \\ 0 \\ 0 \\ \cdot \\ \cdot \\ \cdot \\ 0 \end{bmatrix} + S_{2i}\begin{bmatrix} 0 \\ 1 \\ 0 \\ \cdot \\ \cdot \\ \cdot \\ 0 \end{bmatrix} + \ldots + S_{mi}\begin{bmatrix} 0 \\ 0 \\ 0 \\ \cdot \\ \cdot \\ \cdot \\ 1 \end{bmatrix}$$

vectors in Figure 12•1 could equally well be described by a set of m mutually-orthogonal vectors from the same vector space (Figure 12•2). The trick is, therefore, to choose a basis set such that the basis vectors $\boldsymbol{u}_j$ describe the shapes within the set of spectra, very much as the synthetic line-shapes might do when curve fitting a profile, and allow the coefficients c_{ji} (Figure 12•2) to provide quantities characteristic of the surface.

FIGURE 12•2

$$\begin{bmatrix} S_{1i} \\ S_{2i} \\ S_{3i} \\ \cdot \\ \cdot \\ \cdot \\ S_{mi} \end{bmatrix} = c_{1i}\begin{bmatrix} u_{11} \\ u_{21} \\ u_{31} \\ \cdot \\ \cdot \\ \cdot \\ u_{m1} \end{bmatrix} + c_{2i}\begin{bmatrix} u_{12} \\ u_{22} \\ u_{32} \\ \cdot \\ \cdot \\ \cdot \\ u_{m2} \end{bmatrix} + \ldots + c_{ni}\begin{bmatrix} u_{1n} \\ u_{2n} \\ u_{3n} \\ \cdot \\ \cdot \\ \cdot \\ u_{mn} \end{bmatrix}$$

Consider the following example. Figure 12•3 on page 240 illustrates a relatively small data set of artificial data created from idealised Gaussian/Lorentzian peak positions, FWHM and relative intensities based on PMMA and PVC C 1s spectra. The pure PMMA and pure PVC data envelopes are the top and bottom spectra displayed in the active tile in CasaXPS. Between these two spectra is a sequence of curves generated by linear interpolation using the pure shapes, so the set of curves gradually blend from PMMA to PVC. The spectra represent six vectors in a 322-dimension vector space, where the numerical value 322 is the number of acquisition bins (channels) for each spectrum. Although the number of data bins per spectrum is 322, it is easy to realise that six vectors can describe at most six dimensions and therefore the Cartesian representation uses many more unit vectors than are required to describe the data in Figure 12•3. Simply from the description for the construction of the data set, it is clear the two spectra for pure PMMA and pure PVC contain all the shape information within the sequence and therefore the set of six vectors must all lie in a plane. Further, the best choice for the two basis vectors for this plane would be precisely the two pure spectra normalised. In this case, all the quantitative information would lie in just two co-ordinate values with respect to these two unit vectors. What is more, a linear least squares procedure using the PMMA and PVC spectra as target vectors would fully characterise the variations designed into the data set. The key point here is that the set of six vectors displayed as spectra in Figure 12•3 only need two vectors and six sets of co-ordinates to characterise the entire data set.

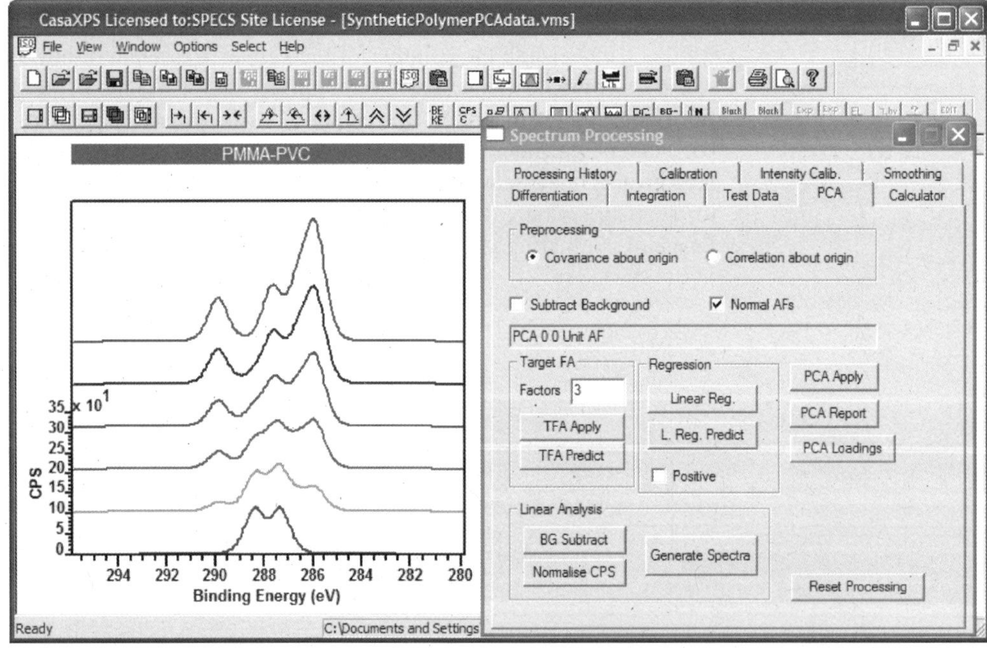

FIGURE 12•3

The problem of characterising a data set by inspection without prior knowledge is a little more difficult. The procedure fundamental to determining dominant shapes within a data set is related to linear-least-squares, where the mechanism by which a linear least squares analysis using the pure PMMA and PVC spectra can also be used to determine a basis for the vector space spanned by the six spectra in Figure 12•3. Consider the linear-least squares problem: given two basis vectors corresponding to the PMMA and PVC spectra, the linear-least-squares procedure constructs a matrix A of size 322 x 2, where the columns of the matrix A are the two vectors for PMMA and PVC. Following through the analysis of linear least squares results in the normal equations $\mathbf{A^TAa = A^Tb}$, where **a** is the least square solution for **b**, an unknown spectral vector. It can be shown that solving these normal equations is equivalent to performing a Singular Valued Decomposition (SVD) of the matrix **A** into three matrices $\mathbf{A = UWV^T}$, where W is a diagonal matrix constructed from the eigenvalues of $\mathbf{A^TA}$ and **V** is the matrix of eigenvectors corresponding to the eigenvalues in **W**. The matrix **U** is the same size as **A** and corresponds to the eigenvectors of $\mathbf{AA^T}$. The least squares solution is computed using the terms in the Singular Valued Decomposition by effectively projecting **b**, the unknown, onto the columns of **U** weighted by the reciprocal of the diagonal elements from **W** and using these quantities to form a linear combination of the vectors from **V**. In summary, the least-squares solution for an unknown spectrum in terms of known spectra is completely determined by the SVD matrices. The eigenvectors for $\mathbf{A^TA}$ form an orthogonal basis for the least squares solution space while the columns of the **U** matrix form an orthogonal basis set for the same subspace spanned by the columns of **A**. What is more, the use of the matrix $\mathbf{A^TA}$ in the algorithm guides the choice of basis towards the variations in the data. By analogy, considering each of the six spectra in Figure 12•3 as a possible basis vector for a linear least squares decomposition of an unknown spectrum. An SVD for the matrix of size 322 x 6, consisting of the six vectors, can be constructed in the image of the matrices shown in Figure 12•4. The result is a similar prescription for

$$\begin{bmatrix} S_{11} & S_{12} & \cdot & \cdot & S_{1n} \\ S_{21} & \cdot & \cdot & \cdot & \cdot \\ \cdot & \cdot & \cdot & \cdot & \cdot \\ \cdot & \cdot & \cdot & \cdot & \cdot \\ \cdot & \cdot & \cdot & \cdot & \cdot \\ S_{m1} & \cdot & \cdot & \cdot & S_{mn} \end{bmatrix} = \begin{bmatrix} u_{11} & u_{12} & \cdot & \cdot & u_{1n} \\ u_{21} & \cdot & \cdot & \cdot & \cdot \\ \cdot & \cdot & \cdot & \cdot & \cdot \\ \cdot & \cdot & \cdot & \cdot & \cdot \\ \cdot & \cdot & \cdot & \cdot & \cdot \\ u_{m1} & \cdot & \cdot & \cdot & u_{mn} \end{bmatrix} \begin{bmatrix} w_{11} & 0 & \cdot & \cdot & 0 \\ 0 & w_{22} & \cdot & \cdot & \cdot \\ \cdot & \cdot & \cdot & \cdot & \cdot \\ \cdot & \cdot & \cdot & \cdot & 0 \\ 0 & \cdot & \cdot & 0 & w_{nn} \end{bmatrix} \begin{bmatrix} v_{11} & v_{21} & \cdot & \cdot & v_{n1} \\ v_{12} & \cdot & \cdot & \cdot & \cdot \\ \cdot & \cdot & \cdot & \cdot & \cdot \\ \cdot & \cdot & \cdot & \cdot & \cdot \\ v_{1n} & \cdot & \cdot & \cdot & v_{nn} \end{bmatrix}$$

FIGURE 12•4

determining the set of least-squares vectors within the subspace spanned by the original six vectors. In other words, the SVD offers up the new basis for the vector space spanned by the original spectra. Further, because the matrices **U** and **V** are both orthogonal, their action on any vector preserves the distance metric, therefore, the **W** matrix contains all the information regarding the relative contributions of these eigenvectors to the description of the subspace. If a diagonal element of **W** is too small then this simply means the corresponding eigenvectors should be excluded from any subsequent calculations. The implication of a small eigenvalue is that the original vectors include a linear dependency and therefore the dimension of the subspace was less than the number of spectra used in the **A** matrix. What is more, the diagonal elements of **W** rank the importance of the basis vectors appearing as the column vectors in **U**, hence the commonly used name for the SVD result, namely, Principal Component Analysis. That is, the principal components are those column vectors in **U** for which the corresponding eigenvalue are considered to be significant.

An SVD performed on the six spectra in Figure 12•3 is presented in Figure 12•5 The traces in the active tile (Figure 12•5) are the unit vectors from the **U** matrix ranked by magnitude of the corresponding eigenvalue. The tables appearing in the dialog windows in Figure 12•5 list the eigenvalues and related statistics, plus the co-ordinates of the original spectra with respect to the unit vectors from the **U** matrix. These results are entirely consistent with the interpretation of the data set given above, namely, only two eigenvalues are non-zero and therefore only two linearly independent basis vectors were found. The outcome of the PCA is therefore that the entire set of six spectra is completely described by two vectors and six pairs of co-ordinates. The Cartesian description (Figure 12•1) required 322 unit vectors and 322 co-ordinates per vector. The PCA has therefore performed a significant data reduction. However the outstanding question of how to interpret the reduced number of characteristic values still remains.

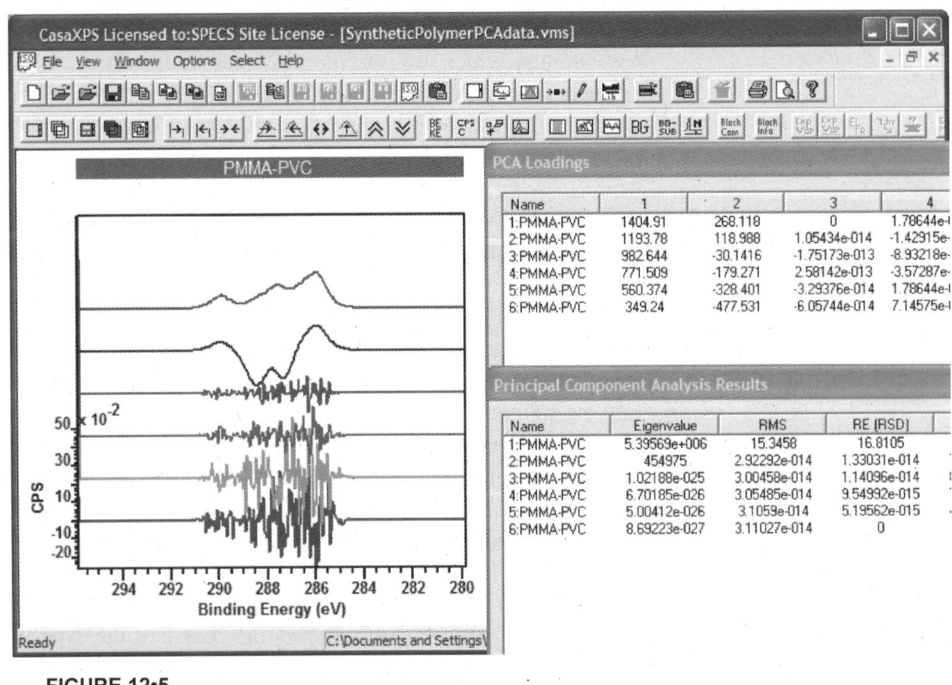

FIGURE 12•5

The unit vectors displayed in Figure 12•5 are referred to as abstract factors (AF).

These abstract factors encapsulate the shape information from the set of six spectra, but have been chosen in such a way that the two significant abstract factors account for the maximum contribution to the entire set of spectra subject to the constraint that the abstract factors are mutually orthogonal. The co-ordinates are often referred to as the loadings for the data with respect to the abstract factors and, for this example, do follow a trend from pure PMMA to pure PVC; however the numerical values relate to the proportions of the abstract factors rather than a direct connection with the pure spectra. The abstract factors are the appropriate linear combination of the PMMA and PVC spectra so that if the data set, as a whole, is approximated by the abstract factors weighted using the corresponding loading, the root mean square error monotonically decreases with the addition of each new abstract factor ordered by eigenvalue. Thus, the reduced set of characteristic values does not clearly identify the chemical information but rather indicates a transition between two states has occurred.

Some Cautionary Notes on PCA

It is clear from the above discussion that PCA is a mathematical tool offering a data oriented perspective for a large data set. The relationship to the linear-least-squares procedure provides some comfort to those wishing to dissect an experiment; however the results of PCA are not always as easy to interpret as might be thought at first sight. The following are a list of cautionary notes, where the intention is to expose these pitfalls rather than condemn PCA as a technique. Indeed, in the right context, PCA is a powerful tool and for certain types of multi-dimensional data set, applying the theory of PCA as part of an analysis yields remarkable results. The first note relates to curve-fitting and the ability of PCA to assist when constructing a peak model. Once again, consider the data set in Figure 12•3 on page 240. It is worth emphasizing the fact that the set of six spectra represent a movement between two data envelopes, however even though each envelope is in turn constructed from individual Gaussian/Lorentzian peaks, the PCA analysis has no way of identifying the underlying peaks, since the variation in the data set was achieved by varying these underlying peaks as a unit. PCA therefore indicates the number of synthetic models required to describe the changes in the data set, not the number of synthetic peaks required to construct the synthetic models. As a result PCA makes a rather weak statement regarding the number of synthetic components in a data set, namely, PCA at best only offers the minimum number of components required and will only deliver the exact number provide all the individual transitions present in a data set vary independently at some point during the course of the experiment. At worst, PCA is inconclusive regarding the true number of trends and is entirely misleading if used to attempt inference of the number of synthetic components in a peak model.

The data set shown in Figure 12•6 is constructed to illustrate possibilities for confusion when interpreting PCA results. The same method for construction of the data shown in Figure 12•3 is employed for the first column of data blocks shown selected in the data set in Figure 12•6. In the case of the data in Figure 12•6, three synthetic polymer data envelopes are blended using linear interpolation, namely, PMMA, PVC and PIB. Three additional columns of data blocks are similar blends of PVC to modified PVC data, where the modified PVC is either energy shifted, intensity scaled or convoluted with a Gaussian to simulate broadening of a spectral feature. All three modifications ("distortions") of an initial line-shape might be expected throughout the course of a multi-layer sputter depth profile, for example. A PCA applied to the data in the first column in Figure 12•6 is displayed in Figure 12•7, where the graphical display plots the loadings for each of the fifteen spectra used in the PCA with respect to the three most significant abstract factors. The axes in the 3-D plot represent the first three abstract factors, which as expected, account for all the shape information in the data set. The abstract factors are also plotted in Figure 12•7 and the coordinates for each of the original spectra used in the PCA are listed.

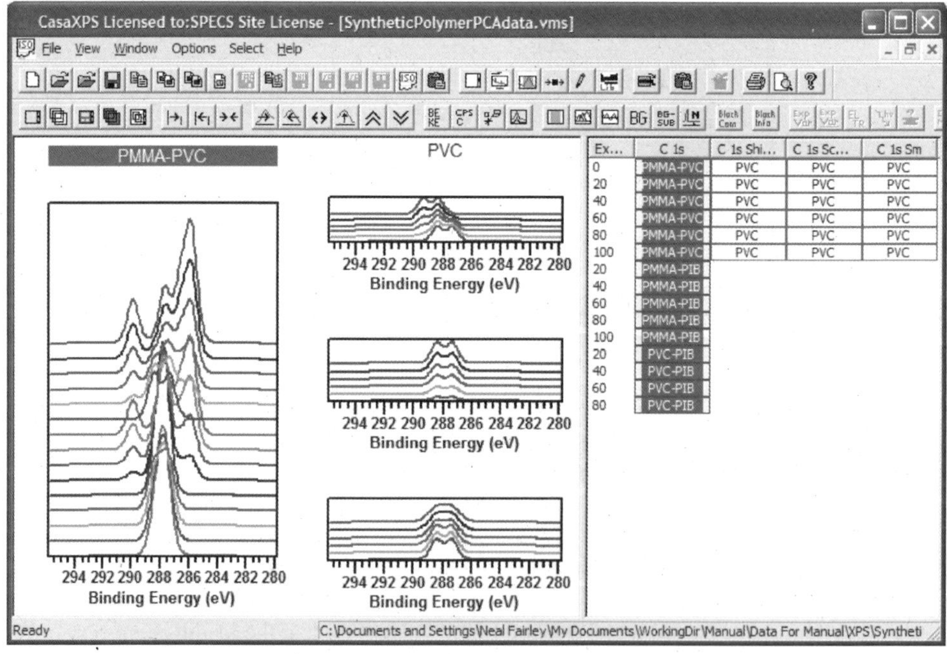

FIGURE 12•6

The first observation based on the data in Figure 12•7 is the importance of both parts of the PCA to the description of the data set. That is, the loadings clearly show an orderly pattern for the spectra when projected onto the first three abstract factors, but the meaning of this pattern is only discerned, if at all, by examining the abstract factors. Now consider a PCA performed using the same data as that used in Figure 12•7 but with the addition of the set of PVC spectra in the third column of data blocks in Figure 12•6. These new additions to the PCA calculation are merely the PVC spec-

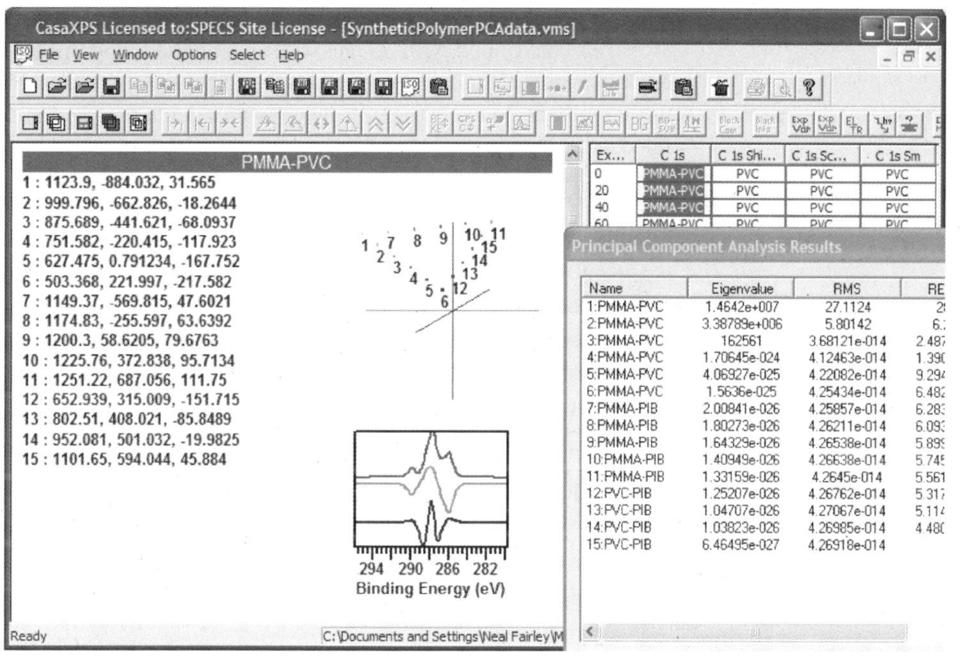

FIGURE 12•7

trum scaled by various constants. Clearly, since PVC is already present in the original data set, scaling the PVC spectrum and including these data will not add any new shape information to the data set. Thus performing a PCA on the new set of spectra results in the information displayed in Figure 12•8, where the PCA result dialog window indicated there are still only three significant (non-zero) eigenvalues, as expected. The addition of these scaled PVC spectra to the PCA calculation has, however, significantly altered the abstract factors. Comparing the factors between Figure 12•7 and Figure 12•8 it is clear the loadings are very different and if the loadings are taken in isolation, the inference would be a dramatic difference had occurred between the two data sets. An initial response might be such differences are good and offer an indicator of changes between samples. However consider a depth profile through an oxide layer in silicon. If the same material is profiled, but using a different number of etch-cycles, then the two data sets would be very similar to the situation in this example. Extra etch cycles through the metallic silicon layer would bias the data set towards the metallic signal and therefore the two experiments might easily generate two different profiles based on the loadings alone. The important thing to note is that the PCA is characterising changes in not only the sample, but also the experimental parameters used to perform the measurement. Changing the acquisition parameters can introduce major changes in the abstract factors and therefore the loadings.

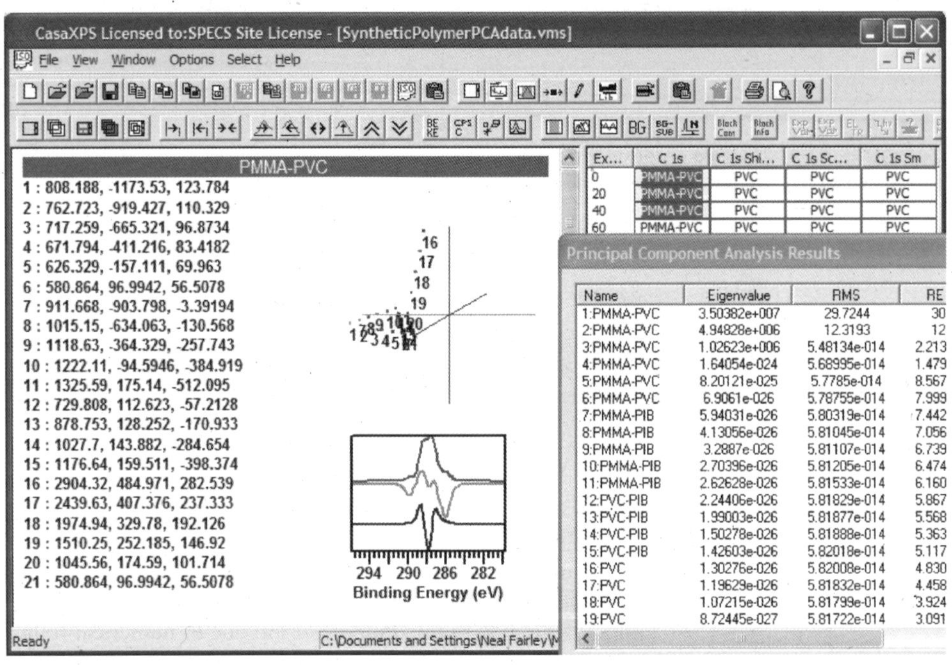

FIGURE 12•8

Two further sequences of PVC spectra are included in Figure 12•6. The consequence of a peak shifting in energy is an additional significant abstract factor. Similarly, a peak broadening also introduces a new abstract factor so when these two sets of essentially PVC spectra are added to the data from the first column, a PCA for these data results in five significant non-zero eigenvalues and therefore abstract factors (Figure 12•9 on page 247). At this point it is worth noting that the synthetic data used in these tests does not include noise, hence the eigenvalues, while small, are still very distinct from the eigenvalues genuinely equal to zero. The point of introducing these modifications to the PVC spectrum is to show that PCA, when confronted with commonly found variations in XPS spectra, indicates an increased number of trends in the data set compared with those that might be considered significant with respect to the sample. Bearing in mind that XPS involves other changes to peak shapes resulting from background variations, the prospect of assigning the number of trends within a data set based on the number of significant abstract factors becomes daunting.

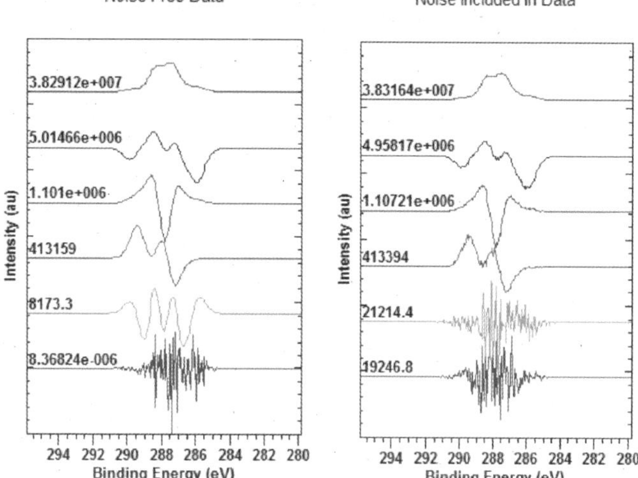

FIGURE 12•9 The first six abstract factors for two equivalent data sets, with and without simulated noise.

Real XPS differs from the spectra in Figure 12•6 in that Poisson statistics cause variations in the signal which can mask underlying trends otherwise identified by PCA. The six abstract factors in the left-hand-side of Figure 12•9 are normalised vectors and while the least significant abstract factor may appear to contain noise, in fact the rapid oscillations in the unit vector are due to numerical round-off errors. The corresponding eigenvalue for the sixth abstract factor is twelve orders of magnitude smaller than the largest eigenvalue and since the eigenvalues contain the size information from a PCA, the sixth abstract factor makes no contribution to the description of the original data set. The same data as seen in Figure 12•6, for which with simulated noise consistent with Poisson statistical variations is added, results in the abstract factors and eigenvalues in the right-hand-side of Figure 12•9. The sixth abstract factor for the noisy data set is as significant in terms of size as the fifth, which itself resembles noise and therefore has lost much of the shape information clearly present in the noise-free PCA abstract factors. The point to take from this example is that information in the data set clearly distinguishable in the absence of noise is dispersed throughout the abstract factors for data with noise content and therefore identifying the number of underlying trends is far less straight forward for real data sets.

The conclusion from these cautionary notes would seem to be that PCA is of limited use for XPS spectra. The data sets, however, have been specially chosen to illustrate potential difficulties and the primary source for these limitations is due to the size of the data set. Statistical procedures are best used when there is difficulty understanding large quantities of data. The test data set are small and very understandable, hence PCA compares poorly to a simple inspection of the data in the light of the methods used to create the spectra. Without prior knowledge and obvious structure in a (larger) data set, the structures pulled from the data by PCA would have been more rewarding. What is more, for genuinely large data sets, reducing the data to a small quantity of information, albeit abstract, does provide an insight into the nature of the problem. With these thoughts in mind, the emphasis in this discussion will switch away from small and less appropriate data sets to large quantities of data for which PCA can be used to great effect.

A Real Application for PCA: Spectra and Images

The name Principal Component Analysis refers to the ability of an SVD to create sets of basis vectors capable of describing the same sub-space as the original set of non-orthogonal vectors defined by the data, where these basis vectors are ranked according to their information content by the magnitude of the corresponding eigenvalues. It is precisely this ability to measure and order the shape information within the data that makes PCA such a powerful tool when handling large data sets with a relatively small number of trends throughout the full set of data. Even for the small data set shown in Figure 12•6, when a PCA is performed on these data with simulated noise added, a PCA concentrates the shape information into the abstract factors with the largest eigenvalues. What is more, the random noise is gathered in the abstract factors with small eigenvalues. A PCA of a data set therefore presents an opportunity to partition the original vector space into useful vectors and noise vectors. The success of this operation will depend on the extent to which the useful information is mixed with noise. The situation most favourable to the separation of noise from useful information occurs when a set of trends is massively over sampled. Happily, these types of experiment are becoming more common. To understand the context for information partitioning using PCA, consider the following analogy. Acquisition of a single spectrum is typically performed using a total acquisition time chosen to result in an acceptable signal-to-noise ratio for the ensuing data. Logically the spectrum is constructed from a set of spectra all assumed to be identical and acquired at smaller dwell-times, all of which when summed equal the desired acquisition-time. Summing these spectra and averaging in time provides a very simple one-dimensional linear-least-squares solution, namely a single spectrum with adequate signal-to-noise. However, the assumption is that all the summed spectra are equivalent. Now consider a variation on this theme, the problem of acquiring data from an inhomogeneous sample, where determinations at different locations on the sample may result in different spectra. This would prevent improving signal-to-noise in the data set by simply summing the spectra from numerous locations on the surface. If every location on a sample surface results in a different spectrum then there would be no alternative but to acquire each spectrum with the dwell-time appropriate for the desired signal-to-noise. On the other hand, if the number of distinct spectra is small compared to the number of locations measured, then PCA offers a means of partitioning the set of spectra from the different locations into a small number of abstract factors representing the shapes in the spectra and a set of abstract factors characteristic on the noise in the entire data set. If linear combinations of the significant abstract factors are used to reconstruct the spectra, then the result is a set of spectra for which the signal-to-noise is characteristic of the acquisition time for the entire set of spectra rather than the dwell-time of the individual spectra. The term, reconstruct the spectra, means perform a PCA, identify the significant abstract factors and only use those abstract factors added together in a sum similar to the sum shown in Figure 12•2 on page 239, where all loadings deemed associated with noise abstract factors are set equal to zero. The result is a set of spectra with signal-to-noise enhanced but each spectrum is associated with a different location on the sample surface and is free to differ to the extent of the information in the retained abstract factors. Further, the reconstructed spectra are a linear least squares approximation to the original data where the basis vectors are the abstract factors containing the shape information.

FIGURE 12•10
Parallel imaging - energy slices

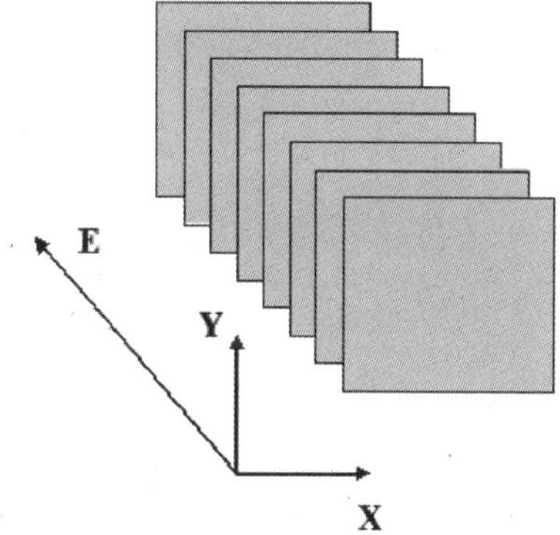

The scenario described above is commonly found for techniques where parallel data acquisition is possible for a set of experimental variables. Two obvious examples are ToF SIMS images, where an ion beam raster over a surface pauses at each pixel to collect a parallel mass spectrum, and stigmatic XPS imaging technology, where parallel images of a surface are acquired at varying energy (Figure 12•10). Both these examples are logically equivalent in that both result in spectra at each pixel of an image. If a set of XPS images are acquired over a sequence of energy steps spanning a photoelectric peak, then potentially numerous images result from a small number of chemical states, background artefacts and instrumental variations; a situation ideal for decomposition using PCA. As stated earlier, the data matrix can be constructed from spectra or images. The SVD proceeds identically for both apparently different, yet mathematically equivalent types of data. The analysis of a stack of images as logically depicted in Figure 12•10 would proceed using the same steps used to process the spectra in Figure 12•6 on page 244. The intention, for the image set, is to identify abstract factor images containing spatially significant information and reconstruct the set of images from only those loadings and abstract factors without noise. Once the images are reconstructed from the significant abstract factors, the spectra at each pixel are extracted with the view to defining quantification regions and/or synthetic component. The quantification items are then used to compute RSF scaled images in preparation for atomic concentration calculations on a pixel-by-pixel basis. The true significance of performing the PCA analysis on the image set lies in the dramatic reduction in noise associated with the spectra at a pixel. Before processing, the signal-to-noise ratio of the spectra leaves much to be desired, however following PCA and omitting the noise components from the reconstruction step, the signal-to-noise is characteristic of the time invested in acquiring the entire set of images. For XPS, the significant gain comes from the quality of the computed background now possible on the spectra at each pixel (Figure 12•11).

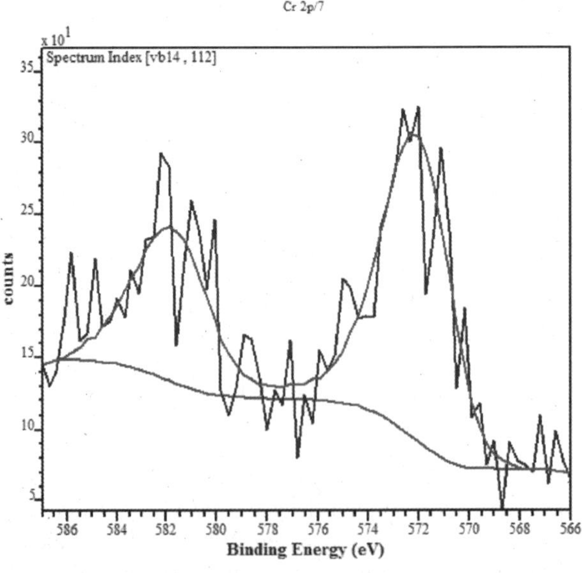

FIGURE 12•11 Spectra from a single pixel

Figure 12•11 shows an example of a spectrum taken from an image pixel, where the two spectra correspond to the unprocessed data and the same data following data reconstruction using the first four abstract factors from a stack of images acquired at 0.3eV energy steps across a Cr 2p doublet. The sample was an MRS-3 (Magnification Reference Standard) where ITO (indium tin (In-Sn) oxide) circles are surrounded by CrO_2. The data was acquired on an un-calibrated Kratos Axis Ultra using a pulse counting delay line detector (DLD). The consequence of using an un-calibrated instrument is that the images contain artefacts due to the instrumental setup and therefore the data set provided an extreme example of mixing information from more than simply chemical state variations.

The image in Figure 12•11 corresponds to the integration of the counts per second between the Cr 2p peaks and background for each pixel in the image set. The variation in intensity in the Cr 2p image is a result of not calibrating the DLD for flat field response before acquiring the data and accounts for one of the four significant abstract factors (Figure 12•12 on page 252) required for the reconstruction step. A second abstract factor can be attributed to a further instrumental artefact, namely, peak alignment over the field of view, again engineered into the data set by design. If the instrument was properly tuned and calibrated, these two factors would diminish, however a third factor corresponding to the background signal contains structure as a consequence of variations due to the In-Sn oxide zones and represent an unwanted mixing of spatial information exposed by PCA. Other possible explanations for these chromium images containing more than one factor might be variations in X-ray flux over the imaged area and topographical effects. Regardless of the source for these additional factors, their inclusion in the reconstruction step is important for a proper move from noise filled data to the smooth curve seen in Figure 12•11.

Interpreting a set of abstract factors as seen in Figure 12•12 is far from easy, as even the most significant abstract factor is determined not just from the most dominant trend in an image set, but is biased by the minor structures too. Visual inspection might lead one to believe the top left abstract factor in Figure 12•12 corresponded to the flat field response of the detection system and excluding this abstract factor from the reconstruction step would lead to a set of calibrated images. The problem is that the abstract factors are determined as a unit and therefore omitting the most significant factor from the reconstruction of the images, while removing a large proportion of the flat field response, will also remove background and subtle chemical information.

The best method for determining chemically meaningful images is to include all the abstract factors of significance, *regardless of source*, and process the chemical information using the spectra at each pixel with standard, tried and tested quantification procedures from spectral quantification. Quantification items from regions and

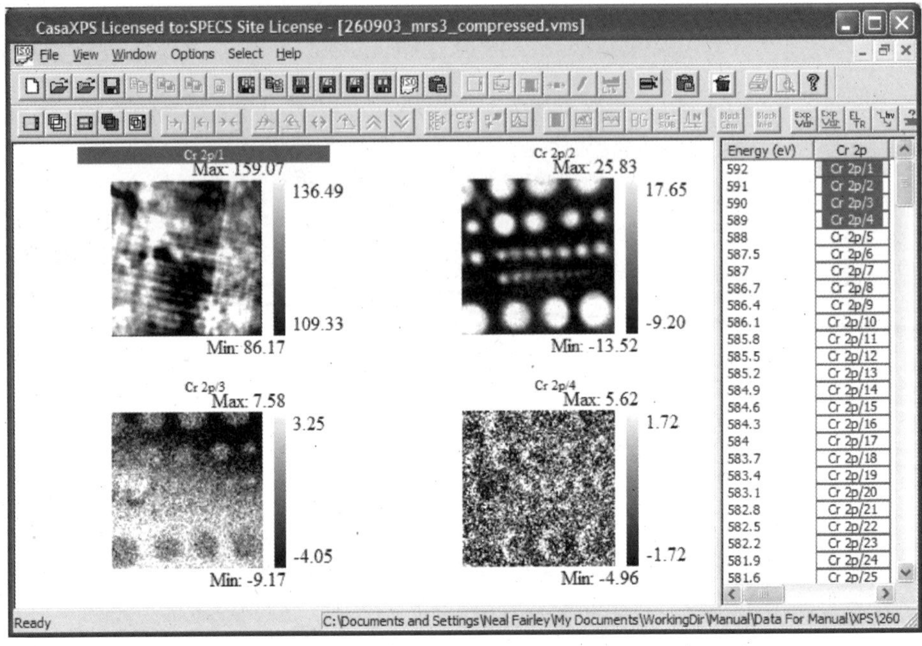

FIGURE 12•12

synthetic components, when combined using the standard atomic concentration formula, produce images free from instrumental, topological and background artefacts. Quantification of the MRS-3 sample using five photoelectric lines and basic quantification regions produces the atomic concentration images seen in Figure 12•13. Compare the Cr 2p image in Figure 12•13 with the image in Figure 12•11 on page 250. The intensity variations due to the DLD flat field response are absent from Figure 12•13 and the image information can be trusted with the same degree as atomic concentration tables generated from conventional spectral data.

The subject of how to determine the number of abstract factors of significance to the data description is yet to be discussed. When creating the Cr 2p spectra in Figure 12•11, four abstract factors where used to approximate each image in the data set. Figure 14 illustrates the linear combination of the abstract factors for one image in the set of images acquired over the Cr 2p doublet energy range.

Included in Figure 12•14 on page 254 are the four abstract factors combined using four coefficients, which are effectively determined in a least squares sense, the image resulting from the least-squares summation, the corresponding raw image and

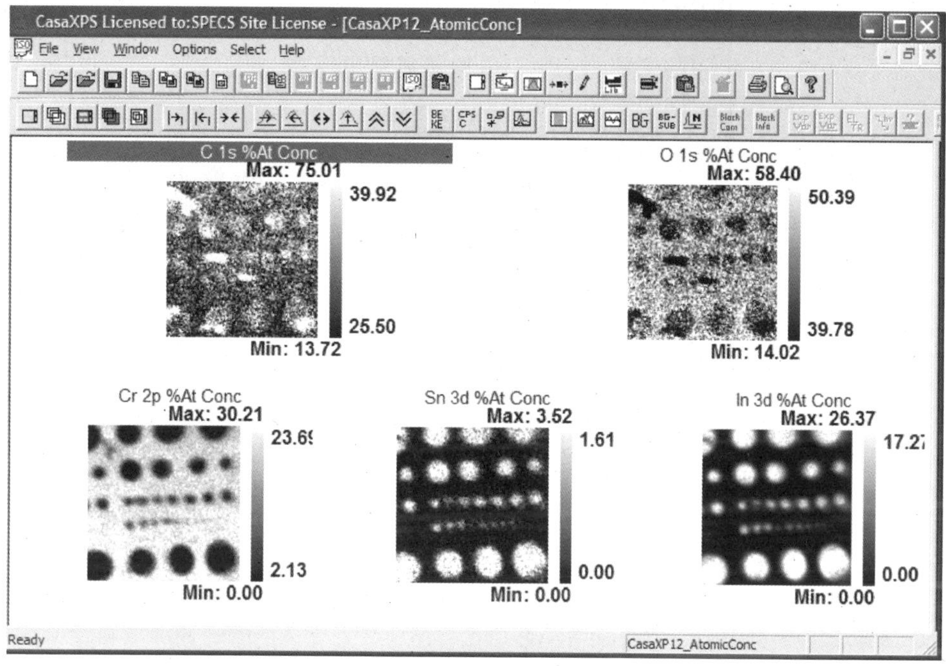

FIGURE 12•13

the fifth abstract factor not used in the reconstruction step. Interestingly, none of the four abstract factors individually bare any resemblance to the resulting image. In fact the characteristics visible in the abstract factors are more easily attributed to instrumental and background artefacts, the chromium image is buried within and only reappears once an appropriate linear combination is prescribed. The main reason for including four abstract factors is the lack of any structure in the fifth abstract factor, also shown in Figure 12•14. The literature of Factor Analysis is filled with statistical methods for assessing the number of significant components: none is foolproof and when the objective is data reconstruction, which is the basis for the application in CasaXPS, assessing the quality of the reconstructed data as a function of the number of abstract factors would seem to be the most reasonable way to proceed. The greatest danger, however, it that minor yet important structure in the data set is blended in with the noise. A consequence of omitting information dispersed throughout the abstract factors representing noise is that the reconstructed spectra contain non-physical features. Indeed, assessing the data using both reconstructed images and spectra provides a consistency check for the procedure in general. Both must continue to make sense form a physical perspective. It is therefore worth experimenting with the number of abstract factors used to reconstruct the data set and observe the development of noise within the reconstructed data with each additional abstract factor.

FIGURE 12•14

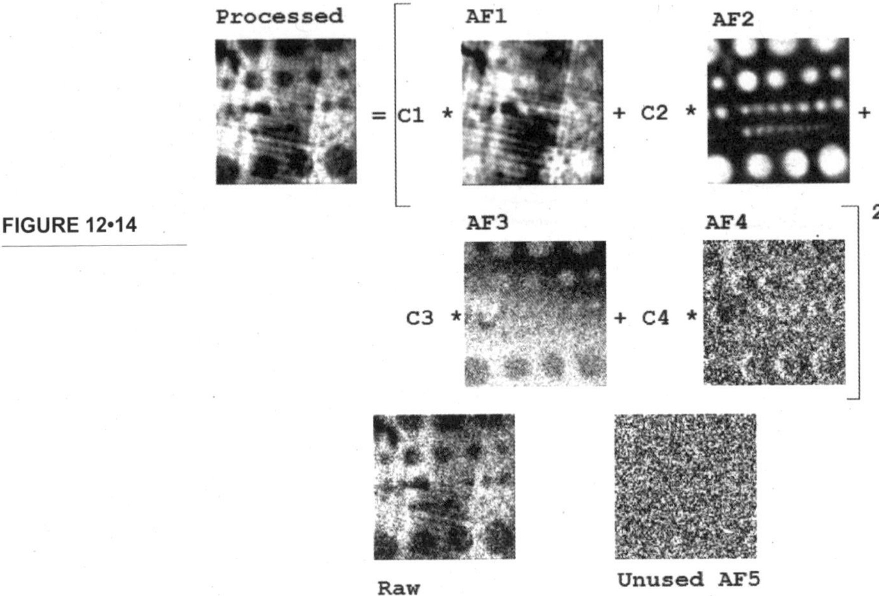

The significance of noise compared to genuine but minor shape information can be important for reconstructed spectra. Figure 12•9 on page 247 provides an example of where the introduction of noise has masked a real structure in the fifth abstract factor. A problem with XPS data is that noise varies in magnitude with the square root of the counts per data channel. For spectral where the peak is significantly more intense than the background, the variation of the noise over an energy interval is non-linear and more significant to a PCA calculation than a broadening or small shifts in a peak position. One way to suppress the variations in the noise as a function of intensity is to divide each channel by the square root of the counts per channel. The act of taking the square root of the counts per channel is equivalent to dividing by the square root and so squaring the result returns the data to the original state. The spectrum in Figure 12•11 on page 250 and the abstract factors in Figure 12•14 are in fact determined for the data following the square root function applied to each pixel and the reconstructed image in Figure 12•14 is therefore the square of the linear combination of abstract factors.

CHAPTER 13

PCA Techniques

PCA and Data Smoothing

PCA creates a set of abstract factors (AF) from the original data such that each AF describes the maximum variation in the data, in a least-square sense, subject to the constraint that the set of AF is mutually orthogonal. These abstract factors represent an orthogonal basis set for the same vector space that is spanned by the original data and therefore each of the original spectra can be equated to a linear combination of the abstract factors. The information content of the original data typically requires fewer AF then is produced by the eigenanalysis and by choosing only those AF with significant structure the noise component of the data can be reduced. Limiting the number of AF is equivalent to projecting the spectra onto the subspace spanned by the first n abstract factors. If the number of AF used in this projection step is chosen appropriately the result is a set of spectra with a reduced component of experimental noise or in other words a smoothed set of spectra.

Reconstruction of Spectra Using a Reduced Set of Abstract Factors

CasaXPS performs the reconstruction step as part of the TFA Prediction option on the PCA property page of the Processing window (see "The Image Processing Dialog Window: A Quick Summary of Features" on page 281).
Figure 13•1 shows a data set [1][2] where quantification regions have been defined for the spectroscopic features of interest. These regions are used during the PCA step

to create the vectors actually used in the eigenanalysis and the region interval also represent the energy range over which the data reconstruction is performed.

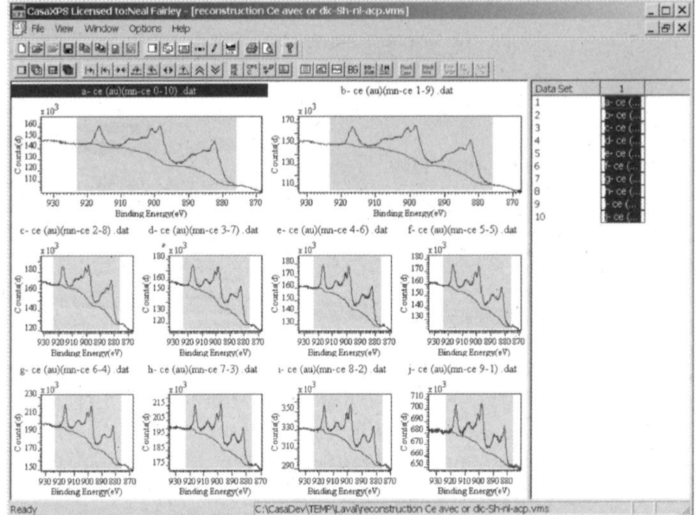

FIGURE 13•1 A data set [1] of related spectra suitable for smoothing by PCA.

Data reconstruction is achieved by a sequence of steps:

1. Load the data set.
2. Before performing any PCA on the data, copy those data for which the data smoothing is to be performed. Two copies of the original data (three identical sets of spectra) allow the results of the analysis to be viewed on completion.
3. Select one of the data sets and perform a PCA. Display the results of the PCA in the active display tile in the left-hand-side of the experiment frame.
4. Using the right-hand-side of the experiment frame, select a second set of the original data and enter the number of abstract factors used to TFA Predict the selected data. Press the TFA Predict option on the PCA property page of the Processing dialog window.
5. Select and display spectra from the reconstruction set. If three copies of the original where created, the predicted spectra can be viewed overlaid with the original spectra and the calculator may be used to compute a residual for the predicted spectra.

Step 1:

Figure 13•1 represents a data set after it is initially loaded into an experiment frame.

In this case a single column of spectra have been entered into an ISO-14976 data file.

Step 2:

Copy the spectra using the right-hand-side Copy/Append dialog window.

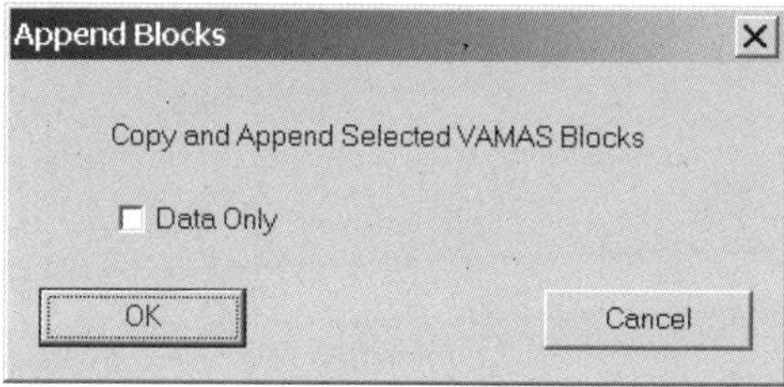

FIGURE 13•2 Dialog for copying the selected data set.

Select the spectra in the right-hand-side of the experiment frame as shown in Figure 13•1. Right-click the mouse over the selected blocks. The Copy/Append dialog window appears. Press the OK button; leave the "Data Only" tick box un-ticked (Figure 13•2). If the original set of spectra is selected a second time and the copy operation performed again then the resulting experiment frame is as shown in Figure 13•3.

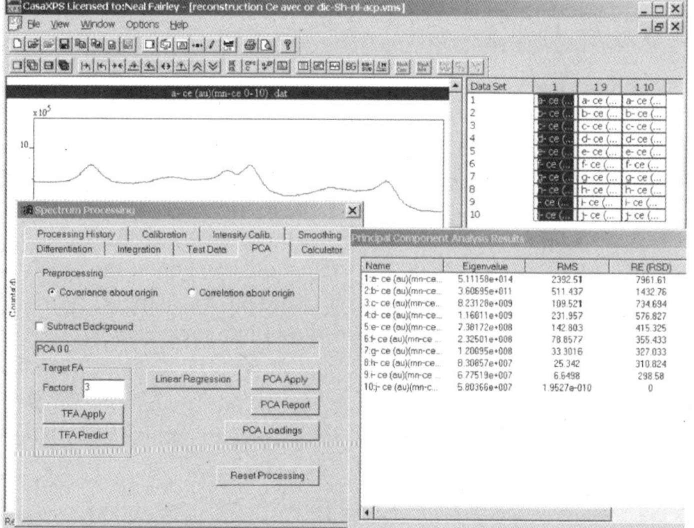

FIGURE 13•3 Data set after two copies of the original spectra have been created. A PCA has been performed on the original column of data.

Step 3

Select the first column of data as shown in Figure 13•3 and display the spectra in the active display tile. Perform the chosen PCA on the displayed data.

Step 4

Whilst still displaying the results of the PCA in the active display tile, select a second copy of the original data as shown in Figure 13•4. Enter the number of abstract factors in the Target FA box on the PCA property page and press the TFA Predict button. The results of the reconstruction step using three abstract factors will be entered into the spectrum blocks selected on the right-hand-side.

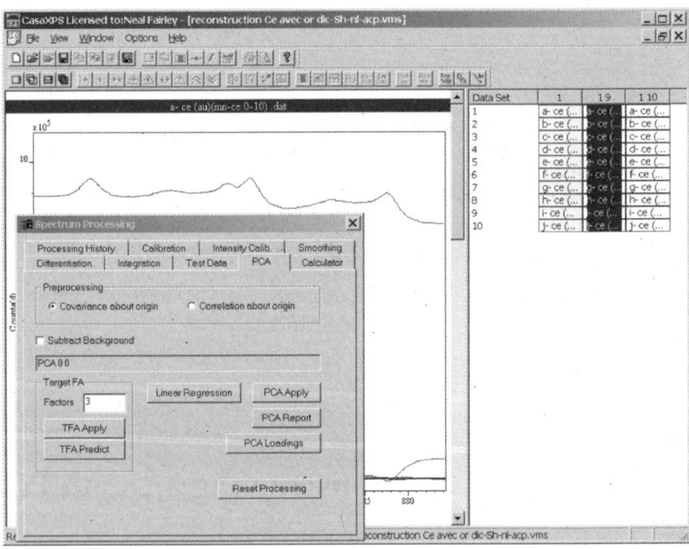

FIGURE 13•4 PCA results are still displayed in the active tile and a copy of the data set is selected in the right-hand-side of the experiment frame.

Step 5

Check the results of the reconstruction step. Figure 13•5 shows how the reconstructed spectra can be compared against the original data. Note that a third copy has been used simply to avoid resetting the PCA column, for example, to avoid losing energy calibration when the PCA is undone. The third column could also be used to compute a residual via the calculator property page.

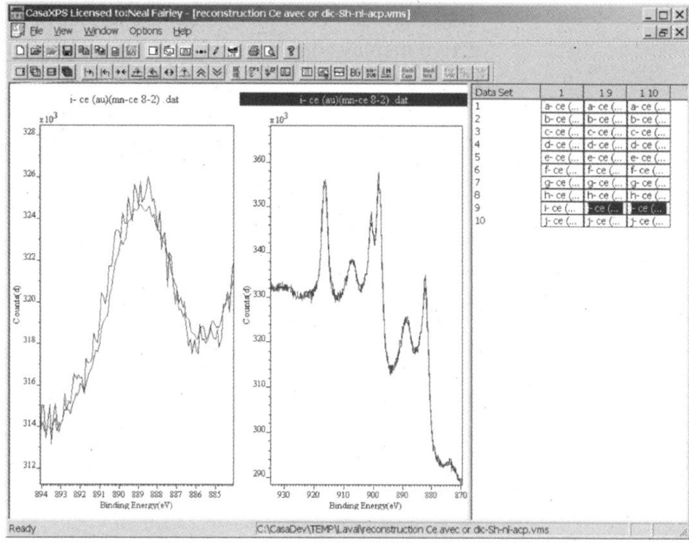

FIGURE 13•5 Overlay the reconstructed spectra and the original data in order to evaluate the consequences of limiting the number of abstract factors.

Constructing a Depth Profile using PCA

Principal Component Analysis (PCA) is best suited to sets of data for which it is meaningful to express the recorded data as linear combinations of fundamental components. When PCA is applied, these fundamental components are constructed from the original dataset using a linear-least-square procedure, where the spectra are considered to be vectors in an n-dimensional subspace and the least-square principle is used to generate an equivalent set of vectors that span the same subspace as the original spectra, subject to the constraint that these new vectors are mutually orthogonal. After analysis, a dataset is simply dissected and then reconstructed using a different perspective. Each spectrum is completely determined by an orthogonal set of n-dimensional axes (abstract factors) and the corresponding co-ordinates with respect to these axes, both of which are generated by the PCA. What this means is that PCA is an essentially mathematical procedure. Individual components generated by the

PCA algorithm may look like chemically meaningful spectra, but it should never be forgotten that abstract factors are indeed abstract. That is to say, the mathematics determines the form of the abstract factors, not the chemistry. Provided the mathematical procedures are recognized as a dominant influence on the outcome, PCA can be usefully employed to analyze depth profiles and other trends within XPS data. Indeed, understanding the mathematical bias in the PCA abstract factors offers a means of manipulating the data to yield characteristic shapes from which peak-fitting can produce the final quantification for a sample.

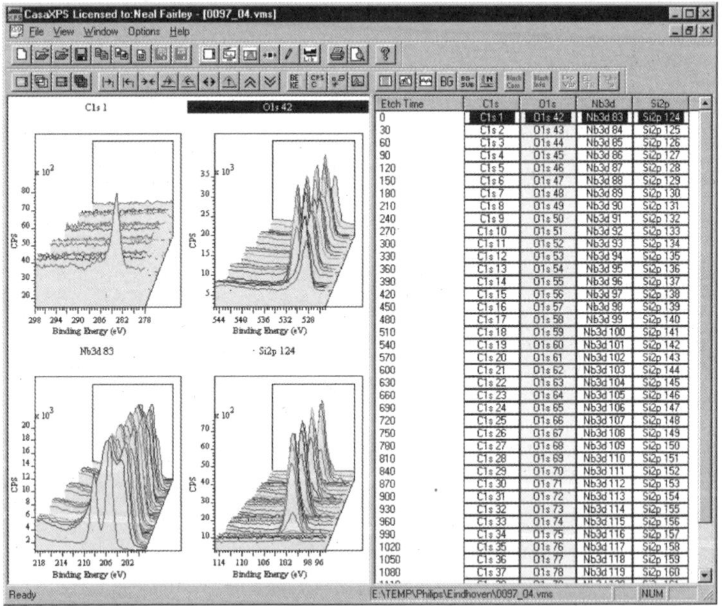

FIGURE 13•6 Multilayer Sample

A Good Dataset

Multilayer materials often produce data well suited to PCA. The dataset shown in Figure 13•6 was acquired on a PHI Quantum 2000 at Philips Eindhoven, where the sample, prior to etching, was an interference filter consisting of alternating layers of SiO_2 and Nb_2O_5.

FIGURE 13•7 O 1s recorded across an interface

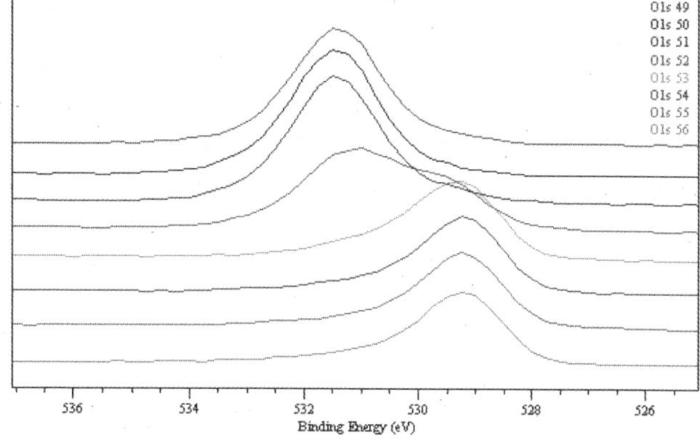

Creating Line-Shapes using PCA

One of the uses for PCA is to sub-sample the data in order to construct line-shapes that emphasize the underlying peak models and aid a systematic approach to fitting the depth profile with synthetic peaks.

FIGURE 13•8 O 1s Abstract Factors corresponding to the data shown in Figure 13•7.

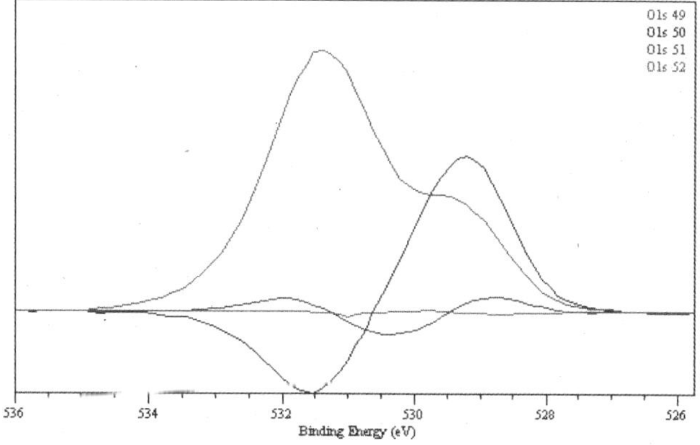

The oxygen peaks shown in Figure 13•7 represent the interface between a SiO_2 and Nb_2O_5 layer. A PCA for the data in Figure 13•7 results in abstract factors shown in Figure 13•8. Three abstract factors apparently account for the variation in the data across the interface however both the size and shape of these abstract factors is strongly determined by the choice of spectra used in the analysis. A second analysis based on the spectra shown in Figure 13•9 results in the abstract factors in Figure 13•10 By choosing the dataset to include additional spectra from the Nb_2O_5 layer, the principal component exhibits a greater proportion of the lower binding energy peak compared to the equivalent computation for the data in Figure 13•7. Clearly the difference between the abstract factors shown in Figure 13•8 and Figure 13•10 demonstrate that chemical mean should not be directly assigned to the results of PCA. This dependence on the data used in the PCA, however, can be exploited to selectively construct spectral shapes from the different environments within the profile.

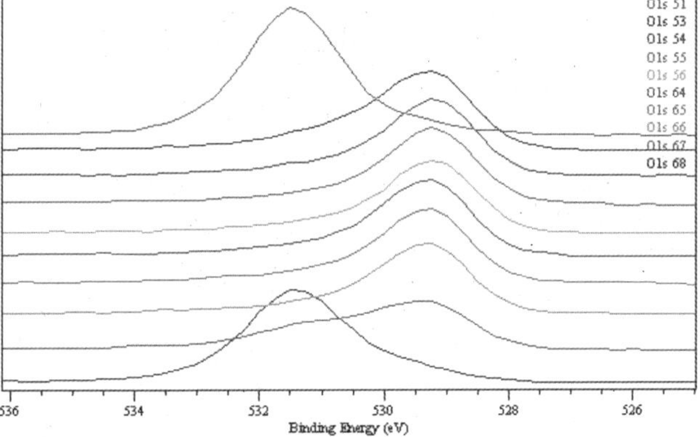

FIGURE 13•9 O 1s sub-sampled across a layer and two interfaces.

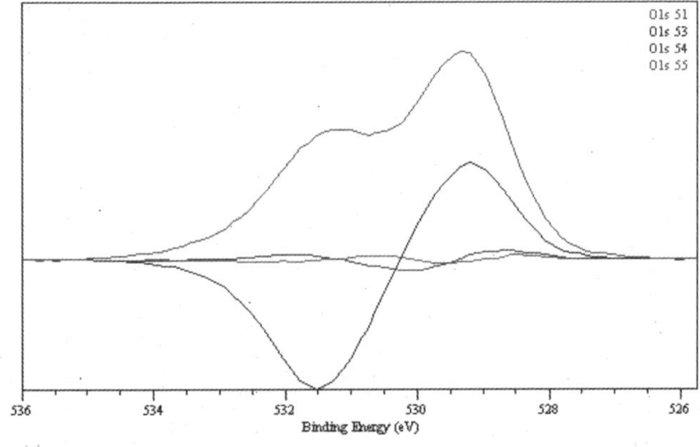

FIGURE 13•10 Abstract factors generated from the data in Figure 13•9.

By sub-sampling the O 1s spectra appropriately PCA can be used to effectively sum the signal taken from equivalent spectra recorded at different depths, whilst offering a measure for how similar the sub-sampled spectra really are to one another. Consider the spectra shown in Figure 13•11. These data are taken from depths for which the O 1s spectra are similar.

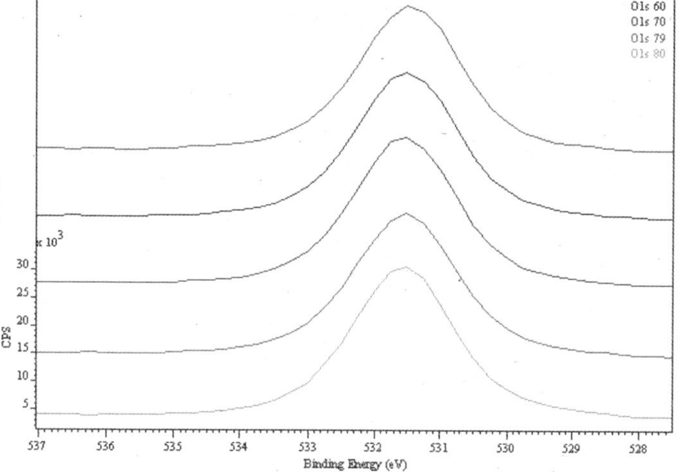

FIGURE 13•11 Subset of O 1s peaks taken from within each layer of SiO_2

The first abstract factor computed by the PCA algorithm represents the maximum variation in the dataset. Thus applying PCA to a set of spectra such as those shown in Figure 13•11 results in an abstract factor that is effectively the sum of the spectra together with a set of vectors representing the deviation from the common shape. Figure 13•12 on page 264 illustrates both the dominance of the first abstract factor and the nature of the remaining information present in the initial set of spectra. Note, the noise within the dataset appears predominantly in the minor abstract factors and, in this case, the major features seen in the inset in Figure 13•12 result from slight shifts in relative peak positions.

Similarly, the shape of the O 1s spectra can also be determined for the Nb_2O_5 layer (Figure 13•14 on page 265). Together, both PCA abstract factors from the two layers provide the basis for synthetic line shapes, which once created, can be used to model the spectra in the depth profile. What is more, PCA can also be used to sample the spectra across the interfaces so that the differences between the interfacial and the non-interfacial regions can be assessed. Figure 13•13 on page 264 shows the first abstract factor computed using spectra taken from near the SiO_2 / Nb_2O_5 boundaries. The synthetic components labelled O 1s 1 and O 1s 2 derive from the abstract factor shown in Figure 13•12 and the equivalent data computed for the Nb_2O_5 layer respectively, but alone these two synthetic peaks when constrained to only adjust in area, do not provide a good model for these interfacial regions.

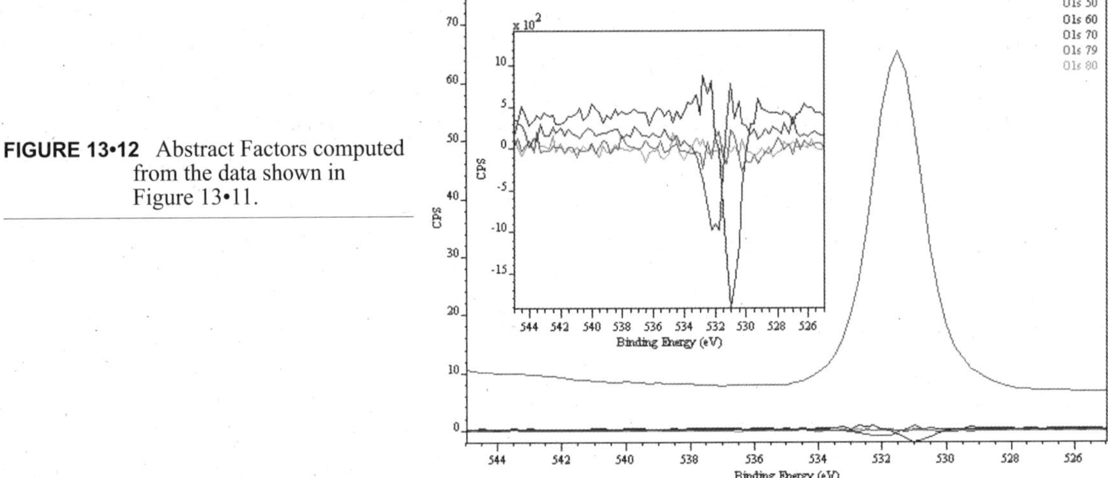

FIGURE 13•12 Abstract Factors computed from the data shown in Figure 13•11.

A third synthetic peak is required before a reasonable fit is produced (Figure 13•13). If the model shown in Figure 13•13 is applied to the O 1s spectra in the dataset and the remaining spectra are analyzed using integration regions only, then the intensity profile shown in Figure 13•15 on page 266 is the result.

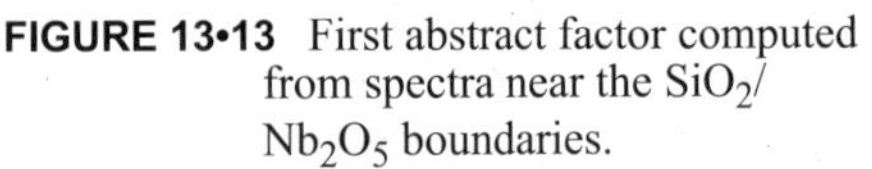

FIGURE 13•13 First abstract factor computed from spectra near the SiO_2/ Nb_2O_5 boundaries.

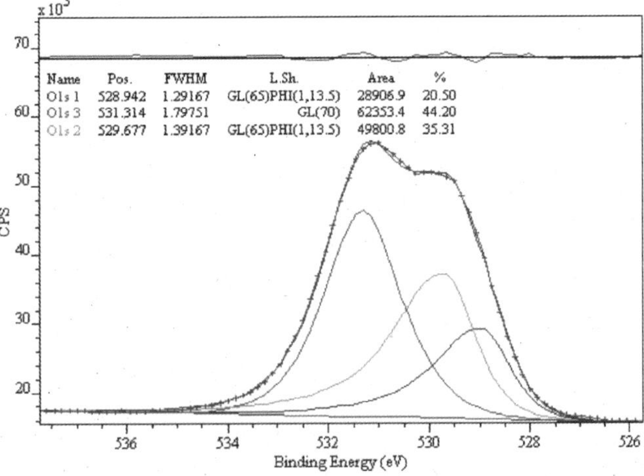

Note how the O 1s peak associated with the Nb_2O_5 layer is modelled using an asymmetric line-shape based upon the abstract factor shown in Figure 13•14. The introduction of asymmetry is based upon an assumption that a single peak should be used to describe the data in Figure 13•14. It is further assumed that the abstract factors determined from the individual layers offer models that have been characterized with respect to relative position and also FWHM. These assumptions cause the third peak to be introduced into the model shown in Figure 13•13. If the entire set of O 1s spectra are used in a PCA, at least six components show significant structure, so the use of three synthetic peaks is possibly under estimating the complexity of the true peak model.

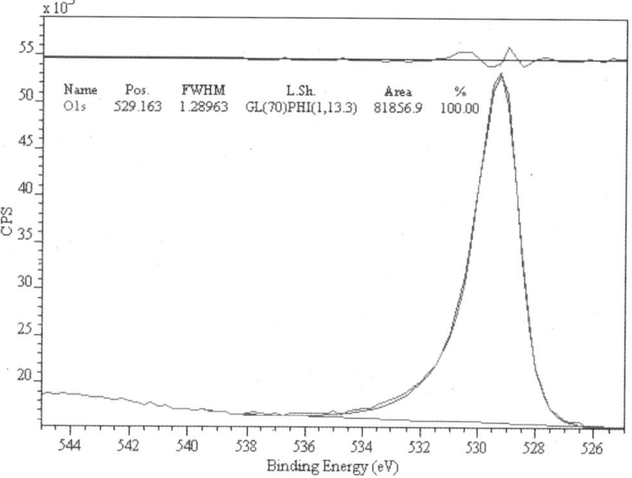

FIGURE 13•14 First Abstract Factor taken from the O 1s spectra within the Nb_2O_5 layer.

Summary

Both PCA and the subsequent peak-fit for the O 1s data support the presence of at least three forms of oxygen, SiO_2, Nb_2O_5 and, possibly niobium silicate. Direct use of Principal Component Analysis to profile XPS data is not always easy to interpret, however, but PCA does offer a means of investigating a dataset and extracting useful information in a form that can be used in more traditional data reduction techniques. The depth profile in Figure 13•15 owes as much to the linear analysis of PCA as it does to the non-linear optimization algorithms used in peak modelling.

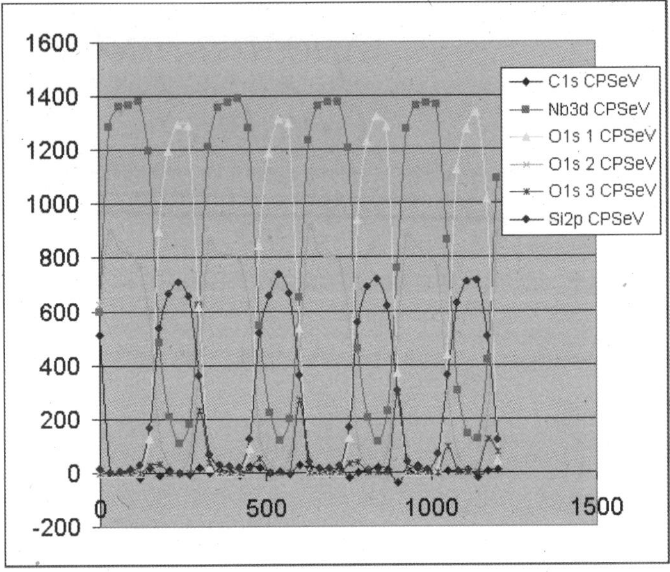

FIGURE 13•15 Intensity profile for the dataset shown in Figure 13•6. The O 1s peak model from Figure 13•12 has been used for the oxygen spectra.

References

[1] ADNOT A., LARACHI F., PIERRE J. (private communication).

[2] "Ce3d XPS Study of Composite $Ce_xMn_{1-x}O_{2-y}$ Wet Oxidation Catalysts", LARACHI F., PIERRE J., ADNOT A., BERNIS A., Submitted to Applied Surface Science – September 2001

CHAPTER 14

SVD Sorting: Images, Spectra and Information Combing

When constructing any analytical experiment, one major aim should be to acquire data for a period just sufficiently long enough to permit the computation of reliable quantitative values. This requirement for adequate precision has to be balanced against efficient use of instrument time and so, since experiments seldom can be prolonged indefinitely, acquisition times are usually a compromise in an effort to optimize the desired outcome. If, however, an experiment can be designed in which the available acquisition time can be used to cover *multiple* variables (energy, position, etch-time, temperature change, and so on), all of which are useful to the analysis, and then by statistical techniques recover the required signal from its noise *en masse - using all the data at once*, this confers a significant advantage over the more conventional measurement which seeks a particular signal-to-noise ratio for a variable simply by devoting large amounts of instrument time to individual measurements.

A paradigm for such an experiment is one in which a set of XPS images is acquired over a sequence of energy values. The result is an image where each pixel offers a spectrum of energy information. Acquisition times which must be employed for such an experiment are usually relatively large as a whole; on an individual spectrum-at-a-pixel basis, however, the acquisition time is small. As a result the signal-to-noise ratio for individual spectra is poor and therefore the data are not suitable for determination of quantitative results. Ignoring the spatial variation in these spectra by simply summing the energy channels across the image produces a spectrum with good signal-to-noise, but without the original spatial information. The problem is therefore to minimise (remove) the noise from the pixel spectra without losing the spatial information originally present in the data set. Such analyses fall within the realms of multivariate statistics and linear algebra, since with the aid of these tools the effective noise in any one spectrum can be reduced to yield remarkable results.

Theoretical Considerations

A Synthetic Data Set

To illustrate the discussion points, a set of synthetic spectra were constructed from five Gaussian line shapes, where spatial variation is simulated by adjusting the area of the Gaussians using sinusoidal functions parametrically dependent on the pixel position. Figure 14•1 shows the relative positions of the five Gaussian peaks with respect to energy; all five synthetic peaks were assigned the same full-width-half-maximum. Using this basic structure, simulated noise consistent with XPS data was used to generate five additional set of image data sets, where the intensity per bin is scaled by a factor of two across the set of five data sets. Since the noise for XPS data varies as a function of the square root of the counts per channel, the signal-to-noise ratio improves throughout the sequence of data sets (Figure 14•2).

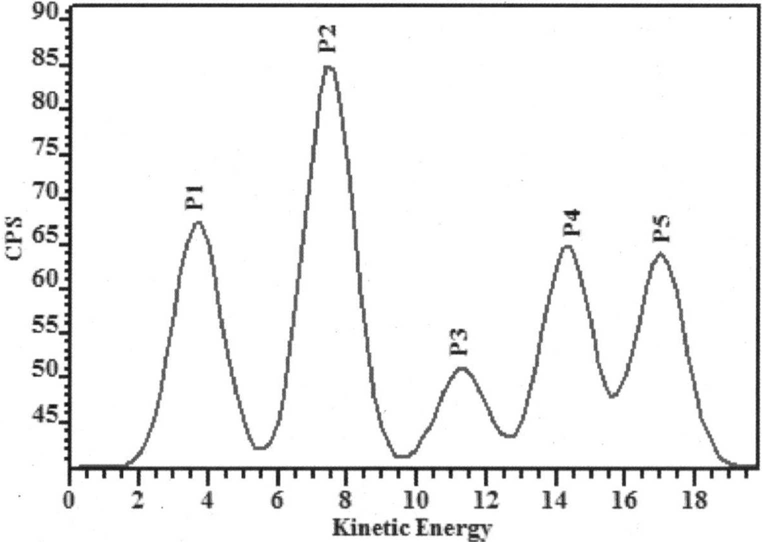

FIGURE 14•1 Synthetic data envelope constructed from Gaussians.

These synthetic image data sets contain a 96 by 96 array of spectra in which each spectrum includes 100 energy channels.

Vector Spaces

While the data shown in Figure 14•2 are intended to look like XPS spectra, the partitioning of such data into useful information and noise is performed using techniques from linear algebra. Therefore the spectra are viewed as vectors in an ***m***-dimensional vector space, where ***m*** is equal to the number of bins (channels) per spectrum. The objective is to determine a set of vectors representative of the useful information, yet belonging to the same ***m***-dimensional vector space as the original set of vectors, and use this subset to model each and every vector in the original data set. The expectation is that the complementary subset contains just the noise component in the data and so by excluding these noise vectors when modelling the raw data, the resulting vectors will be a noise free set of spectra. The partitioning of the vector space is almost never perfect and so the procedure typically yields improved signal-to-noise statistics, not a complete elimination of the noise.

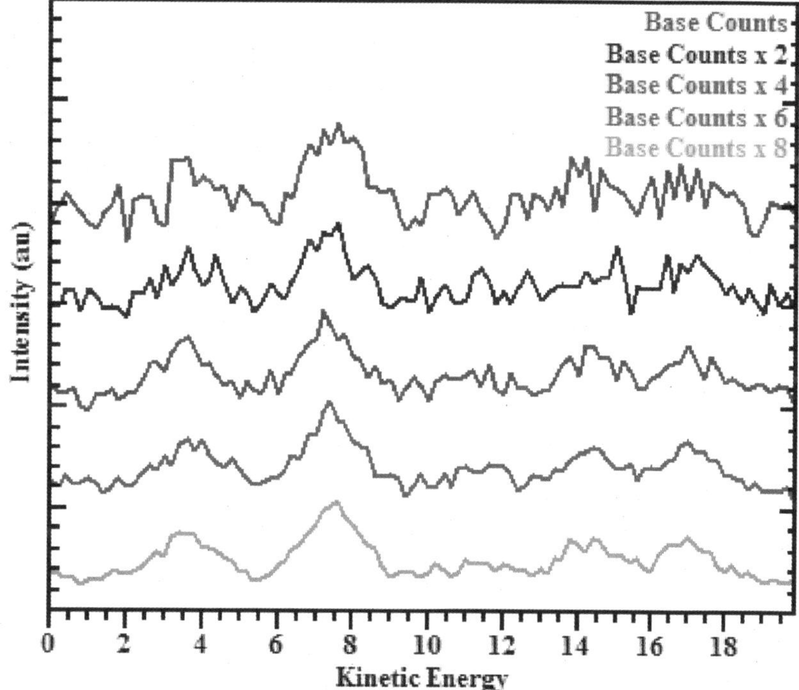

FIGURE 14•2 These spectra correspond to the synthetic envelope in Figure 14•1, where random noise is added on the basis that the intensity per channel varies by a factor of two throughout the sequence bottom to top.

The conventional method for constructing a set of vectors spanning the same subspace as the original vector space is an eigenanalysis based upon the covariance matrix (Equations 1 and 2). The covariance matrix is formed from the dot product of the original vectors and is therefore a real symmetric matrix. The result of an eigenanalysis is an orthogonal set of eigenvectors, where the information content in each of the eigenvectors is ranked by their eigenvalue. This ranking of the eigenvectors allows the significant information to be collected into a relatively small number of vectors from a much larger, but over specified set of vectors. These significant vectors, when constructed using the covariance matrix, are referred to as the Principal Components and hence the decomposition into these new vectors is often called Principal Component Analysis (PCA). What is more, the decomposition of a set of

$$D = \begin{bmatrix} d_{11} & d_{12} & \cdot & \cdot & d_{1n} \\ d_{21} & \cdot & \cdot & \cdot & \cdot \\ \cdot & \cdot & \cdot & \cdot & \cdot \\ \cdot & \cdot & \cdot & \cdot & \cdot \\ \cdot & \cdot & \cdot & \cdot & \cdot \\ \cdot & \cdot & \cdot & \cdot & \cdot \\ d_{m1} & \cdot & \cdot & \cdot & d_{mn} \end{bmatrix} = [\tilde{v}_1, \tilde{v}_2, \ldots\ldots, \tilde{v}_n] \quad \text{.....................Eqn 1}$$

$$Z = D^T D \quad \text{............................Eqn 2}$$

$$D = \begin{bmatrix} u_{11} & u_{12} & \cdot & \cdot & u_{1n} \\ u_{21} & \cdot & \cdot & \cdot & \cdot \\ \cdot & \cdot & \cdot & \cdot & \cdot \\ \cdot & \cdot & \cdot & \cdot & \cdot \\ \cdot & \cdot & \cdot & \cdot & \cdot \\ \cdot & \cdot & \cdot & \cdot & \cdot \\ u_{m1} & \cdot & \cdot & \cdot & u_{mn} \end{bmatrix} \begin{bmatrix} w_{11} & 0 & \cdot & \cdot & 0 \\ 0 & w_{22} & \cdot & \cdot & \cdot \\ \cdot & \cdot & \cdot & \cdot & \cdot \\ \cdot & \cdot & \cdot & \cdot & 0 \\ 0 & \cdot & \cdot & 0 & w_{nn} \end{bmatrix} \begin{bmatrix} v_{11} & v_{21} & \cdot & \cdot & v_{n1} \\ v_{12} & \cdot & \cdot & \cdot & \cdot \\ \cdot & \cdot & \cdot & \cdot & \cdot \\ \cdot & \cdot & \cdot & \cdot & \cdot \\ v_{1n} & \cdot & \cdot & \cdot & v_{nn} \end{bmatrix} \quad \text{............Eqn 3}$$

vectors using the Singular Valued Decomposition shown in Equation 3 is directly related to the linear least squares principle and explains why the data is transformed into an ordered set based on information content. Computing the improved set of spectra, with respect to noise content, is achieved by choosing to set the w_{kk} equal to zero for all k above or equal to some value in Equation 3, where the nonzero w_{kk} below the chosen value of k correspond to the vectors containing the principal components.

The caveat in the previous discussion when applied to the sets of spectra-at-pixels is the sheer quantity of spectra. Given the synthetic data set described above, the vector subspace would be of maximum dimension 100, that is, the number of energy channels per spectrum, while the total number of spectra involved in the calculation is 96^2, or 9216. Forming the covariance matrix from the data matrix prescribed in Equation 2 would result in a matrix of size 9216 by 9216. Alternatively the covariance matrix could be formed from post multiplication by the transpose of the data matrix resulting in a covariance matrix of size 100 by 100. The success of any noise reduction is achieved by a comparison of information in a least squares sense and, as with all least squares calculations, the number of such samples influences the precision of the result. Forming the larger of these two covariance matrices represents the situation where the least squares procedure acts on an energy-channel by energy-channel basis, while the smaller covariance matrix corresponds to the least squares criterion acting on a pixel-by-pixel basis. That is to say, the averaging effect of the least-squares criterion has considerably more data with which to work when using the 9216 by 9216 covariance matrix. However, there are few analysts (at present) with computers capable of performing a direct SVD on a problem of this size. Yet, if the original data are numerous poor signal-to-noise ratio spectra, the use of large numbers of data points for the least-squares procedure is highly desirable. In this sense, an alternative to direct PCA is required.

SVD Sorting

SVD sorting is an algorithm for moving significant vectors to the top of a list of vectors, where the loop invariant when scanning through the set of vectors is the following: the set of vectors all remain within the same subspace spanned by the original set of vectors. The idea is to compare adjacent vectors by computing the eigenvectors and eigenvalues, transforming the vectors in the process, which effectively moves the most significant transformed vector up the list in an analogous way to the Bubble Sort algorithm for ordering keys. The result of stepping through the set of vectors once is that the most significant information is moved up the list in the form of a transformed vector. This procedure employs the same mechanism as SVD to transform the vectors *en route*, however the size of the SVD is always two and therefore, for a single scan through the set of vectors, the computational complexity is $O(n)$, where n is the number of vectors. The number of scans applied to the data set is then dependent on the number of components within the set of vectors and can be chosen for a given application. Bubble Sort is an $O(n^2)$ algorithm; similarly for SVD Sorting, after n scans the vectors would be fully ordered according to this pair wise comparison, however the difference between a full SVD based PCA and SVD Sorting is the final set of vectors are not mutually orthogonal and it is relaxing this mutually orthogonality constraint that yields the reduction in time spent on computation. The fact that the resulting set of vectors is not necessarily mutually orthogonal is only a problem if there are no dominant vector directions; in which case, the PCA would have been of little value anyway. Moving the variation in the vector set to the top of the list allows the useful vectors to be identified and a full SVD can be performed on these data to yield the PCA quantities. Since in general, the number of significant factors is small compared to the overall data set size, the use of a selective PCA via a SVD is time efficient.

The essential steps used in an SVD scan are illustrated in Algorithm 1. Repeated applications of these steps orders the information content in a set of vectors until finally, the important vectors to the description of the subspace, appear at the top of the vector list. Isolating these vectors and applying a standard SVD to only these significant vectors is equivalent to generating the matrices in Equation 3 but where the unwanted w_{kk} are already set to zero.

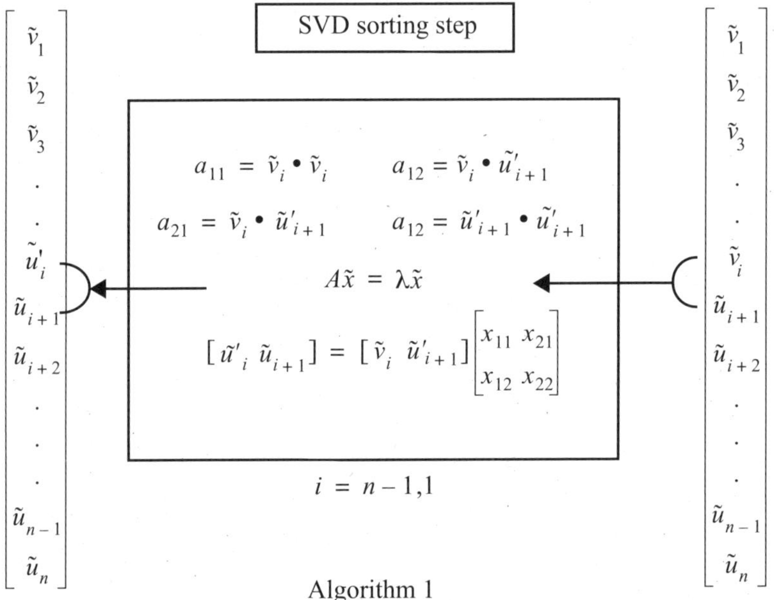

Algorithm 1

Results

Apart from achieving the same result as a massive SVD in a shorter time, the real significance is that the least-squares criterion has been applied in the most advantageous order for removing the noise on an energy channel basis. The consequence of which is that abstract factors containing minor adjustments to the overall description of the spectra will be less likely to include major features due to noise. Thus, more abstract factors can be included in the reconstruction step without reintroducing a significant noise component and so achieve a better approximation to the original data. Figure 14•3 illustrates the results of applying the combination of:

1. SVD sorting followed by,
2. a PCA on a small number of sorted vectors and
3. finally reconstructing the spectra from the five most significant abstract factors.

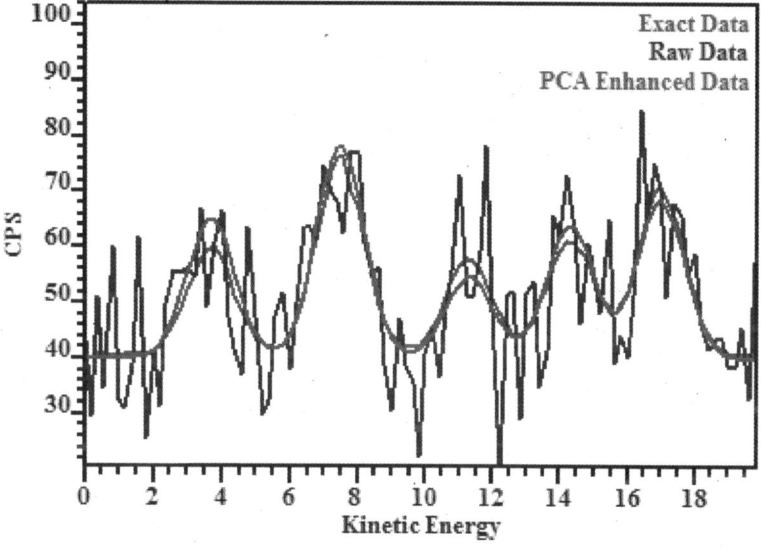

FIGURE 14•3 Application of SVD sort and PCA

These steps have been applied to the image set in which the signal-to-noise is the worst of those shown in Figure 14•2. Once reconstructed spectra are computed for each pixel; simply placing integration regions on the spectra and calculating images using the intensity from these five regions (labelled P1 through P5 in Figure 14•1), a set of spatially resolved quantitative values result (Figure 14•4). The images in Figure 14•4 are computed for each of the five sets of data in which the signal-to-noise varies, plus the PCA enhanced spectra as well as the exact spectra (Figure 14•1). Visual inspection shows the PCA enhanced image is equivalent to the image simulated to have four times the intensity as the data used in the PCA procedure. For further illustration of the fidelity of the method see Walton & Fairley [1]

Summary

For XPS there are two major implications. Firstly, the acquisition time for a given precision can be reduce by roughly a factor of four (provided sufficient information is still present in the data set as a whole) and secondly, the PCA enhanced spectra offer a superior definition for the background to the spectra and consequently the resultant images are free from artefacts often associated from background variations. What is more, the improvements in the background also open up the opportunity to use background information to extract thickness of layered structure on the sample surface. Application of simple techniques proposed by Tougaard in cases where the background above and below a peak is well defined offers alternative image processing resulting in topographical maps.

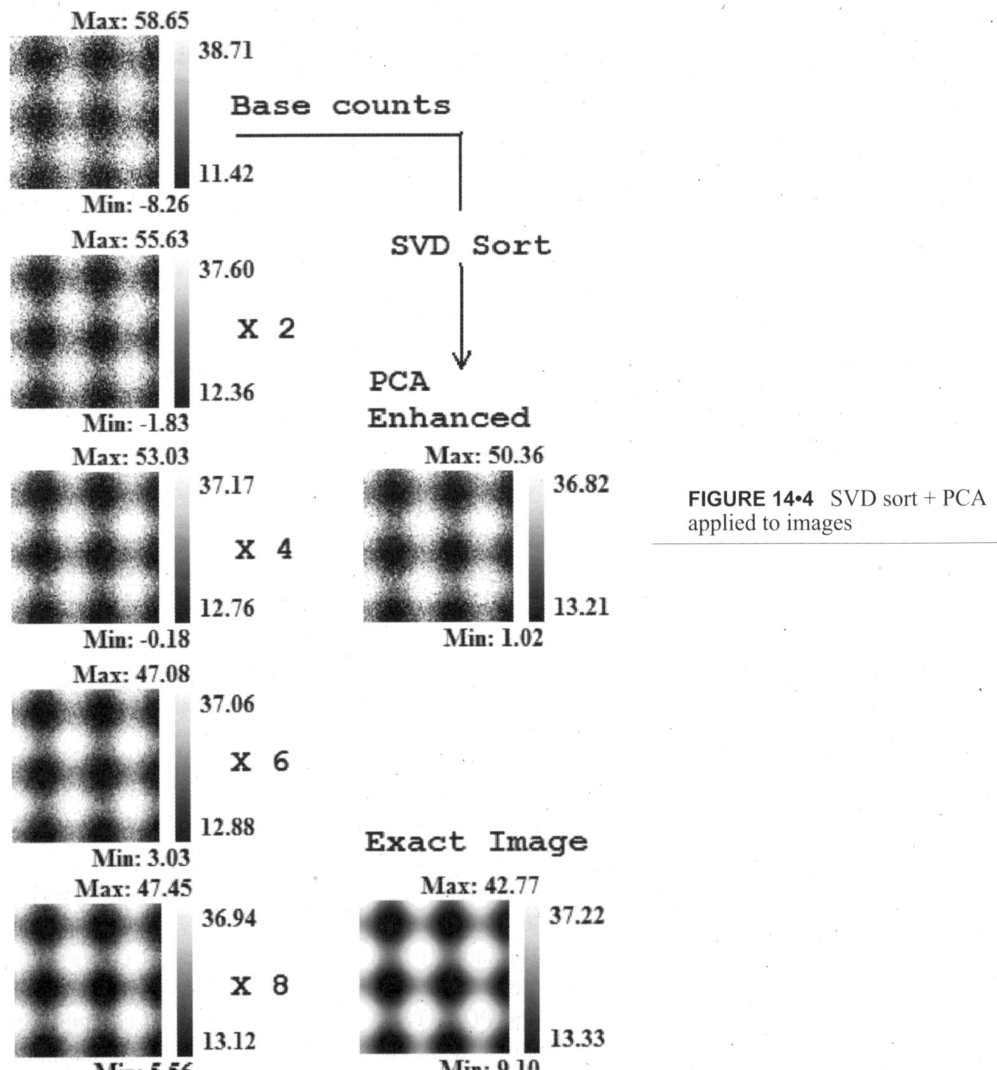

FIGURE 14•4 SVD sort + PCA applied to images

Image Processing with SVD sort - a practical example

The advantages of the approach described above are perhaps best illustrated by examining in detail the steps required to proccss data from a specific type of experiment in which a sequence of images is acquired at regular steps in energy. The resulting data are effectively an image where each pixel within the image contains a spectrum from which spatially resolved quantitative information can be extracted, and although the initial description relates to a specific acquisition system, the processing is entirely general and may be applied to any suitably acquired set of images.

Data Acquisition (using Kratos Vision 2.x)

The Acquisition Manager in Vision 2.x offers a means of defining a sequence of images, each image acquired at regular steps in energy. The mechanism by which the set of acquisition regions is defined is facilitated by use of the Interpolate option available for acquisition regions when in Imaging Mode.
The procedure is as follows:

1. Switch the (Vision) menu to Imaging Mode. When in imaging mode a panel below the Regions Definition table will be visible offering a text-field for entering an increment and a button labelled Interpolate.
2. In the Regions Table enter two acquisition regions. The first region represents the start energy for the sequence of images and the second region defines the end energy for the last image in the sequence. Ensure these two regions are entered with identical acquisition times.
3. Ensure the high-light marker is over the second of the two regions and enter the desired energy interval into the text-field labelled Increment; then press the Interpolate button.

The result of these steps is a set of acquisition regions in the Region Table. There is a limit on the number of images defined in this fashion (about 200); if more images are required, then a second acquisition object will need to be defined and entered into the Vision Manager flow chart. Run the experiment and save the images to a single ISO file.

Processing the Image Set

The signal enhancing steps help to reveal the true information within the image set. These enhancements are achieved using PCA and PCA related algorithms to partition the useful information from the noise and therefore create a set of images with reduced noise content. Figure 14•5 on page 276 shows a set of images before and after processing using PCA. The two images in the first column are displayed after PCA has been applied, while the second column contains the same two images before processing.

There are two ways to approach enhancing a set of energy stepped images. The PCA procedure can be applied in either the spatial domain (that is, treating each image as vector) or, with the help of SVD Sort, in the energy domain, where the raw images are converted to spectra and the spectra are treated as the vectors. The data shown in Figure 14•5 has been processed in the spatial domain, however in general, applying the noise partitioning in the energy domain offers a greater opportunity to suppress the noise in the set of data.

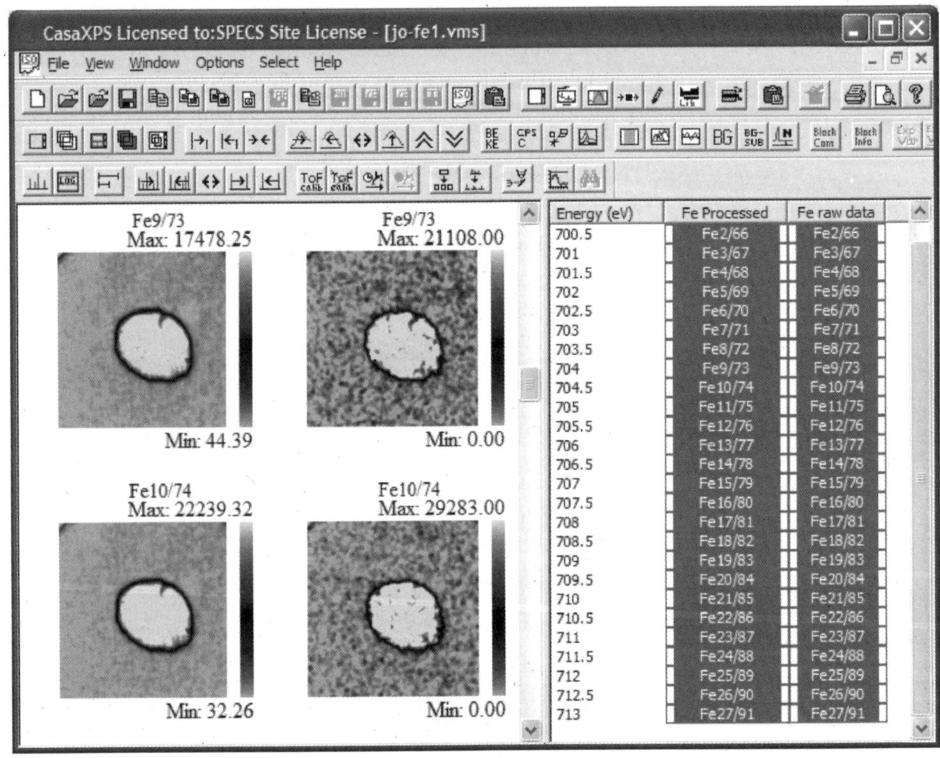

FIGURE 14•5

SVD Sort and PCA Noise Reduction

The following is a description of how to reduce the noise in a data set using the energy domain. The steps described below are intended to provide an understanding for the method as well as the mechanics of performing the analysis. Once a dataset has been assessed, the all-in-one buttons under the TFA Predict section greatly reduce the sequence of steps required for an analysis. See the section detailing the Image Processing dialog window.

The starting point for an analysis is a set of images as shown in the second column in Figure 14•5. These images represent a sweep of energies (kinctic energy appears as the experimental variable in the right-hand-side of the Experiment Frame), where in this case the step size is 0.5eV. From these images, the data are converted to an equivalent set of spatially resolved spectra, from which a single image will be created, where the intensity at each pixel is determined by integrating the signal between a background and the signal defined on each and every pixel in the image. In this example, the advantage of performing this type of experiment is that the final image is without artefacts due to sample charging (albeit small shifts in energy) and the computed intensity will be unaffected by variations in the background itself.

The processing steps are as follows:

1. Convert the images to an equivalent set of spectra. Overlay the set of raw images in the Active Display Tile. Click on the Active Tile to enable the toolbar buttons and select the Image Processing dialog window (Figure 14•6) from the Options menu.

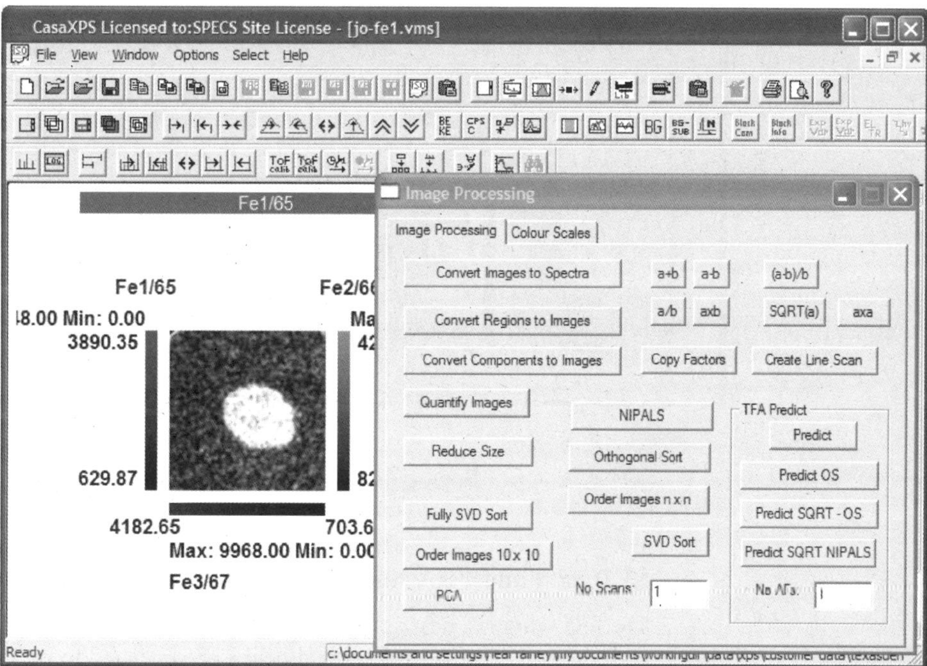

FIGURE 14•6 Image Processing Dialog Window

2. Press the button labelled Convert Images to Spectra on the Image Processing property page. The images in the Active Tile must all correspond to regular energy increments. Given that the images do correspond to a sequence of regular energy intervals, a new Experiment Frame is created containing a column of data blocks, one for each row in the original image, where each data block contains a spectrum for each column in the original image. Figure 14•7 shows the spectra from the 61st column (index starts at zero) overlaid in the Active Tile. To move through the columns of spectra, use the Control Key and the Page Up/Page Down/Home/End keyboard buttons.

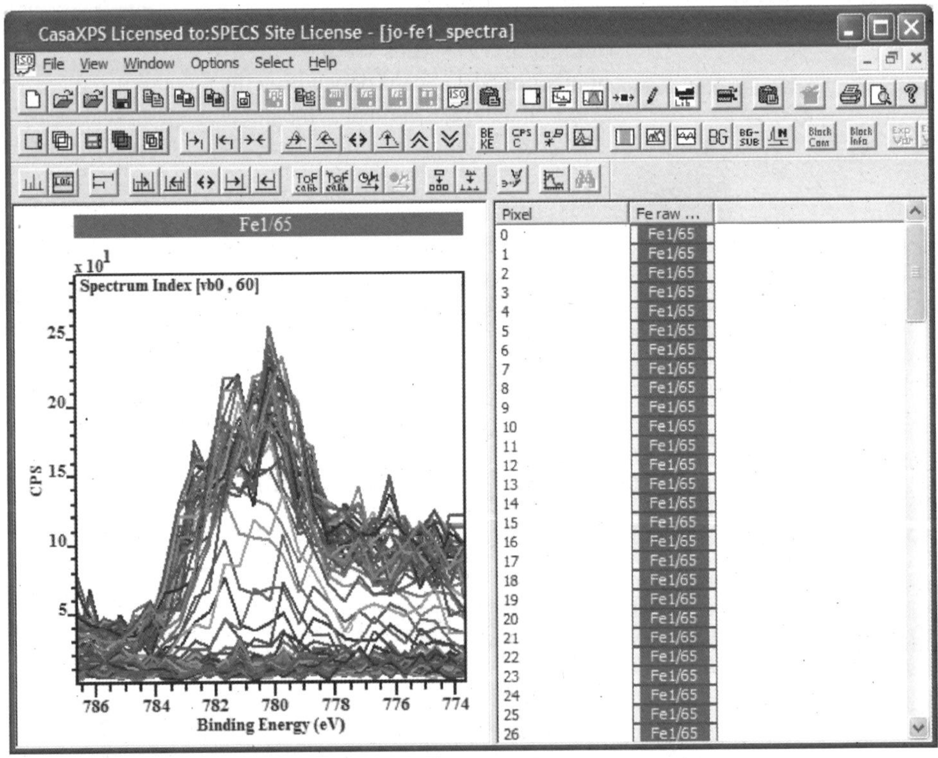

FIGURE 14•7

3. To extract the useful information from the entire set of 128 by 128 pixels requires the use of SVD Sort. SVD Sort is a method for combing large data sets (in this case 16384 spectra) for the useful information in that data and results in a small number of vectors from which all the original data can be approximated in a least-squares sense. Of course, for SVD sort to be successful the original set of vectors must be over the range specified by the original set of data and this is

clearly the case for the spectra shown in Figure 14•7. To perform an SVD Sort on a set of spectra, these data must be overlaid in the Active Tile (as shown in Figure 14•7), the number of SVD scans entered in the text-field on the Image Processing property page and the SVD Sort button pressed. The action of the SVD Sort button is to perform the operation on each spectrum within a data block (in this example there are 128 spectra per block) and then follow the process through each of the data blocks. The result of each invocation is the set of most significant vectors are moved to the top of the list and appear in the first corresponding variable in each block (that is, the spectra visible when Ctrl-Home is pressed). The SVD Sort button can be pressed multiple times or the number of SVD scans set to the desired number of operations, where the chosen number of SVD Scans depends on the number of vectors exhibiting information rather than noise. In this example, ten scans results in the information content shuffling into the top twelve data blocks (Figure 14•8).

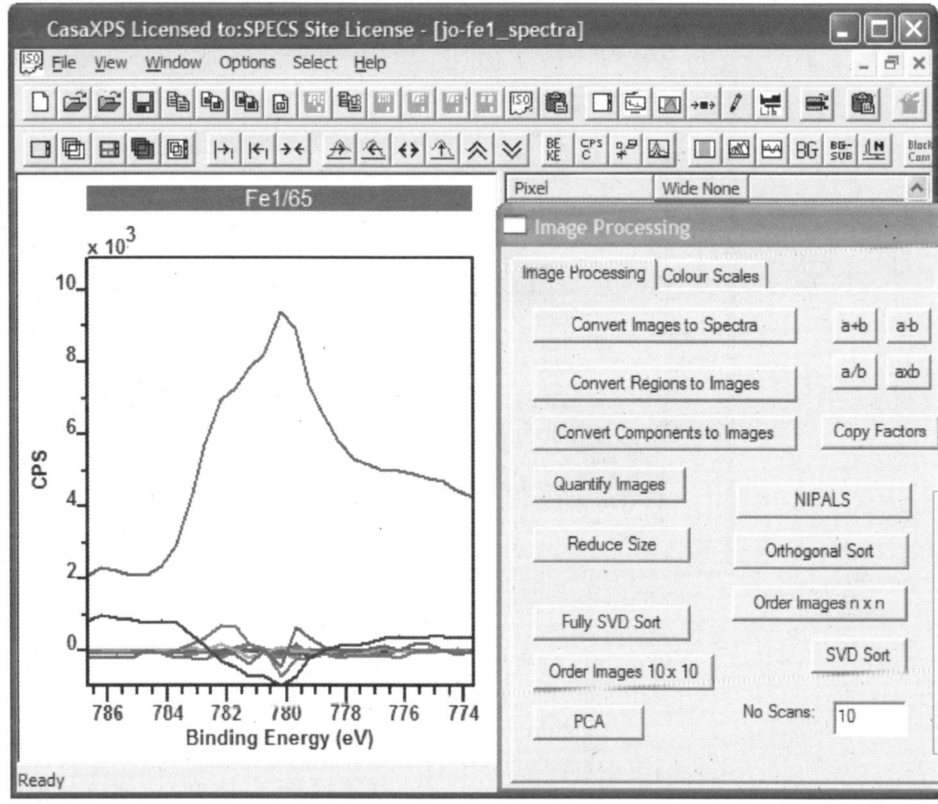

FIGURE 14•8

4. Next step is to extract the useful vectors into the same Experiment Frame and then finish the data organization by performing a PCA. In the right-hand-side of the Experiment Frame, select the top set of data blocks containing the useful information. These are essentially those blocks for which the transformed data does not look like noise, plus a couple more for luck. Press the Copy Factors button on the Image Processing dialog window, check the list of blocks to be copied and press OK. The Processing Data Only tick-box should be ticked. A new column of data blocks will be created in the Experiment Frame.
5. Select the newly created data blocks and overlay these data in the Active Tile. Then press the PCA button. The data previously ordered with respect to the SVD Sort will now be adjusted to ensure all the vectors are mutually orthogonal and in a state which can be used to recreate the raw data using only the most significant abstract factors.
6. Return the original column of data blocks to their unsorted state. This is accomplished by overlaying the first column in the Active Tile, invoking the Spectrum Processing dialog window and pressing the Reset All button on the Processing History property page.
7. Next, determine the number of factors required to represent the data. Typically this involves inspecting the abstract factor and rejecting those after the last abstract factor for which a recognizable shape is present. The abstract factors rejected are considered to be due to noise.
8. Display the data blocks containing the abstract factors in the Active Tile and select in the right-hand-side the raw data blocks. Enter the chosen number of abstract factors considered to be representative of the data (in this example the chosen number is two) in the TFA Predict section and press the Predict button. The raw spectra are replaced by predicted spectra, where the predicted spectra are a linear combination of the chosen number of most significant abstract factors displayed in the Active Tile. Figure 14•9 shows a comparison of the raw and the predicted spectrum for one pixel.

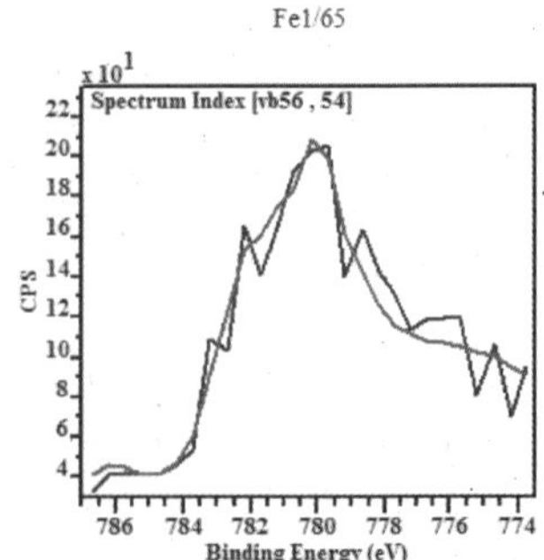

FIGURE 14•9 (left)
Overlay of raw (noisy) data and predicted spectrum

FIGURE 14•10 (right)
Final processed image

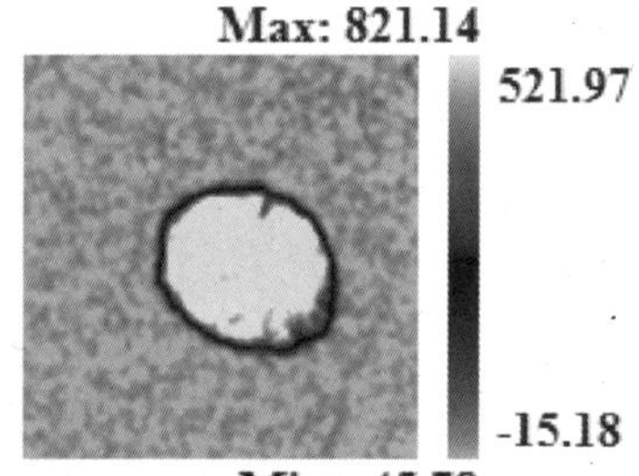

9. To create the final processed image(s) (Figure 14•10), define quantification region(s) on each block containing a spectrum from a pixel, overlay these spectra in the Active Tile and press the Convert Regions to Images button on the Image Processing property page.

The Image Processing Dialog Window: A Quick Summary of Features

Convert Images to Spectra

The Convert Images to Spectra button creates a set of data blocks containing spectra corresponding to each pixel from a set of energy stepped images. Overlay a set of images in the Active Tile and press the Convert Images to Spectra button. The images must be assigned sequential energy values defined by the experimental variable and these energies must be part of a sequence defined using a common energy increment.

Convert Regions to Images

Given an image data set in which pixel information is represented by spectra, where a column of data blocks, each holding a row of pixel spectra, define the number of rows within the image, then new images are created from Quantification Items defined on these spectra using Quantification Regions by pressing the Convert Regions to Images button. At least one Region must be defined on each data block and those blocks used to create the image must be overlaid in the Active tile. The Tag field in the region determines the type of quantification information used to create the image. Keywords in the Tag field indicate the required image type (all in lower case characters):

area	RSF adjusted peak area CPSeV
height	RSF adjusted peak height above background CPS
position	Position of the peak maximum relative to the background
centroid	Position of the peak centroid relative to the peak background

If the Tag field in the Quantification Region is none on the above keywords, the image is created from the RSF and transmission corrections area CPSeV (where ever possible).

Convert Components to Images

Given an image data set in which pixel information is represented by spectra, where a column of data blocks, each holding a row of pixel spectra, define the number of rows within the image, then new images are created from component intensities in CPSeV. A peak model must be defined on each data block and the blocks displayed in the Active Tile. On pressing the Convert

Components to Images button, the peak model is fitted to each spectrum in the image and an image is generated for each component in the peak model, plus a figure-of-merit image.
Keywords in the Tag field indicate the required image type (all in lower case characters):

position	Position of the peak maximum relative to the background
fwhm	The Full Width Half Maximum from each component

If the Tag field in the Component is none on the above keywords, the image is created from the RSF and transmission corrections area CPSeV (where ever possible).

Quantify Images

Given a set of images overlaid in the Active Tile, each image is divided by the sum of all the images and normalised to 100%. If the images are generated from quantification items such as Regions or Components, then the result of the Quantify Images button is a set of Atomic Concentration images displayed in a new Experiment Frame.

Reduce Size

The Reduce Size button sums sets of four pixels in squares to produce images with half the original pixel dimension. Overlay a set of images in the Active Tile and press the Reduce Size button. A new Experiment Frame is created in which the reduced size images appear.

Create Line Scan

To create a line scan from one or more images, overlay the images in the Active Tile, mark out a line on the image display using the left-hand mouse button and a drag action, then press the Create Line Scan button. The dimensions of the line scan are in pixel coordinates. The X-axis will display distance when the display mode in not binding energy and, if desired, these distance dimensions must be calibrated elsewhere.

Copy Factors

The Copy Factors button is a means of copying the chosen SVD sorted vectors to a new column in preparation for use to TFA predict the original data.

Order Images 10 x 10

An alternative SVD Sort strategy in which the vectors are ordered using a ten dimensional subspace filter.

Order Images n x n
No Scans: 1

An alternative SVD Sort strategy in which the vectors are ordered using a subspace with dimension n equal to twice the value of the text-field.

Fully SVD Sort

A complete ordering of a data set using the SVD Sort algorithm, in which the set of n vectors is scanned n times. Generally, it is quicker and just as accurate to use the SVD Sort with fewer SVD scans.

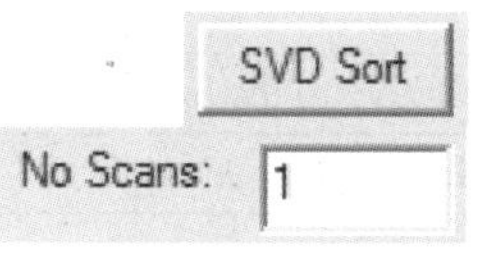

The text field is used to input the number of SVD Sort scans required when the SVD Sort button is applied. The number of SVD Sort Scans should be more than the number of expected abstract factors for a given data set. If the expected number of factors is four (say) then a safe number of SVD Scans might be ten.
The important thing to monitor is the convergence of the vectors, where all vectors excluded from the final PCA have converged to exhibit only noise like structure.
To perform an SVD Sort, enter the desired number of scans in the text-field, overlay the data in the Active Tile and press the SVD Sort button. When spectra are displayed in the Active Tile, the SVD Sort will work on all spectra stored in the corresponding variables.

Orthogonal Sort

An extension of the SVD Sort option, where each scan through the set of vectors produces a vector and this vector is then projected out of all the vectors in the list before applying the next SVD Sort. The result is a set of orthogonal vectors spanning the original vector space. The number of vectors identified and projected out of the data set as a whole is specified in the No Scans text-field.

NIPALS

Nonlinear Iterative Partial Least Squares (NIPALS) method for computing the significant abstract factors is applied to the data displayed overlaid in the Active Tile. The number of vectors sought is specified in the No Scans text-field. A one stop enhancement button using the NIPALS method is also available on the TFA Predict panel.

PCA

Perform a PCA, via a Singular Valued Decomposition on those data displayed in the Active Tile. The PCA is performed directly on the data channels without shifting or other adjustment available on the Spectrum Processing PCA property page.

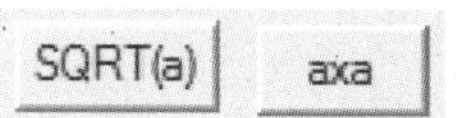

Data can be pre-processed prior to SVD Sorting and/or PCA using these options. Dividing data by the square root of the counts per channel suppresses the influence of noise on the structure in an image or spectrum. These buttons represent a simple, yet effective means of reducing noise artefacts from the abstract factors by offering the forward and inverse transformation steps.
To operate on multiple data blocks, the data from the blocks should be overlaid in the Active Tile before pressing either of these buttons.

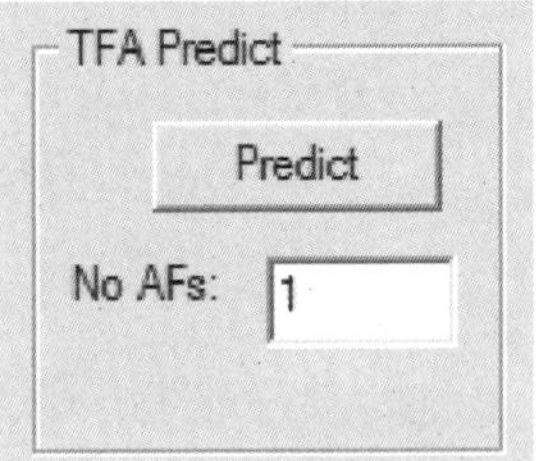

Enter a valid PCA abstract factor in the Active Tile and use the right-hand-side of the Experiment Frame to select those images/spectra for which the TFA prediction is to be applied. Enter the number of Abstract Factor used in the TFA Prediction step in the text-field and press the Predict button.

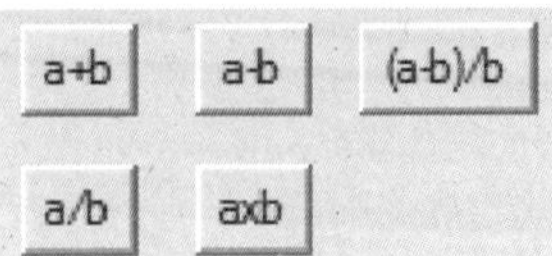

These buttons represent direct computation on a pixel-by-pixel basis. The operands are defined by overlaying two images in the Active Tile. The first of the two images selected in the right-hand-side will correspond to a; the result of the computation appears in the data block corresponding to a. The Spectrum Processing dialog window can be used to reset the processed data to the original raw data.

One Stop Data Enhancement

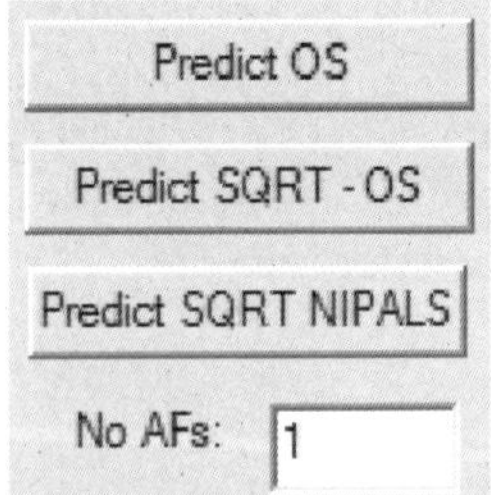

These buttons offer a means of enhancing the signal-to-noise in a data set at a single press of a button. Once the number of significant abstract factors are determined and the number of entered into the given text-field (No AFs), the set of data currently overlaid in the Active Tile is processed into significant vectors and reconstructed from the number of significant abstract factors specified in the text-field. The buttons labelled with SQRT will perform the operation using the recommended noise suppressing SQRT/square transformation. The resulting spectra calculated directly or indirectly from images are ready for quantification.

Practical Advantages

- for the PCA / SVD sort approach include:

- Given pulse counted images, the resulting images are suitable for “proper” quantification.
- Chemical state images are possible.
- Background subtraction is performed in analogous fashion to normal quantification of spectra. Image artefacts due to background variation are removed.
- Statistical techniques can be applied to the data set as a whole to offer alternative views of the data and enhance the signal-to-noise in resulting images.
- The background information can be used to offer layer thickness information in image format.

CHAPTER 15

More Numerical Methods - Savitzky-Golay and Linear Analysis

Savitzky-Golay Methods

Three options on the Spectrum Processing dialog window utilise a simple but effective means of extracting information from data containing noise. Two of these options are commonplace, where smoothing of experimental data and the determinations of derivatives are performed using the algorithm proposed by Savitzky and Golay [1],[2]. The third option utilises the same principle behind the smoothing and differentiation algorithm to perform integration on a set of data and represents an extension of the Savitzky-Golay approach.

Theory

Given a set of data containing both signal and noise, the initial objective of the Savitzky-Golay method is to replace the raw data by a smoother set of data representing the true signal responsible for the record intensities. For a constant underlying signal, the most natural means of estimating the true signal from a set of measurements would be to average the values. The act of averaging a set of values is in fact one example, and possibly, the simplest applications of the linear-least-squares principle. Spectral data, on the other hand, typically contain peaks superimposed on a background signal and therefore a more subtle use of averaging is required if the essential structure in the data set is to be retained. One way to use the averaging process, but to maintain information relating to the variation in the intensities, is to perform a local averaging for each energy channel within a spectrum, for

example, each data bin (channel) could be replaced by the average of three bins - itself and the two bins either side. The averaging operation could be applied to a data set via a digital convolution of the data bins with a convolution kernel consisting of the values {1/3, 1/3, 1/3}. These simply, yet often used concepts are at the basis of the Savitzky-Golay method, which in essence applies the least-squares principle to determine an improved set of kernel coefficients for use in a digital convolution, where these improved coefficients are determined, in the least-squares sense, using polynomials rather than, for the case of averaging, simply assuming a constant value determined from a sub-range of data bins. Indeed, the Savitzky-Golay method could be seen as a generalisation of averaging data, since averaging a sub-range of data corresponds to using a Savitzky-Golay polynomial of degree zero.
To illustrate the Savitzky-Golay method, consider the specific example in which five data bins are used to approximate a quadratic polynomial. The polynomial can be expressed in the form:

$$p(x) = a_0 + a_1 x + a_2 x^2$$

Where the coefficients a_0 , a_1 and a_2 are determined from the simultaneous equations in which the abscissa x is the index for the data bin; the origin is always placed at the central data bin and so the abscissa values corresponding to each of the data bins are {-2, -1, 0, 1, 2}:

$$\begin{bmatrix} 1 & -2 & 4 \\ 1 & -1 & 1 \\ 1 & 0 & 0 \\ 1 & 1 & 1 \\ 1 & 2 & 4 \end{bmatrix} \begin{bmatrix} a_0 \\ a_1 \\ a_2 \end{bmatrix} = \begin{bmatrix} d_{-2} \\ d_{-1} \\ d_0 \\ d_1 \\ d_2 \end{bmatrix}$$

Or

$$A\underline{a} = \underline{d}$$

Where the evenly spaced data bins $\{d_{-2}, d_{-1}, d_0, d_1, d_2\}$ are selected with the target of replacing the value for d_0 with the value for the polynomial at $x = 0$ or $p(0) = a_0$. Since there are five equations and only three unknowns, the coefficients to the polynomial must be determined in the least-squares sense, where the linearly independent basis functions are $1, x$ and x^2 , therefore the normal equations yield:

$$A^T A\underline{a} = A^T\underline{d}$$

Since $A^T A$ is a square symmetric matrix of rank three, the coefficient vector $\underline{a}$ is determined from $[A^TA]^{-1}A^T$, the top row of which yields the prescription for computing the value of a_0, namely:

$$a_0 = [s_0, s_1, s_2, s_3, s_4]\begin{bmatrix} d_{-2} \\ d_{-1} \\ d_0 \\ d_1 \\ d_2 \end{bmatrix}$$

Thus, for each set of five such data bins, the central bin can be replaced by the value determined for a_0, or in other words, a digital convolution using the five point kernel $\{s_i\}$ and the raw data bins results in a smoothed set of data bins, where a linear least squares quadratic polynomial is used to model the data, five channels at a time. Similarly, the derivative of a spectrum can be computed using the Savitzky-Golay polynomial. Again the intention is to approximate the derivative at a given point in the spectrum using the derivative of the polynomial at $x = 0$. Since $dp(0)/dx = a_1$, the second row of the matrix $[A^TA]^{-1}A^T$ yields a second convolution kernel for computing the derivative of the spectrum and, apart from the difference in the kernel values, the computation of the derivative proceeds in an analogous fashion to that of the smoothing calculation.

The variation on a theme occurs with the integration algorithm. Here, the polynomial is again used in the approximation to the integral of the spectrum, however rather than evaluating the polynomial at the point $x = 0$ the integral between $x = -1$ and $x = 0$ is computed. Completing the integration over a range of data bin is then achieved by summing the set of sub-intervals computed using a convolution kernel determined from all three rows the matrix $[A^TA]^{-1}A^T$:

$$\int_{-1}^{0} p(x)dx = a_0 - \frac{a_1}{2} + \frac{a_2}{3}$$

Therefore

$$\int_{-1}^{0} p(x)dx = [i_0, i_1, i_2, i_3, i_4]\begin{bmatrix} d_{-2} \\ d_{-1} \\ d_0 \\ d_1 \\ d_2 \end{bmatrix}$$

Where the i_j in the convolution Kernel are the linear combination of the three terms from each of the rows of $[A^T A]^{-1} A^T$ prescribed by the expression above. To compute the integral with respect to energy requires the multiplication of the convolution integrals by the energy step-size.

The use of the Savitzky-Golay polynomial to evaluate the derivative is essential for noisy data, since an interpolating polynomial would produce an inferior result to that of a least-squares approximation. On the other hand, there is an academic nature to using Savitzky-Golay integration, since simply summing the data bins includes the averaging mechanism required to reduce the influence of random noise on the computed values; however the symmetry of using the same algorithm to compute both the derivative and its inverse is somewhat pleasing.

The theory behind the Savitzky-Golay options, whilst illustrated using a five-point quadratic approximation, can be applied for any odd number of points applied to any degree of polynomial, provided the number of data bins over-specifies the information required to compute the polynomial coefficients. CasaXPS offers quadratic, quartic and linear polynomials, while the number of data bins used to process the spectra is specified in a text-field on both the Smooth and also the Differentiate property pages on the Spectrum processing dialog window. Integration is always performed using a five-point quadratic Savitzky-Golay approximation. One further point regarding integration is that integration cannot recover constant offsets lost during differentiation, so there is an option (Zero Adjust) on the Integration property page, which when ticked, ensures the data is all positive following an integration calculation.

Linear Analysis of Spectra

In working with blends of polymers it is sometimes useful to analyse the spectrum from a blend of two or more materials in terms of the spectra from the basic polymers. Linear Analysis is a technique for estimating the proportions of these basis spectra found in the spectrum taken from the composite material. By way of example, consider the spectra shown in Figure 15•1, where three possible basis spectra are used to analyse the blend.

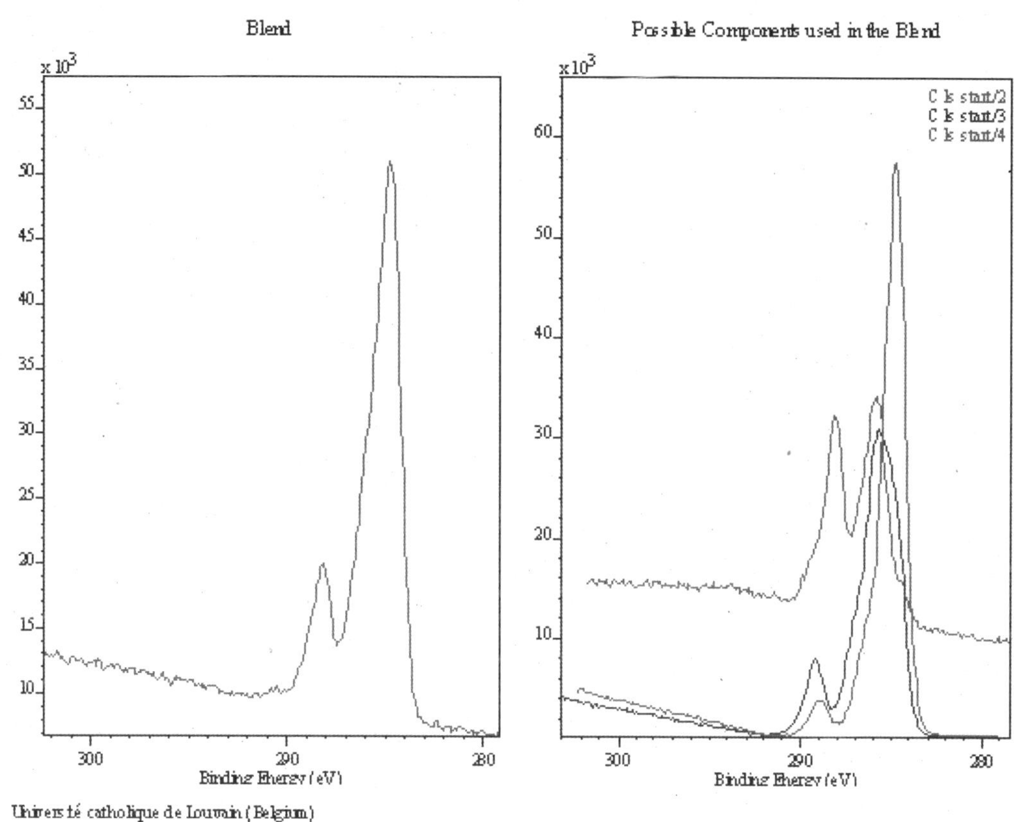

FIGURE 15•1 Polymer blend (left) together with spectra from the basic polymers (right).

The procedure to determine the composition of the polymer blend is as follows:

1. Transform the basis spectra into energy-corrected, background-subtracted normalised spectra.
2. Background-subtract the spectrum to be analysed (Figure 15•2) Load the transformed basis spectra into the active display tile as overlaid traces.
3. Select the spectra to be analysed in the right-hand-side of the Experiment Frame (Figure 15•3).
4. Press the Generate Spectra pushbutton in the Linear Analysis section of the PCA property page on the Spectrum Processing dialog window.

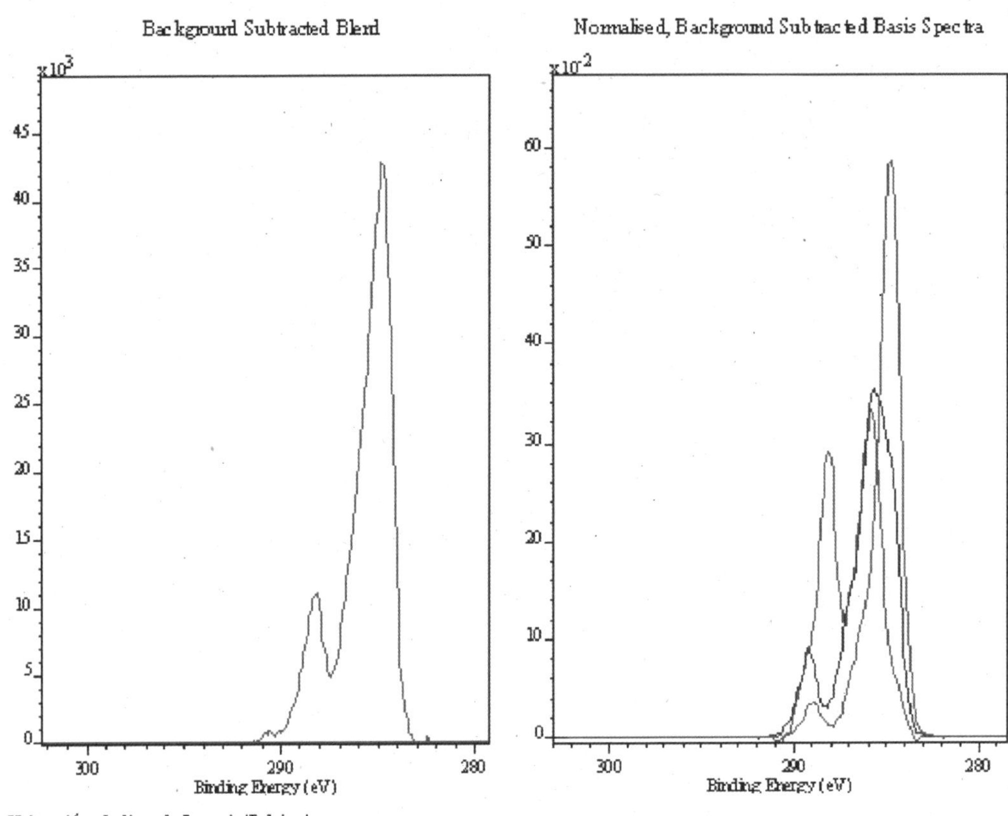

FIGURE 15•2

A new Experiment Frame in created containing the original spectra analysed, plus each of the weighted basis spectra together with the sum of all the weighted basis spectra. If overlaid in the same tile, this result is not dissimilar to the display of a peak-fitted spectrum using synthetic components. The relative contributions to the polymer blend can be determined using integration regions on each of the basis spectra, while the coefficients determined by the linear analysis are entered into the block comment fields for each of the spectra in the new Experiment Frame.

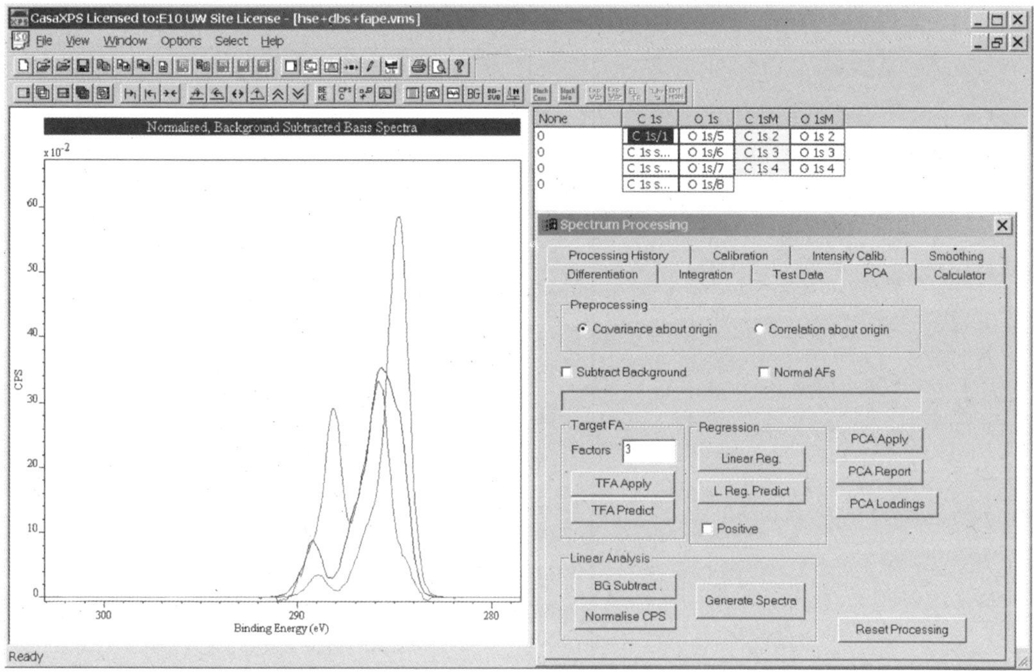

FIGURE 15•3

What is Linear Analysis?

Linear analysis is simply a linear least square solution where the loadings for each of the basis functions are constrained to be non-negative. The non-linear least square method used to fit synthetic components allows line-shapes to move relative to one another via the position parameter and alter in shape via the FWHM parameter. While such parameters are necessary in general peak modelling scenarios, in the case of well-defined basis spectra these additional degrees of freedom are undesirable since there is an opportunity to adjust the line-shapes beyond the realms of reality. By energy-calibrating the basis spectra the positional information is fixed and so only the relative intensities of the basis spectra found in the unknown spectrum are determined using the linear analysis option. The equivalent calculation could be performed using line-shapes determined from the basis spectra, however once the line-shapes have been extracted, the curves become abstract shapes open to optimisation without direct reference to the information in the original basis spectra. The linear analysis retains all the information about the basis spectra and the resulting spectra can be managed as if they were ordinary spectra.

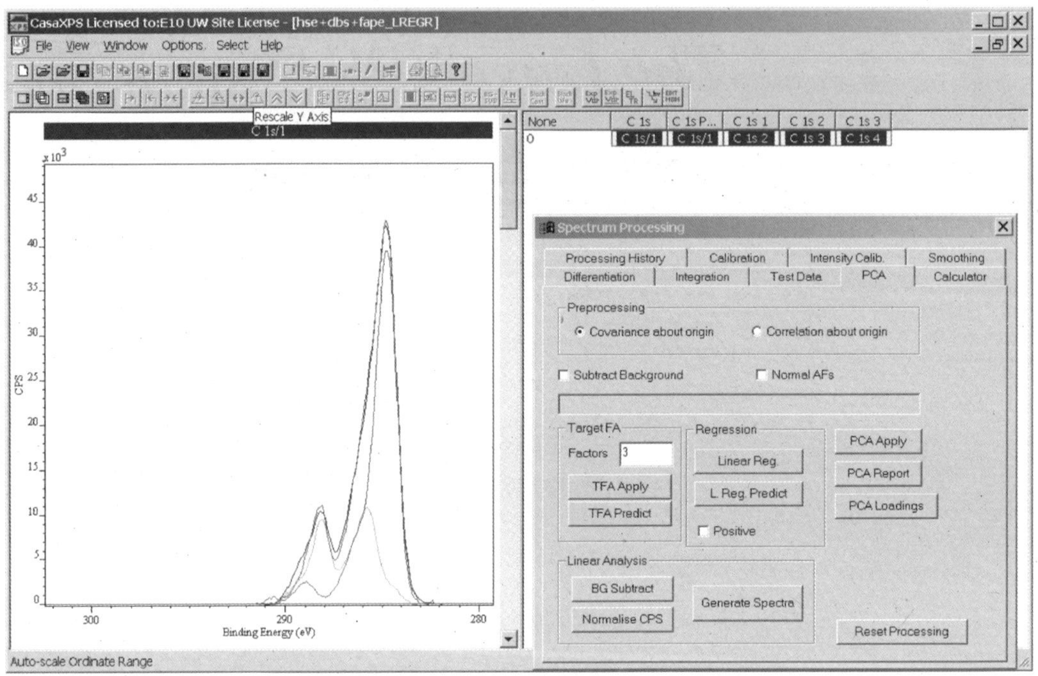

FIGURE 15•4

References

[1] Savitzky, A. and Golay, M.J.E.
"Smoothing and differentiation of data by simplified least squares procedures"
Analytical Chemistry, **36** (8), 1627 -1639 (July 1964)
(the original)

[2] Steiner, J. Termonia, Y. and Deltour, J.
"Comments on smoothing and differentiation of data by simplified least squares procedures"
Analytical Chemistry, **44** (11), 1906 -1909 (September 1972)
(correcting some of the coefficients in the original)

Section 4 - The CasaXPS System and some background to XPS

CHAPTER 16

Converting File Formats - using different systems

In order to further enhance the "cross platform" approach inherent in ISO 14976, CasaXPS provides several file conversion routines for the "native" file format of systems which either do not have any "VAMAS" output capability or which do not implement the specification fully. Brief descriptions of procedures for these conversions are given in the following sections.

PHI MultiPak files

MultiPak Binary files

PHI MultiPak binary-file with extensions .spe/.pro/.ang/.map/.sem/.pho can all be converted to ISO 14976 file format by simply selecting the files with these extensions through the Convert to VAMAS File dialog window. The conversion operation creates a new ISO file leaving the original PHI binary-file unaffected.

MultiPak ASCII files

PHI MultiPak writes an ASCII format. Data in this format can be converted to ISO format through CasaXPS.

The ASCII files may contain single regions, so called multiplexed regions, or depth profile data. To convert a set of files in any of these formats into a single ISO file, the .asc files must be collected into a sub-directory that contains only valid .asc data files and then a new file name must be entered with an extension of .qua (Figure 16•1) An ISO file containing all the regions from the files in the directory with an .asc extension will be created.

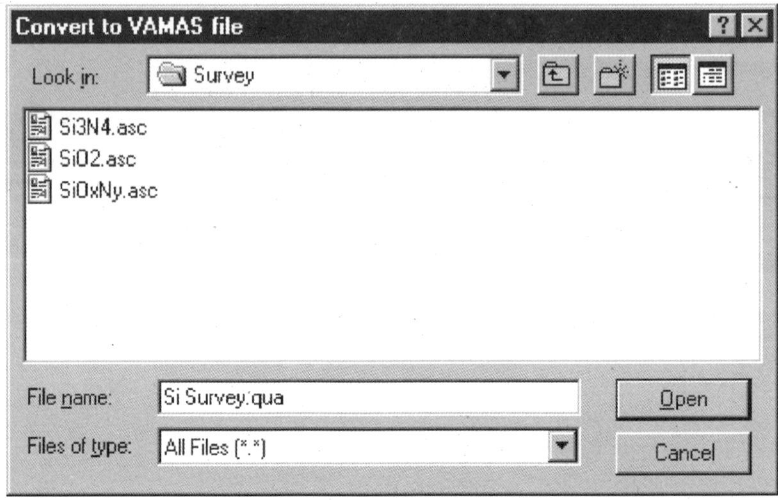

FIGURE 16•1 MultiPak ASCII files are converted by entering a file name with an extension of .qua.

VG Eclipse Files

VG Eclipse files are binary files, one file per spectral region, organised in directories where files with common acquisition characteristics appear in the same directory or inside sub-directories within that directory. CasaXPS will convert these directory structures into a single ISO 14976 file where experimental information and relationships are maintained. Quantification in CasaXPS can be performed in an identical way to the Eclipse data system; transmission correction is accounted for by CasaXPS whenever the information is included in the original files.
Individual spectral regions are stored in files with a .dts extension. Sets of regions are stored in individual files all located in the same sub-directory. Older versions of Eclipse may have .col extensions associated with these data directories, but later versions of Eclipse generate such directories without this naming conversion. Similarly, depth profiles or angle resolved experiments appear as sets of directories, one per acquisition region, where each sub-directory contains the .dts files for one of the ac-

Post Publication Addendum.........

A selection of ".vms" data files used in the preparation of examples and figures in "The Casa Cookbook", including some of the Chapter 2 Case Studies is now available for download from the CasaXPS web site at:-

http://www.casaxps.com/berlin/

ACOLYTE SCIENCE (Publishers)
2, Kinderton Close, High Legh, Knutsford, Cheshire WA16 6LZ U.K.
Tel. / FAX: +44 (0) 1925754064 / 755451 E-mail: sales@ acolytescience.co.uk

quisition regions involved in the experiment. Again, older versions of Eclipse use a naming convention where each sub-directory was given a .mle extension.

There are four ways to convert VG Eclipse data.

1. If an individual file is required then selecting the .dts file via the "Convert to VAMAS" file dialog window will result in the creation of an ISO 14976 file containing the spectrum from that .dts and no others.

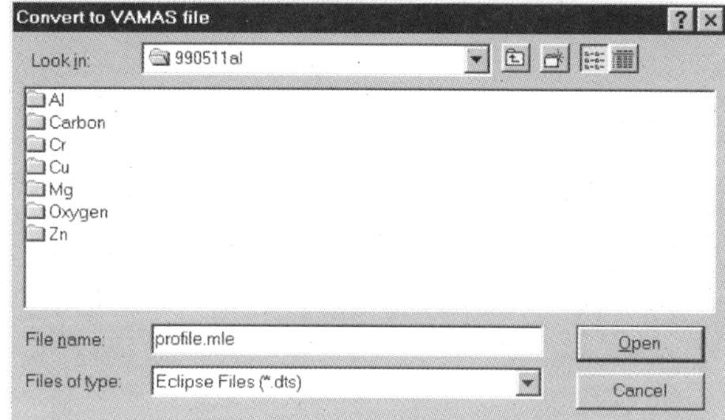

FIGURE 16•2 "Convert to VAMAS" File Dialog Window.

2. If, on the other hand, all the .dts files are required in the same file, the "Convert to VAMAS" file dialog window should be used to find the directory containing the spectral regions. The name of the new file entered in the dialog window must be specified in the text-field, but with a .col extension. It is the extension given to the filename that flags the type of conversion required, however the new file created by CasaXPS will replace the entered extension by .vms.
3. The third method converts depth profile type file structures. Using the "Convert to VAMAS" file dialog window, move to the directory that contains the sub-directories containing the spectral regions. Whilst at the level of the sub-directories enter the name of the new filename and add a .mle extension. The .mle flags that CasaXPS should read each sub-directory within the current directory and convert all the .dts files found within the sub-directories. The new filename will replace the .mle extension with .vms. Figure 16•2 shows the dialog window where the visible directories are about to be converted using the .mle flag. The filename profile.mle indicates that each of the directories form part of a depth profile and should be interpreted as the names for the regions found inside these directories. The result of converting these directories is shown in Figure 16•3.
4. A fourth conversion filter appears in v2.1.0. The .all flag allows data held in separate sub-directories in a similar structure to the .mle directory grouping to be converted into a single ISO file. The sub-directory names are used to label the rows of spectra as seen within CasaXPS when viewed in Edit Mode.

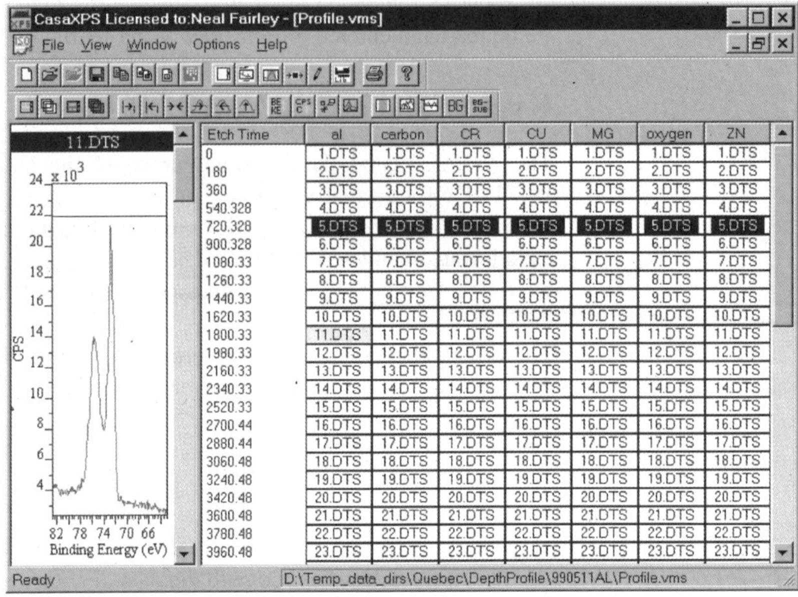

FIGURE 16•3 VAMAS File Block Structure from a VG Eclipse Depth Profile.

VG Avantage Files

The VG Avantage binary files must be converted via an Avantage utility program (DataSpace_BatchDump.exe) to ASCII equivalent files. These ASCII files are created with the file extension .avg and can be converted to ISO 14976 (VAMAS) format using the Convert to VAMAS dialog window:

1. Using the VG utility program DataSpace_BatchDump.exe, convert all the .vgd binary files for an experiment to .avg ASCII files.
2. Select the Convert to VAMAS dialog window in CasaXPS either using the toolbar button or via the Convert menu item on the File menu.
3. Move to the directory containing the .avg ASCII files; the DataSpace_BatchDump.exe program offers the user the choice of where the ASCII files should be created.
4. Enter a name for the new VAMAS file using the file extension .avg and press the Open button. For example, if you wish to call your new file sample001.vms, then you should enter sample001.avg into the file name field on the Convert to

VAMAS file dialog window. Each .avg file found in the current directory will be read and entered into the new VAMAS file named sample001.vms.

Kratos Vision 1.x/2.x ASCII files

The Kratos Vision data sets are probably the closest in logical design to the ISO 14976 file format. A single binary file maintains the context for an experiment, where an experiment may include multiple acquisition regions characterized by, so-called, state changes. These state-changes are equivalent to the experimental variable fields used in the ISO 14976 standard and the Vision Objects are equivalent to the ISO blocks.

The principal problem with converting the Vision 1.x/2.x data sets is the internal data-structures used to maintain the experimental information. The file structure is a link-list of file records that hold a hierarchical description of the data. Fortunately there is a utility called dump_dataset (available from Kratos), which converts the binary format to an ASCII version and it is the ASCII version that CasaXPS converts to the ISO 14976 standard.

To create an ASCII version of a Kratos Vision data set do the following.

On the Unix workstation move to the data directory and type

```
dump_dataset filename.dset all > new_filename.kal
```

The extension .kal is recognized by CasaXPS and tells the conversion routine to parse the data using the Kratos Vision option.

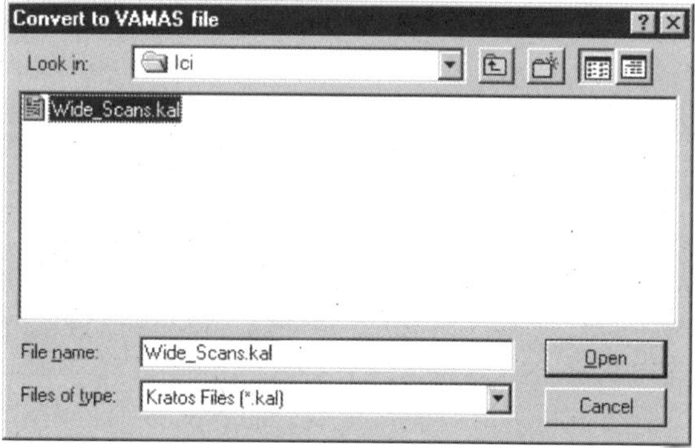

FIGURE 16•4 Convert to VAMAS file Dialog Window. The file display filter is set to show only .kal files.

Selecting a .kal file via the "Convert to VAMAS file" dialog window causes an ISO 14976 file to be generated with the same name, but with a .vms extension added. The spectra will be converted and other experimental information extracted so that the ISO 14976 file may be quantified to produce identical results to those from the Vision systems. That is, transmission correction is included as part of the ISO 14976 file.

Element library files from the Vision 1.x system can also be converted to CasaXPS format and Vision 2.x element libraries can be converted on request.

"Dayta" System Files (University of Bristol IAC)

Spectra acquired by the Bristol "Dayta" system are stored as sets of ACSII files with .SP_ file extensions. These files may be connected to a particular experiment through an associated .seq file. Sets of .SP_ spectrum files are listed in the .seq file together with the any experimental information (e.g. etch times.)
There are two methods for loading Bristol files.

- The simplest method is to convert files that are listed in a .seq file. All that is required is for the .seq file and all the .SP_ files listed within that file are located in the same sub-directory, then the .seq file is selected through the "Convert to VAMAS file" dialog window. If for any reason the set of .SP_ files does not match the set listed in the .seq file, then the .seq will need to be edited to remove or add entries as necessary. The ISO file that results will include an experimental variable for each spectrum loaded.

- The alternative method for loading the .SP_ files is to collect a set of these files in a sub-directory and then enter a new file name in the "Convert to VAMAS file" dialog window, but add an extension of .bri to the specified name. All .SP_ files in the directory will be read and appear in a single ISO file. The default action is that no experimental variable it defined for each spectrum. As a consequence spectra read will all appear in the same row in CasaXPS (i.e. in a format suitable for quantification of high resolution spectra). If on the other hand it would be more convenient to separate spectra by an experimental variable then the new file name entered in the dialog window should be followed by a –col string e.g. "newfile.bri –col" (omit the quotation marks). The experimental variable is the index number determined form the order in which the files are read.

VGX900 (Ron Unwin)

Ron Unwin developed the VGX900 system and so all the various forms of Ron Unwin's files can be converted in the same way.
Files generated by the VGX900 system are ASCII files and each file may contain

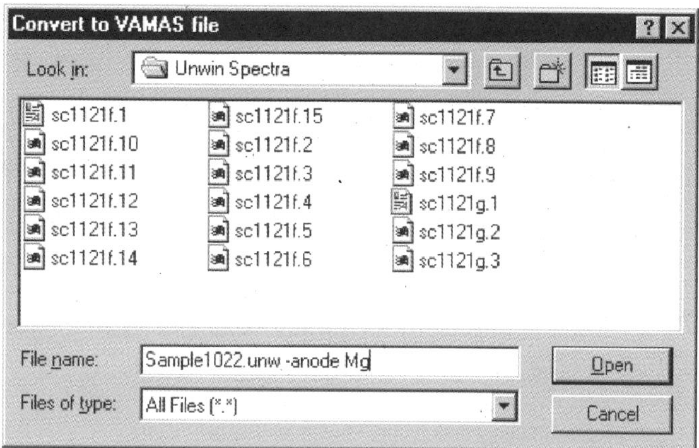

FIGURE 16•5
Convert to VAMAS file dialog window. Note the .unw extension used to indicate that the set of files contain spectra acquired using Ron Unwin's data system. The –anode Mg is only used to force the X-ray source to be Magnesium instead of the default setting. The latest release of Ron Unwin's system includes this information.

one or more spectral regions. The method for loading these files into CasaXPS is to collect a set of the files in a sub-directory and then enter a new filename in the "Convert to VAMAS file" dialog window, but add an extension of .unw to the specified name (Figure 16•5). All files in the directory will be read and appear in a single ISO file. Since the files have no characteristic file type it is essential that only files generated from the VGX900 system are present in the sub-directory and no others.

Some formats of the VGX900 system do not include information necessary for a proper description of the data (in the VAMAS sense), but the system does allow the writing of ISO 14976 files based on a user specified template. This allows data to written in a form directly readable by CasaXPS, but the user must ensure that the correct fields have been used to describe the spectra.

Alternatively, the conversion option in CasaXPS can specify some of the missing information using flags. The X-ray anode defaults to Aluminium and the energy-scale defaults to Binding Energy. Appending the following strings to the file name at conversion time can alter these two options:

-anode Mg	specifies Magnesium anode
-energy KE	specifies that the scans were performed using an increasing KE sweep.

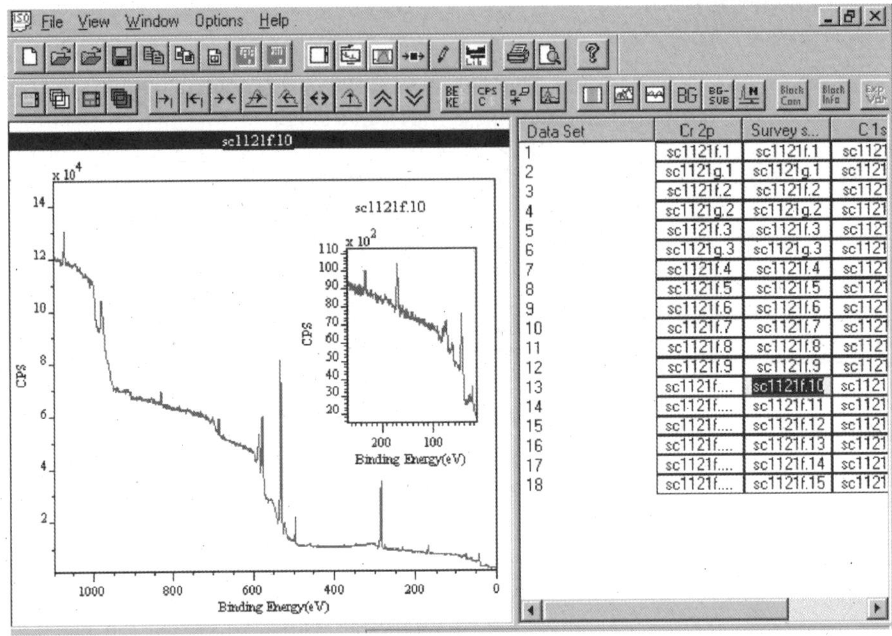

FIGURE 16•6 The set of spectra contained in the sub-directory shown in Figure 16•5 are offered in CasaXPS using the browser and spectrum display frames

SSI M-Probe Files

SSI M-Probe files may exist in an ASCII format where multiple regions appear in the same file or alternatively, the data is described by a control file with the spectra stored in separate .reg files in a sub-directory where the sub-directory has the same name as the base-name of the control file.

There are two methods for loading M-probe files.

- ASCII format files must have an .mrs extension and may be opened directly through the Convert Dialog window.
- The binary format consists of an .mrs file together with a directory of the same name as the base-name of the .mrs file. To load the information held in this binary format the base-name of the .mrs file must be specified but with an .ssi extension replacing the .mrs extension

RDB Text Files

RBD data files are in a binary format, however these files can be exported as ASCII files which can be converted through CasaXPS to the ISO 14976 standard file format. The RBD ASCII files may contain multiple spectral regions. To convert one or more of these files into a single ISO format, collect related data files exported from the RBD software into a sub-directory and ensure that the only .txt files in that directory are files in the RBD export-format. Using the File Dialog Window under the Convert toolbar option, go to the directory containing the RBD files. Enter a name for the new file to be created by CasaXPS and add a file extension of .rbd. All .txt files in that current directory will be read and entered into the new .vms file. Note that the files and therefore the spectral regions will be entered into the ISO file in alphabetical order with respect to the names of the original .txt files.

Kratos DS800 Binary Files

Kratos DS800 binary files can be directly converted through CasaXPS.
Use the "Convert to VAMAS file" dialog window as above and set the "Files of type" filter to allow only .mpa files to appear in the listing. Simply select a DS800 file and press Open or double-click the desired file. CasaXPS will convert the binary file into the ISO 14976 file format, where the new file will have the same name as the original .mpa file but with a .vms extension added.

Scienta DOS Binary Files

The Scienta DOS binary files are organized in the sub-directories SPECIFIC, REGIONS and RESULTS. To convert a spectrum stored across these directories, simply select the Convert option in CasaXPS and move to the SPECIFIC directory, then open the required file within the SPECIFIC directory using the Convert to VAMAS File Dialog window. A new VAMAS file will be created in the SPECIFIC directory.

Scienta IGOR .pxt files

Simply select the .pxt file via the Convert to VAMAS dialog window.

Specs SpecsLab I .exp files

Simply select the .exp file via the Convert to VAMAS dialog window.

X/Y Column Data Files

Columns of abscissa/ordinate pairs must be stored in files with a .dat file type. Any files that need to be loaded into the same ISO 14976 file should be collected into a sub-directory that contains only those ASCII files that are suitable for conversion through CasaXPS. Select the Convert toolbar button and move to the sub-directory containing the .dat files. Type a filename for the new ISO file and add a .dat file extension to the new file name before pressing the Open push-button on the File Dialog window. Note that each file may contain more than one spectral region, where the regions are concatenated using a blank line to separate the spectra.

The default input mode assumes the abscissa is in binding energy and that the X-ray anode is Aluminium. To change these settings additional flags may be specified after the file name (Figure 16•7) where "–anode Mg" specifies a Magnesium anode for the X-ray source and "–energy KE" informs CasaXPS that the abscissa is kinetic energy rather than binding energy. All the .dat files in the directory shown in Figure 16•7 will be entered into the same ISO file. Names, element/transition, comments and other ISO 14976 fields may be modified after the data has been loaded into CasaXPS via toolbar buttons (see below).

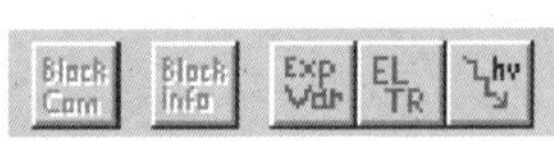

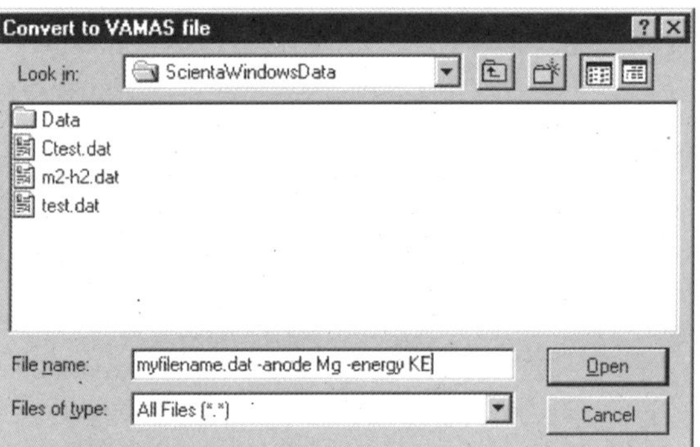

FIGURE 16•7
Loading a set of .dat files through the Convert Dialog Window. Note that the default settings for the energy scale (BE) and X-ray anode (Al) are over-ridden by explicitly entering "–anode Mg" and "–energy KE" after the file name.

CHAPTER 17

Auger Considerations and Casa Auger Features

Auger Imaging and the Advantages of Spectra at Pixels

Auger images measured using a peak maximum are often difficult to interpret. Fluctuations in the signal as a function of position on the sample cause variation of intensity independent of the chemical information desired from the surface. The difficulty arises because Auger images, collected in direct mode at a peak maximum, are often taken in isolation; background intensity variations are sufficient to hamper the interpretation of such images. In an attempt to reduce these effects, images are often also acquired at an energy representative of the background - and sometimes at a second background energy - with the view to estimating the background contribution to the final image thereby improving the measure of peak height at each pixel. A common calculation for an Auger analysis is the so called "topographical correction" computed from the peak maximum N_1 and a background image N_2, namely $(N_1-N_2) / N_2$. A variation on a theme is to compute an image from $(N_1-N_2) / (N_1+N_2)$. The principal reason for measuring Auger images using simple peak and background intensities is the time penalty involved in doing more, but the benefits of applying spectroscopic techniques to image determination may out-weigh the time penalty of the extended acquisitions, when this enables determination of spectra at each pixel. The case examined below (data provided by Pennsylvania State University, USA), is a simple experiment in which a gold grid is placed over a clean silver surface, and a set of Auger images acquired at unit eV step size using a silver Auger peak. As luck would have it, a weak gold Auger line is also included in the energy range over which the images were recorded and so the data offer a means of illustrating a simple, yet effective procedure for creating "deconvoluted" elemental Auger images. Rather than using direct spectra, following the conventional wisdom for Auger work, data are differentiated and peak-to-peak measurements made for each pixel in the image. Simple quantification regions are defined for the spectra, one for the silver Auger line and one for the weak gold Auger line, then two images of the peak-to-peak intensities are constructed, and these *images* in turn are quantified using the standard procedure of quantifying *spectra*. The results are two images (Figure 17•2), where the silver image is free from intensity variations other than the elemental surface composition. In many ways, the improved quality of the gold and silver images is reassuring, since for years it has been assumed that Auger spectra taken from a single point on a sample can be compared by determining atomic concentrations from the relative intensities of differentiated Auger peaks. The images in Figure 17•2 were determined from the peak-to-peak intensities without application of relative sensitivity factors, but provided the appropriate values are available, proper atomic concentration images are possible and such images then show surface composition on the same basis as that normally associated with spectra.

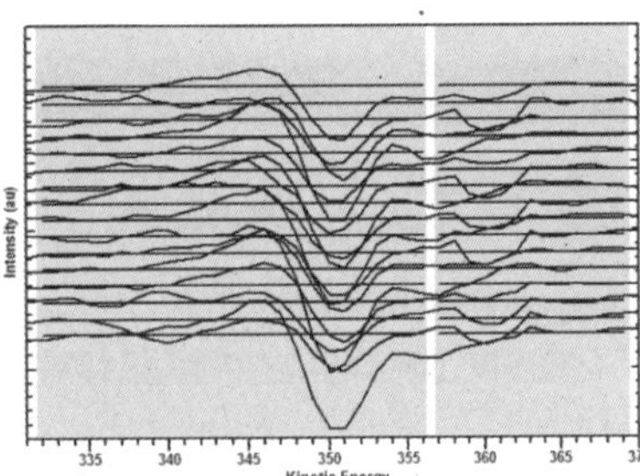

FIGURE 17•1
(above) A sample of spectra taken from individual pixels constructed from the original set of images.

FIGURE 17•2 (right)
Top two images are generated from peak-to-peak intensities from spectra and regions similar to those shown in Figure 17•1. Bottom two images, while representing an improvement over the peak maximum image, still include artifacts of the measurement process.

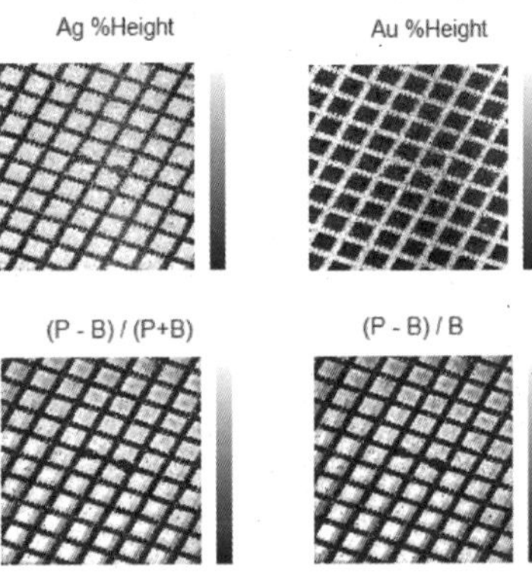

Analysis of Auger Images in CasaXPS

The essential sequence of steps required to process the image data is as follows:

1. Convert the image set to spectra at each pixel.
2. Differentiate all the spectra.
3. Define quantification regions for each transition and propagate these quantification regions to each block containing a row of spectra.
4. Convert the region intensities to images.

Within these steps there are potentially sub-steps and these sub-steps are detailed below based on the original data for the given example.

The data in Figure 17•1 and Figure 17•2 were acquired on a PHI Auger system. Auger images were acquired, in this particular case, using two data files in which the saved images correspond to a sequence of unit energy steps over an interval spanning 370 eV to 331 eV. Since the data are in two separate files, the first job is to assign the correct energy to the experimental variable for each image and then move the images into a new Experiment Frame.

Ordering the Images using the Experimental Variable

Initially, when converted through CasaXPS, the PHI .map files each contain half the images for the entire data set. The images are assigned the same experimental variable and the species/transition block fields are all different, therefore the data blocks appear in the right-hand-side of the Experiment Frame as a single row (Figure 17•3). To convert a set of images to spectra at each pixel, it is necessary for each image to appear in the same column and the experimental variable for each image to be assign the value for the energy at which the image was acquired. Given data in the format seen in Figure 17•3, the first step is to assign the species/transition fields for each block in the file:

FIGURE 17•3

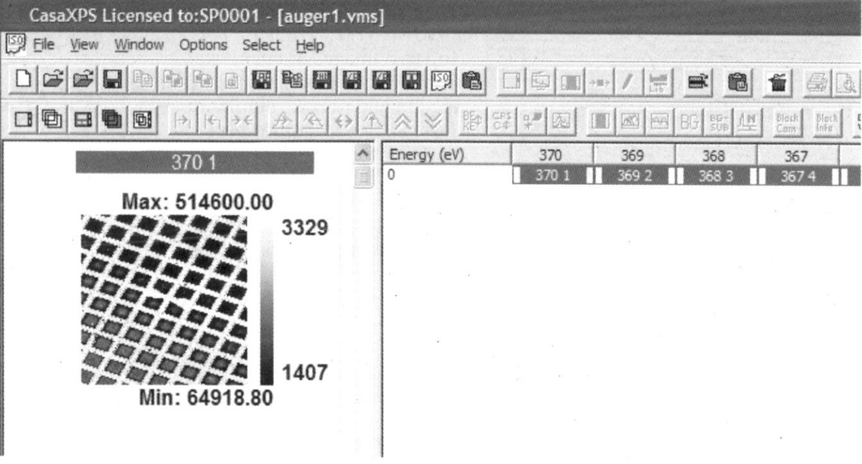

1. Select the entire set of blocks in the Experiment Frame. Either click on the single experimental variable value to select the row of blocks or click once on the button heading the column of experimental variable values. Both these actions will result in the full set of blocks being selected. The former action relies on all the data appearing in a single row, whereas the latter selects the entire set of data blocks in the Experiment Frame.

2. Press the "ElTr" button on the second toolbar and enter new Element/Transition strings on the resulting dialog window. When the OK button is pressed, all the selected data blocks are assigned the pair of strings entered on the dialog window and as a consequence, the right-hand-side of the Experiment Frame is reordered so that

all the blocks appear in a single column.

Once the blocks appear as a single column in the Experiment Frame browser, the values for the experimental variable are adjusted to the energy at which each image was acquired. The two "ExpVar" toolbar buttons provide a means of assigning the experimental variable value on bulk; however the left-most of this pair of buttons offers a means of specifying a range of values for the experimental variable, where each row of data blocks will be assigned intermediate values. Thus for the data in Figure 17•3 following rearrangement into a single column, are assigned the appropriate experimental variable values by specifying the range 370 – 351 eV. Similarly the second file can be adjusted into a column of data blocks by assigning the same element/transition strings to each block as was used in the first file and then the

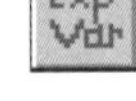

"ExpVar" toolbar button used to specify the range 350 – 331 eV. Having made these assignments for the experimental values to both files, the two sets of blocks can be merged into a new Experiment Frame. Select all the data blocks in one Experiment Frame, click on the second Experiment Frame to make it the active window, then press the "Add Blocks" toolbar button. The action of this toolbar button (page 56) copies the selected blocks from all the open Experiment Frames to the active Experiment Frame (Figure 17•4).

FIGURE 17•4

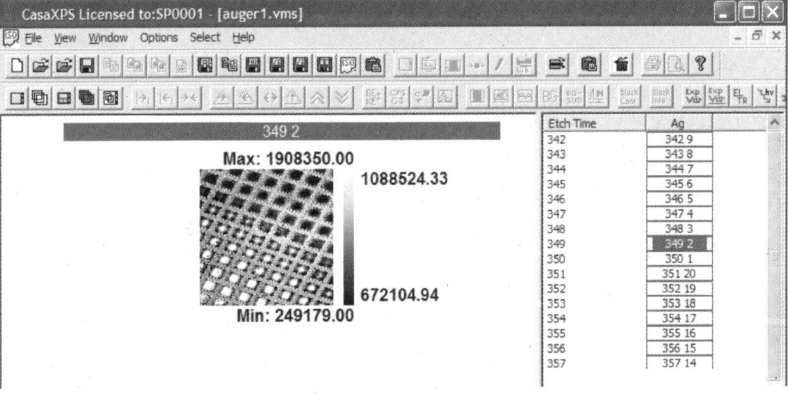

Although the data blocks now appearing in the same Experiment Frame are all in one column and the map energy is assigned to the experimental variable, there is still one possible problem. As seen in Figure 17•4, the images are ordered from high kinetic energy to low and so the spectra, when generated, will be assumed to be XPS data rather than AES, as is the case. It is also possible that the merged files appear back to front, from the energy scale perspective, and so the true order of the images, with respect to energy, is yet to be realized. The functionality in the second button for assigning the experimental value is required to force the appropriate reordering. The

dialog window invoked by the right ExpVar toolbar button allows the assignment of the experimental variable for a selected set of blocks. Further, once the assignment for the selection is made, the blocks are reordered with respect to the new set of experimental variables. To effect a reordering of the image, it is simply the case of selecting a single block and pressing the OK button on the dialog window. Although no new value was actually assigned, the reordering will still take place and the images then appear in the Experiment Frame ordered with respect to the map energies.

Converting Images to Differentiated Spectra

Given the set of images, now ordered by the energy at which the data are acquired, the analysis proceeds by converting the data set into one viewed as a set of spectra at each pixel in an image. The conversion from images to spectra is achieved using the Image Processing property page on the Image Processing dialog window. To invoke the Image Processing dialog window, click on the displayed image to ensure the Options menu items are active, and then select the Image Processing menu option. Overlay the images in the Active Tile and press the Convert Images to Spectra button (Figure 17•5) on the Image Processing property page. It is important that all the images are acquired using the same acquisition time and that the experimental variable represents a sequence of evenly spaced energies. If the step in energy between the images deviates from a constant difference, then an error message will appear and no conversion to spectra takes place.

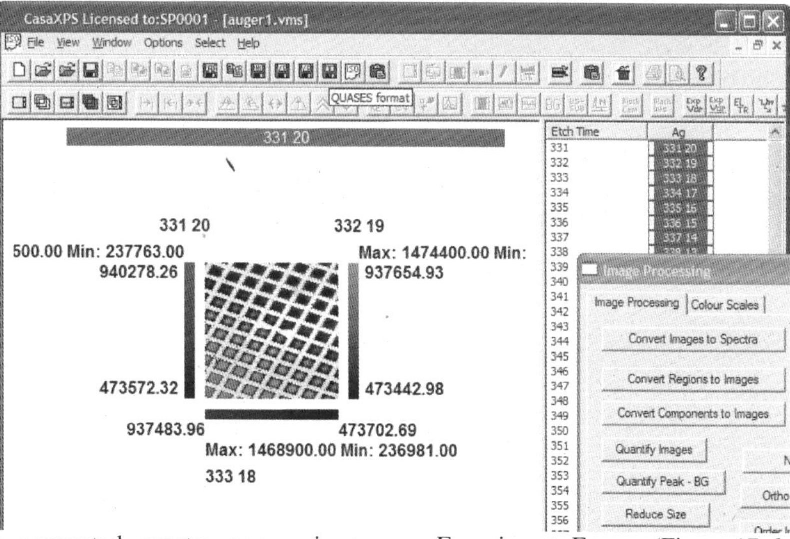

FIGURE 17•5

The converted spectra appear in a new Experiment Frame (Figure 17•6 on page 308), where the experimental variable is labeled pixel and the values for the experimental variables represent row indices of the pixels in the original images. Each data block contains an entire row of spectra, the spectra being stored as the corresponding variables in each block. As a result, a new processing option on the Differentiation property page of the Spectrum Processing dialog window (Figure 17•6) specifies that the operation should be applied to all the corresponding variables in a block, rather than simply to the corresponding variable currently displayed in the Active Tile. The data in all the data blocks must be differentiated, therefore once the data in the Active Tile is differentiated, the propagate mechanism should be used to process all the blocks similarly. To propagate the differentiation: select the entire set of blocks in the Experiment Frame Browser and right-click the mouse button over the Active Tile containing the differentiated spectrum. On the Browser Operations dialog window, tick the Processing tick-box within the Propagate section and press the OK button. All the spectra within the targeted data blocks will be differentiated. Since the spectra are stored as multiple corresponding variables and only one corresponding variable from a given block can be viewed at a time, it becomes necessary to step through the corresponding variables to inspect the results of the differentiation operation. To step through the corresponding variables in a block, the Control + Page-Up and Control + Page-Down keys are used. The corresponding variable index in each block within an Experiment Frame is adjusted by the Control + Page-Up and Control + Page-Down keys. Similarly, Control + Home and Control + End move the corresponding variable index to the beginning and end indices for each block in the Experiment Frame. To adjust the index for a corresponding variable on a block by block basis, using the Page-Up, Page-Down, Home and End keys whilst holding down the Shift key causes only the blocks in the Active Tile to be adjusted.

FIGURE 17•6

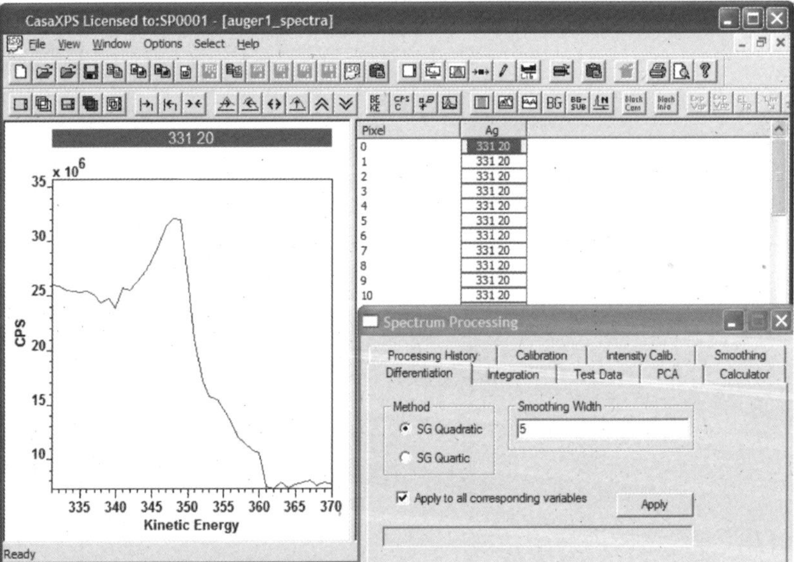

Quantifying the Spectra at Pixels to Produce Images

Examples of the differentiated spectra are shown in Figure 17•1. The quantification of Auger spectra is typically performed using the peak-to-peak metric to measure line intensity. In this example, two Auger lines are evident in the energy range 331 – 370 eV and so two quantification regions can be defined on each spectrum as indicated by the grey bands over the data in Figure 17•1. These quantification regions are defined on the Quantification Parameters dialog window via the Regions property page. The regions for use with the spectra-to-images options are identical to those used for general spectral quantification, but with one slight difference. The default action of the Convert Regions to Images button on the Image Processing property page is to create images using the integrated intensity based on peak area. For differentiated AES data, the peak-to-peak intensity is required and to create images appropriate for the data in Figure 17•1, the Tag field in the region specification table must be assigned a keyword string, namely, peak to peak. The Tag field is also used to switch between other regions parameters such as fwhm, position and centroid; by using these various region outputs, the surface mapped by the images can be viewed with respect to peak broadening or peak shifts. Again the Tag field is entered with the appropriate keyword string e.g. fwhm, position or centroid. Figure 17•7 shows the Regions property page where two quantification regions have been defined on a spectrum in which the Tag field are set to the keyword for generating images using the peak to peak intensities.

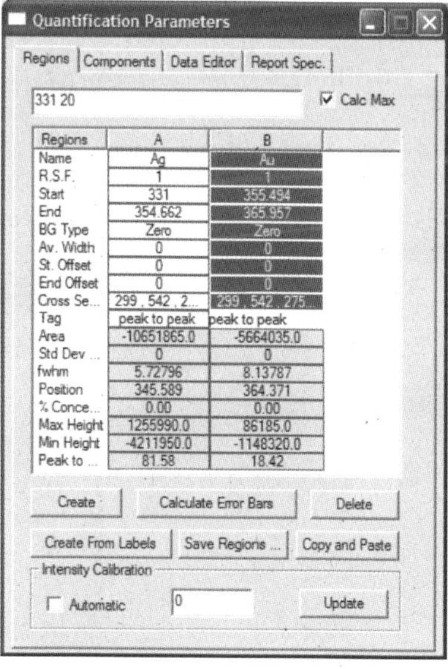

FIGURE 17•7

Once quantification regions have been propagated to each block in the spectrum file, the corresponding images are generated by overlaying all the spectra from the blocks in the Active Tile and then pressing the Convert Regions to Images button (Figure 17•5 on page 307).

Ideally, images for each element identifiable on the sample surface should be measures and a similar analysis to the Ag and Au data performed for each element. In this example only two elemental images were recorded, but nevertheless, proceeding to the quantification step based on the two available images has merit in that the resulting images (Figure 17•2 on page 304, top) are normalized with respect to each other across the field of view. Figure 17•5 on page 307 shows a section of the Image Processing property page, on which the button labeled Quantify Images can be used to perform the operation. The Quantify Images button assumes the images used in the calculation are generated from either quantification regions or synthetic components, both of which are in units of CPSeV (for area based intensities) or CPS (for peak-to-peak measurements), therefore the new images are generated simply using the formula $I_j / (I_0+I_1+ \dots +I_n)$.

Note, a second button labeled Quantify Peak – BG offers a second means of quantifying a set of images. The procedure in this case does not involve images preprocessed using spectral regions or components and therefore adjustments for time will be included in the calculation. This is neither appropriate nor desirable for data processed as described in this section, but is available for use with raw images, where a less sophisticated approach is adopted.

Converting AES Direct Spectra to Differential Form

The steps required to convert an entire data-set to differential format are as follows:

1. Select a spectrum into the Active Tile.
2. Enable the Processing menu/toolbar option by left-clicking the Active Tile, then invoke the Spectrum Processing dialog window using the Processing option on the Options menu.
3. Using the Differentiation property page on the Spectrum Processing dialog window, apply a five point Savitzky-Golay, quadratic polynomial differentiation to the data (Figure 17•8).

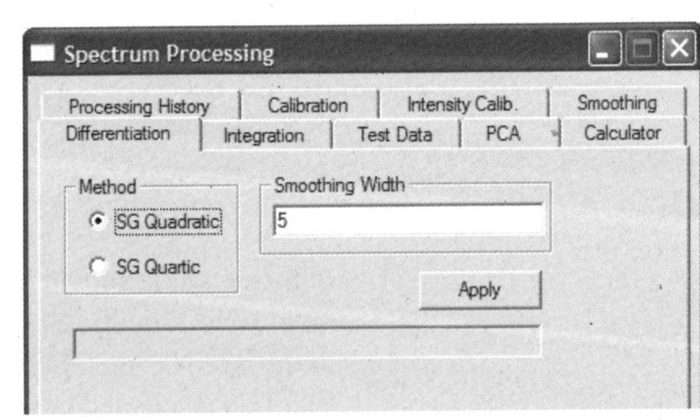

FIGURE 17•8

4. Select all the data blocks in the Experiment Browser and propagate the processing to all the selected spectra (Figure 17•9).

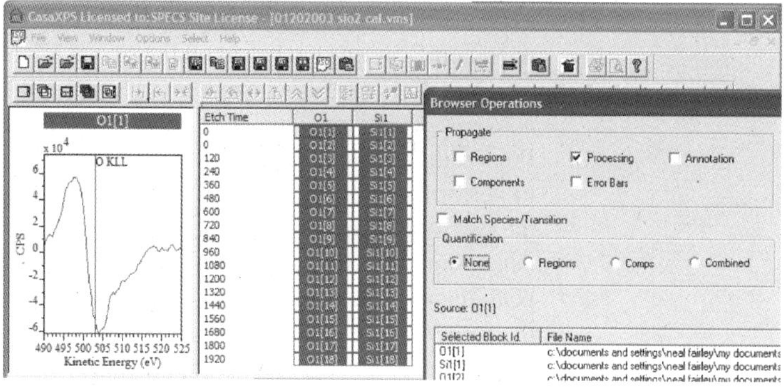

FIGURE 17•9

Linear Least Squares Analysis of Chemical States

Chemical State information within AES differential spectra can be separated using linear least squares (LLS) provided representative spectra are available for the desired chemical states. The LLS option is available on the PCA property page on the Spectrum Processing dialog window (Figure 17•10).

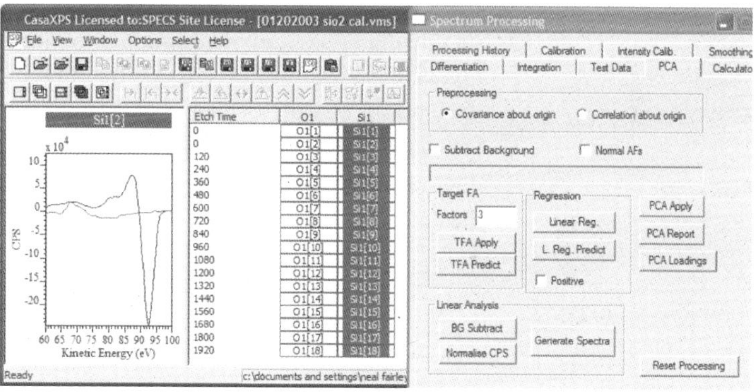

FIGURE 17•10

In the example shown in Figure 17•10, two states have been selected to represent the Oxide and Metallic Si signal and are displayed in the Active Tile. These two spectra in the Active Tile represent the basis set used as part of the LLS procedure, while the set of data selected in the right-hand-side of the Experiment Frame are the unknown data to be analysed by LLS. That is to say, when the button, labelled Generate Spectra on the PCA property page shown in Figure 17•10, is pressed the appropriate linear combinations of the two spectra in the Active Tile are determined for each of the spectra selected in the right-hand-side and a new Experiment Frame is created. The contents of the new Experiment Frame are: the original spectra, a spectrum computed from the linear combination and the scaled basis spectra determined from the LLS calculation. Assuming the LLS can properly account for the states within the unknown spectra, the scaled basis spectra can be used to quantify the relative proportions of the known chemical states. It will be necessary to move the scaled basis spectra into the same Experiment Frame as any other spectra also part of the analysis, such as the oxygen region also shown in Figure 17•10. Quantification can then proceed as described in the next section.

Quantifying Spectra Using Peak-to-Peak

Quantification of AES differential spectra is typically performed using peak-to-peak intensities. The Regions property page on the Quantification Parameters dialog window (Figure 17•11) displays both the maximum/minimum intensities within a region and also the Peak-to-Peak atomic concentration (although the term atomic concentration only applies when an elemental analysis is performed on a survey spectrum). The background type is irrelevant to the computation of these intensities, provide a background can be computed, however from an aesthetic point of view a choice of mean or zero background types would be appropriate.

FIGURE 17•11

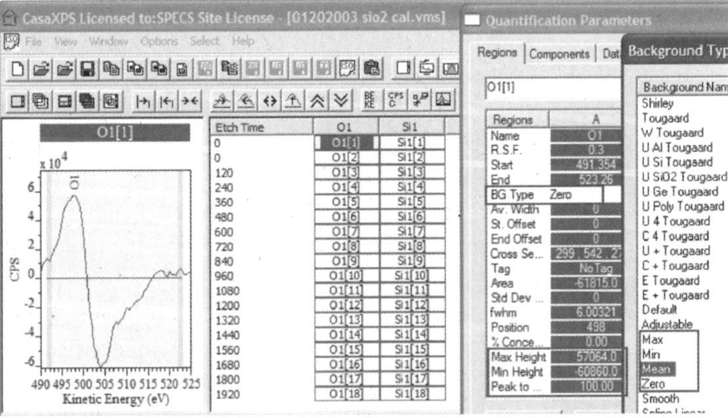

Quantification for sets of narrow scan data is achieved using the Report Spec property page on the Quantification Parameters dialog window. The Standard Report can be customized to display only the Peak-to-Peak concentration results, while the, so called, Custom Report interface offers a button for generating a profile using peak-to-peak intensities. The essential mechanisms described elsewhere for XPS spectra also apply to AES quantification.

Quantification: An Example

In this example, an AES depth profile acquired using the RBD Auger Scan software is analysed using LLS to separate metallic Si from Si oxide. The profile is generated from quantification items defined on each of the LLS components together with the oxygen spectra.

Figure 17•12 shows the direct spectra before any processing is applied.

FIGURE 17•12

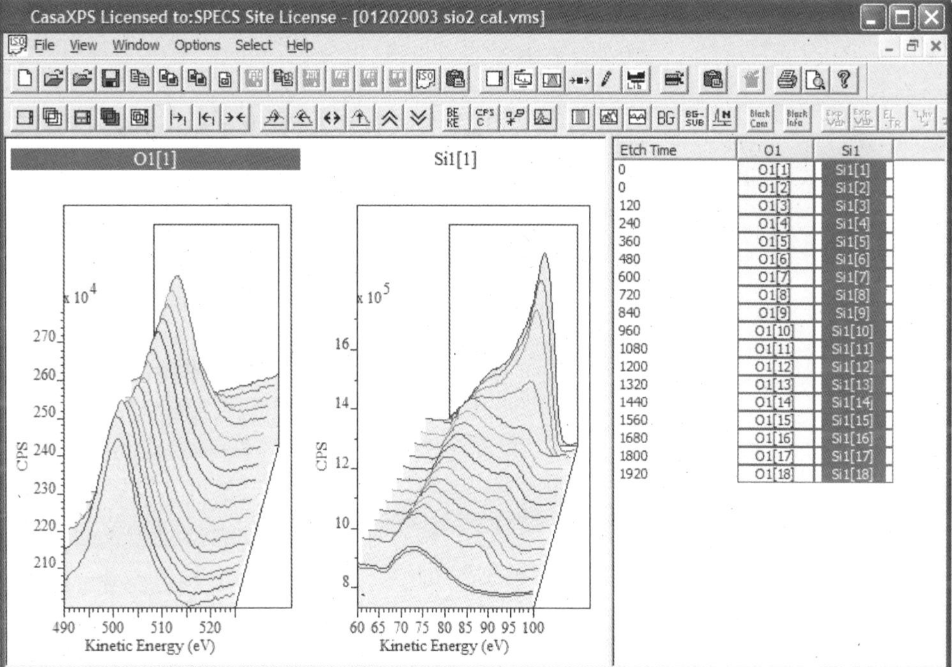

Step 1: Differentiate Spectra

AES spectra are typically quantified using peak-to-peak intensities taken from the differential of the direct spectra. The spectra must be all converted to differential form using the procedure described above.

Step 2: Separate the Metal and Oxide Silicon States

A LLS procedure using three basis spectra taken from the profile provides the means for estimating the contributions from the metal and oxide present in the silicon spectra. A representative set of spectra after the LLS has been applied as shown in Figure 17•14 on page 315. The three basis spectra scaled by the LLS procedure are displayed in the second column labelled Surface Si, Si in SiO_2 Layer and Si Metallic.

FIGURE 17•13

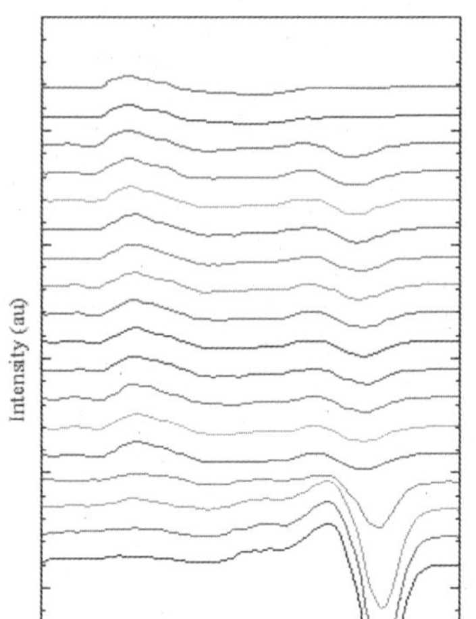

Combining these three scaled Si spectra results in trace 1 in the top left-hand tile, while trace 2 is the original spectrum modelled by trace 1.
Note, the O1 spectra have been copied into the Experiment Frame created when the LLS procedure was applied to the Si spectra. This allows quantification regions to be assigned to each of the Si basis spectra and also the O1 data. The profile shown in Figure 17•15 on page 316 is the result of quantifying the entire data set using the Height Report from the Custom Report section of the Report Spec property page on the Quantification Parameters Dialog Window.

Three basis spectra were used in the LLS, one taken from the surface regions, one taken from the metallic region and the third created from a Si spectrum within the oxide layer after first subtracting the surface spectrum. Such a procedure is necessary, from a numerical perspective, to allow reasonable accuracy between the raw Si spectra and the reconstructed spectra. The trace labelled "Si?" in Figure 17•15 on page 316 could be viewed as part of the metallic signal and therefore these two intensities could be summed using the Custom Report (Figure 17•17 on page 316). The result of creating a Height Report using the definition shown in Figure 17•16 is offered in Figure 17•17.

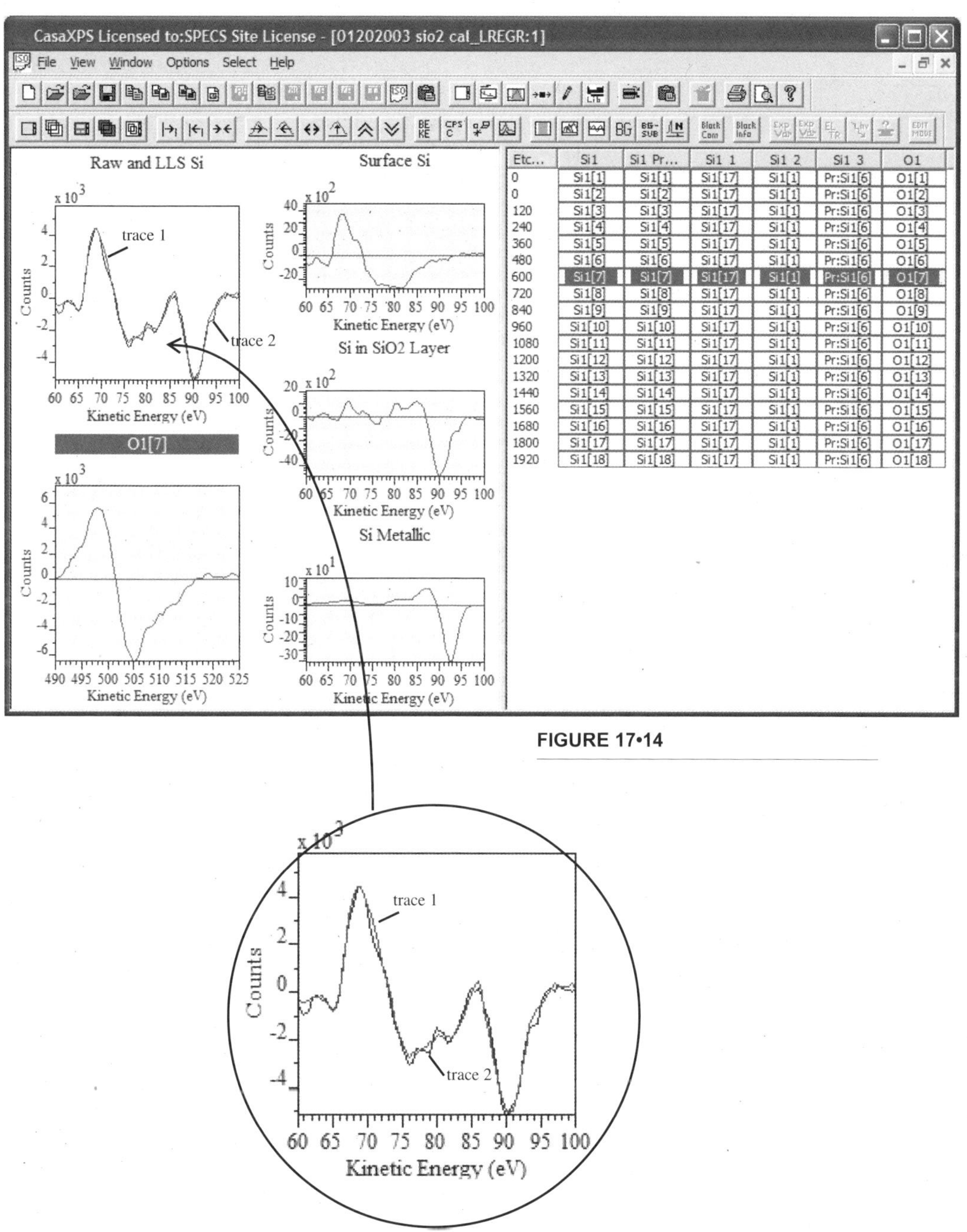

CasaXPS Licensed to:SPECS Site License - [01202003 sio2 cal_LREGR:1]
File View Window Options Select Help
Raw and LLS Si
x 10^3
trace 1
trace 2
Counts
Kinetic Energy (eV)
O1[7]
Surface Si
Si in SiO2 Layer
Si Metallic
Etc... | Si1 | Si1 Pr... | Si1 1 | Si1 2 | Si1 3 | O1
0 | Si1[1] | Si1[1] | Si1[17] | Si1[1] | Pr:Si1[6] | O1[1]
0 | Si1[2] | Si1[2] | Si1[17] | Si1[1] | Pr:Si1[6] | O1[2]
120 | Si1[3] | Si1[3] | Si1[17] | Si1[1] | Pr:Si1[6] | O1[3]
240 | Si1[4] | Si1[4] | Si1[17] | Si1[1] | Pr:Si1[6] | O1[4]
360 | Si1[5] | Si1[5] | Si1[17] | Si1[1] | Pr:Si1[6] | O1[5]
480 | Si1[6] | Si1[6] | Si1[17] | Si1[1] | Pr:Si1[6] | O1[6]
600 | Si1[7] | Si1[7] | Si1[17] | Si1[1] | Pr:Si1[6] | O1[7]
720 | Si1[8] | Si1[8] | Si1[17] | Si1[1] | Pr:Si1[6] | O1[8]
840 | Si1[9] | Si1[9] | Si1[17] | Si1[1] | Pr:Si1[6] | O1[9]
960 | Si1[10] | Si1[10] | Si1[17] | Si1[1] | Pr:Si1[6] | O1[10]
1080 | Si1[11] | Si1[11] | Si1[17] | Si1[1] | Pr:Si1[6] | O1[11]
1200 | Si1[12] | Si1[12] | Si1[17] | Si1[1] | Pr:Si1[6] | O1[12]
1320 | Si1[13] | Si1[13] | Si1[17] | Si1[1] | Pr:Si1[6] | O1[13]
1440 | Si1[14] | Si1[14] | Si1[17] | Si1[1] | Pr:Si1[6] | O1[14]
1560 | Si1[15] | Si1[15] | Si1[17] | Si1[1] | Pr:Si1[6] | O1[15]
1680 | Si1[16] | Si1[16] | Si1[17] | Si1[1] | Pr:Si1[6] | O1[16]
1800 | Si1[17] | Si1[17] | Si1[17] | Si1[1] | Pr:Si1[6] | O1[17]
1920 | Si1[18] | Si1[18] | Si1[17] | Si1[1] | Pr:Si1[6] | O1[18]

FIGURE 17•14

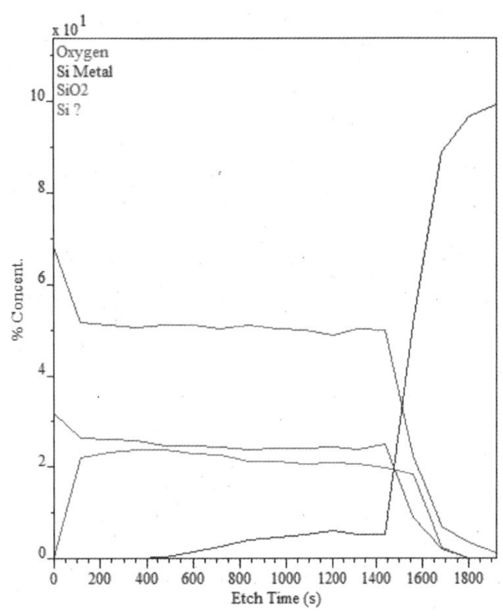

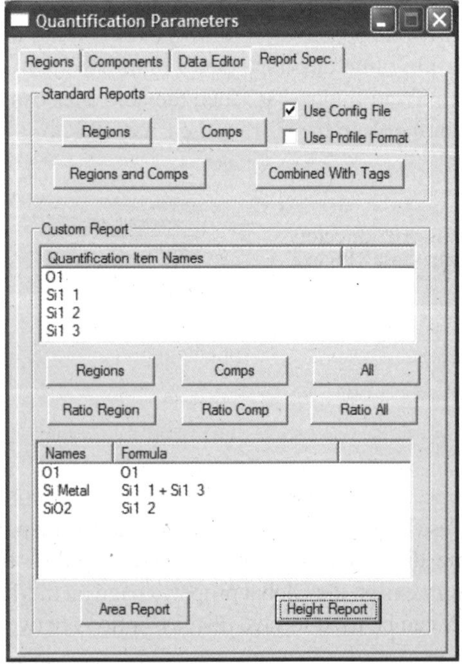

FIGURE 17•15 (top)

FIGURE 17•16 (bottom)

FIGURE 17•17 (below)

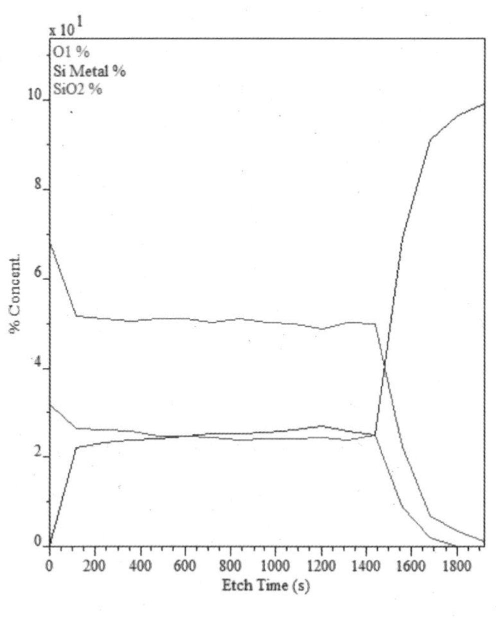

CHAPTER 18

The CasaXPS File System

Help Files

The CasaXPS Help system is a set of HTML files. To view these HTML documents it is necessary to ensure that the CasaXPS executable (CasaXPS.exe) and the help folder (CasaXPS.HLP) are located in the same directory (see Figure 18•1). If the files are structured as seen in Figure 18•1 and Figure 18•2 then the help files can be viewed by selecting the Help option on the Help menu from the CasaXPS program window.

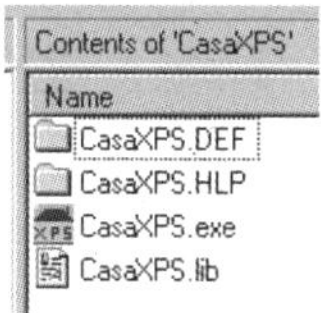

FIGURE 18•1 Contents of the default CasaXPS directory showing two directories, the CasaXPS.exe file and a CasaXPS element library file.

Default Settings

A set of files containing default settings, such as colours, fonts and display options are maintained in a directory called CasaXPS.DEF. This directory, like the help directory, must be located in the same directory as the CasaXPS executable file CasaXPS.exe. Once the default directory exists, the Global property page on the Tile Parameter dialog window in CasaXPS can be used to save display options between

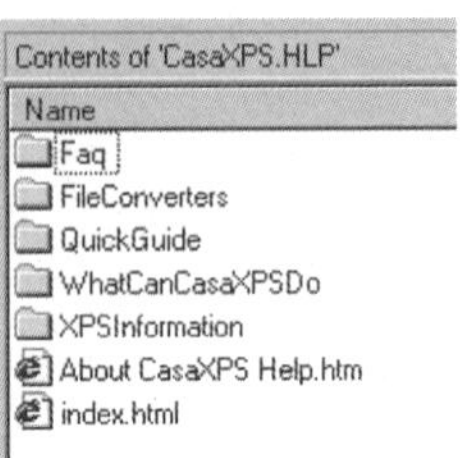

FIGURE 18•2 Contents of CasaXPS.HLP. The index.html file is the top level HTML file and may be loaded into an Internet Explorer as one means of viewing the help information.

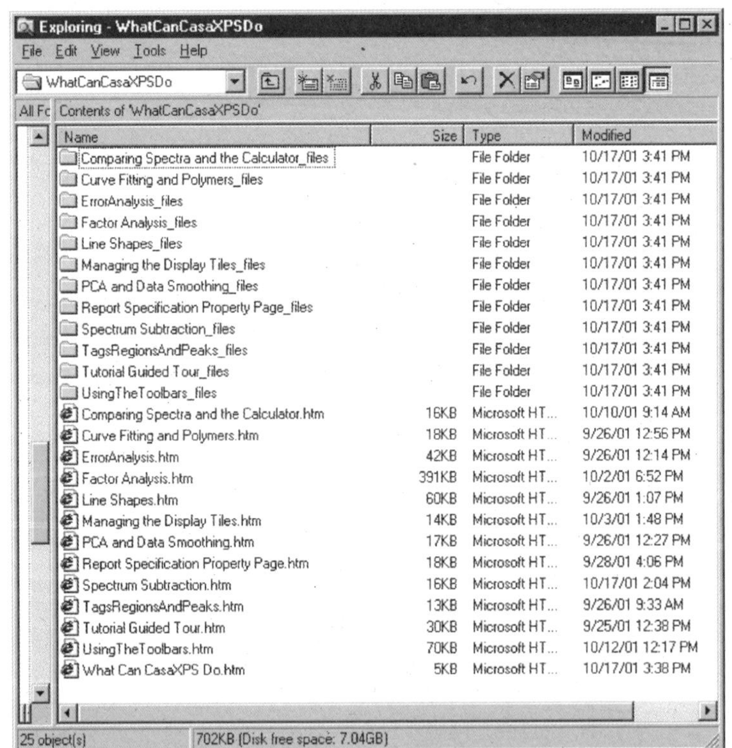

FIGURE 18•3 Contents of the directory WhatCanCasaXPSDo. Note that the images for the html files are held in sub-directories.

sessions of CasaXPS. Note, the default directory on the CD is empty and the save options on the Global property page of the Tile Parameter dialog window, when used, will create a set of files containing the new default settings for CasaXPS. Configuration files are also located in the CasaXPS.DEF directory:

ParameterFile.txt

TABLE 18•1. ParameterFile.txt

Optional Entries in File	Action
mean free path =<real number>	Defines the default exponent displayed on the Regions property page on the Quantification Parameters Dialog Window
order by position	Causes quantification reports to be ordered by energy
number of decimal places concentration = <int>	Number of decimal places used to print the concentration fields in a quantification report
number of decimal places fwhm = <int>	Number of decimal places used to print the fwhm field in a quantification report
number of decimal places area = <int>	Number of decimal places used to print the area field in a quantification report
number of decimal places position = <int>	Number of decimal places used to print the position fields in a quantification report
number of decimal places stdev = <int>	Number of decimal places used to print the stdev fields in a quantification report.
rsf table = source angle 54 44	Angular distribution correction to RSF entries should be applied
hp text bug	Compensate for vertical text misalignment in HP print-driver.
emf text bug	Compensate for vertical text misalignment in Word 2000 EMF
bg = shirley	Initial default background type is Shirley rather than linear
clip concent	Restricts the values in the Quantification Reports to lie in the range [0,100].

PrintFootNote.txt

An ASCII text files. Each page printed by CasaXPS includes a text string printed at the bottom of the page. If a string is entered into the PrintFootNote.txt file then the string, so defined, is printed instead of the default message. Version 2.1.0 and above will also print the string in each EMF or Bitmap created by CasaXPS. To remove the footnote, simply enter a space character into the PrintFootNote.txt file.

ComponentQuantTable.txt and RegionQuantTable.txt

ASCII text. Entering predefined strings into these two files configures standard quantification reports (see detailed description, page 143 ff.)

Element Library Files

The CasaXPS element library (discussed in greater detail in "Qualitative Analysis and the Element Library" on page 81) is also an ASCII file. The first line in the file is a version number, which may be '0' or '1' depending on whether the entries appear on separate lines or are separated by TAB characters, respectively. The TAB spaced format (shown in Figure 5•2 on page 83) supersedes the 'Type 0' format to improve the ease with which the library data may be edited in spreadsheet programs. For a large number of changes, a spreadsheet is the best means of constructing a CasaXPS library, but for a few alterations the Element Library property page (see Figure 5•3 on page 84) offers an integral dialog window based adjustments capability. Figure 5•2 shows library entries from two different excitation source strings, "Al" and "Mg". A single library file may contain any number of entries corresponding to different excitation source strings. These strings are matched to the excitation source string in the displayed data block of the Active Tile and only those library entries for which a match occurs are displayed in the Element Table property page on the Element Library dialog window, so excitation source strings should be edited with care.
At start-up CasaXPS loads the element library file called CasaXPS.lib. This file must be located in the same directory as the main executable file (CasaXPS.exe). For version 2.1.0, an additional file called CasaXPS_scofield.lib also must be located in this same directory. CasaXPS_scofield.lib is also loaded on start-up and provides the relative peak intensities used as part of the automatic peak identification algorithm (Peak Find button available on the Element Library dialog window).
Creating quantification regions from element markers displayed on the spectrum via the Element Library is performed by selecting the transition with the largest RSF for a given element in the active element table. If the transition corresponding to the largest RSF for an element is not the required quantification region, then element library entries, one per element per anode type, may be collected in a further file CasaXPS_quant.lib, which must be located in the same directory as the CasaXPS.lib and CasaXPS_scofield.lib files. The transitions held in the file CasaXPS_quant.lib determine the preferential creation of quantification regions, but RSF and other quantification information is still read from the active element table, i.e. the entries from the CasaXPS.lib file read on start-up or subsequently.

CHAPTER 19

Background to the XPS Technique

The Nature of X-ray Photoelectron Spectra

X-ray Photoelectron Spectroscopy involves irradiating a sample with X-rays of a characteristic energy and measuring the flux of electrons leaving the surface. The energy spectrum for the ejected electrons is a combination of an overall trend due to transmission characteristics of the spectrometer, energy loss processes within the sample and resonance structures that derive form electronic states of the material under analysis. The instrumental contribution is an unwelcome fact of the measurement process, but the background and resonance peaks offer information about the sample surface.

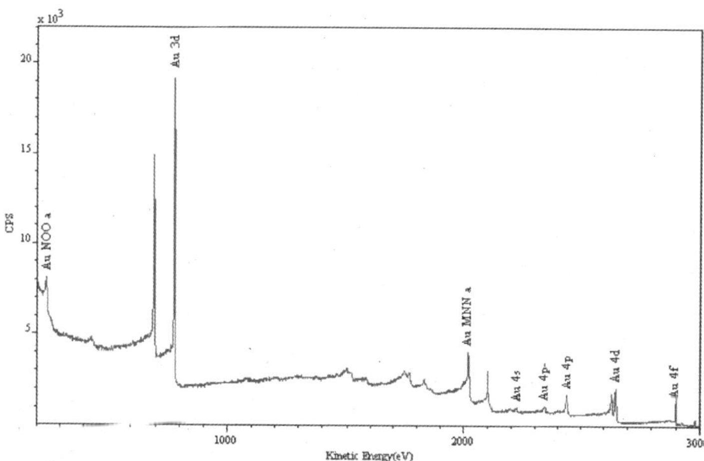

FIGURE 19•1 Wide scan energy spectrum acquired using a silver anode on a Kratos Axis ULTRA.

An energy spectrum from a clean gold sample using monochromatic silver X-rays is shown in Figure 19•1 on page 321. The peaks in the spectrum correspond to two distinct but related processes showing electrons ejected due to the photoelectric effect and emission due to the Auger effect.

Photoelectric Peaks, Auger Peaks and Energy Scales

Photoemission results in electrons of a given kinetic energy where the recorded energy is determined from a combination of the binding energy of the electron (with respect to some molecular form for each element) and the characteristic energy of the X-ray source. The dependence of the energy spectrum on the anode used to produce the X-rays means that the photoelectric lines move in position when the same sample is analyzed using a different anode to produce the X-rays. Since these lines provide the bulk of the chemical information found in the spectra, and it is chemical information that makes XPS such a powerful technique, it has become common practice to display XPS spectra using binding energy for the abscissa. The kinetic energy scale is reported relative to the photon energy of the excitation source and so photoelectric line positions with respect to a binding energy scale become independent of the X-rays used to excite the sample, while Auger line positions are invariant with respect to the X-ray anode only when plotted against a kinetic energy scale.

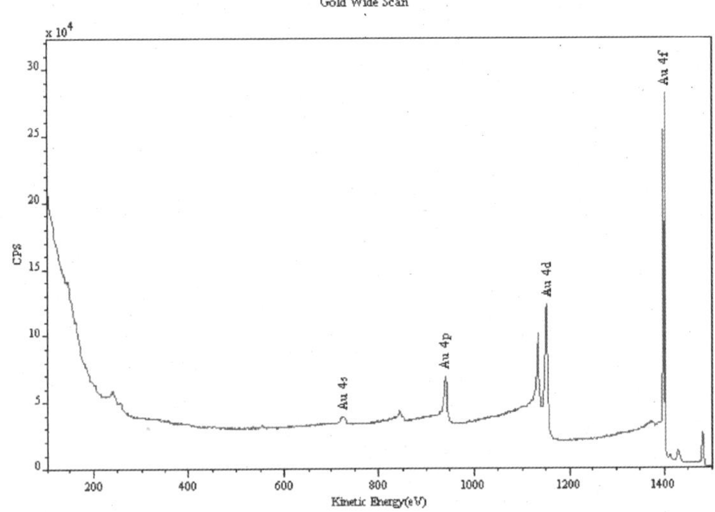

FIGURE 19•2 Clean gold sample spectrum - monochromatic Al x-ray source

Auger electrons are emitted with kinetic energies that are only dependent on the electronic state of the element responsible for the ejected electron. That is to say, unlike the photoelectric lines, changing the X-ray characteristic energy does not alter the position of the Auger lines in the recorded spectra with respect to a kinetic energy scale. Auger electrons are produced by complex multi-stage processes which include radiative de-excitation, auto-ionization, and other mechanisms. The collision processes involved with XPS result in electronic excited states with energies that lie above the ionization threshold. Some of these states are meta-stable with respect to radiative transitions within the neutral atom, but may couple with the continuum states of the ionic form to produce a radiation-less transition where the excess energy is transferred to the emitted electron. For this reason an electron spectrum may include Auger features that, when viewed using a binding energy scale, appear to move as a function of the X-ray source energy. In reality the Auger peaks always appear at the same *kinetic* energy and it is the photoelectric lines that "move" when a different excitation source is used.

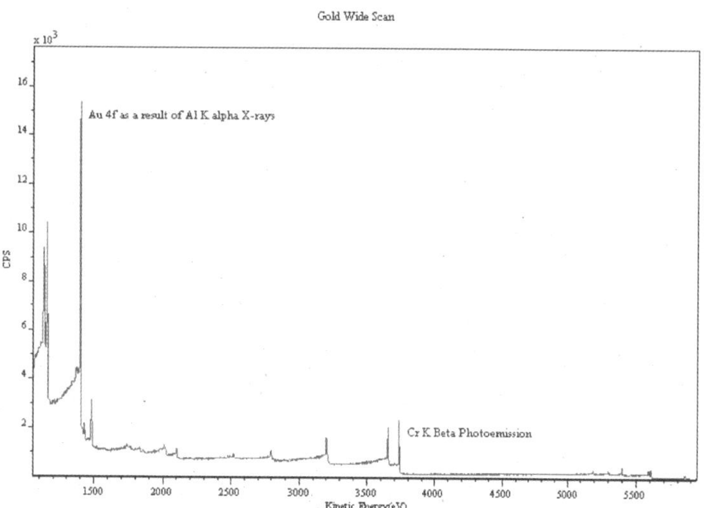

FIGURE 19•3 X-rays from both Cr and Al targets may be monochromated using quartz crystals. Photoelectric lines due to excitation from Cr Kβ and Al Kα X-ray sources are visible in this gold spectrum.

Although binding energy is the natural scale to use from a chemical perspective, the kinetic energy of an electron is more significant from an instrumental point of view - kinetic energy is usually the "independent variable" (the scanned parameter) for most instruments. Instrumental transmission characteristics change with kinetic energy and these variations can be quite dramatic. The spectrum in Figure 19•1 on page 321 has been recorded using a monochromatic silver X-ray source with photon energy about 2984.2eV The intensities of the 4f, 4d, 4p and 4s photoelectric lines are very different from the same peaks measured by the same instrument with an aluminium X-ray source (1486.6eV Figure 19•2). The change in transmission efficiency of the analyzer for larger kinetic energies accounts for part of the change in intensity, but the major cause of change in relative size of these lines when excited by the different photon energies is the difference in photo-ionization cross-sections for these transitions, which also contributes to the overall reduction in signal for the silver X-ray induced spectrum compared to the aluminium Kα. Note that in both Figure 19•1 on page 321 and Figure 19•2 on page 322 the abscissa is kinetic energy to emphasize the different energy of the electrons form these same states, but excited with different photon energies.

Electrons generated either by the photoelectric effect or by the mechanisms elucidated by Pierre Auger, leave the surface with a characteristic energy provided they have not undergone some energy loss process. The probability of these electrons emerging from the surface without some energy loss is related to the inelastic mean-free-path, which is in turn a function of the kinetic energy of the ejected electrons. The sampling depth for a given photoelectric line recorded with different X-ray anodes is therefore different, while the sampling depth for the same Auger line is unaffected by similar considerations.

Photoelectric lines for a given element and given X-ray anode are governed by the same sampling depth variation as a function of the ejected electron energies, that is lines with higher kinetic energy (lower binding energy, in general) will sample deeper. Variations in the relative intensities for lines from a given element may exist due to differences in the depths sampled by the individual lines.

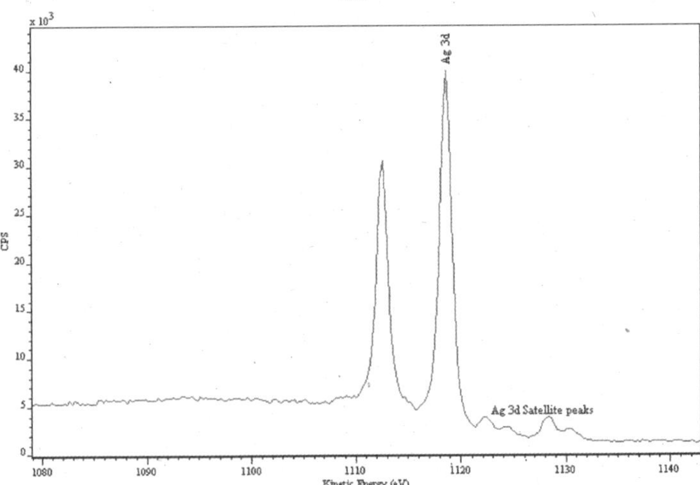

FIGURE 19•4 Satellite lines appear with higher kinetic energy values than the dominant Al K alpha induced photoelectric line

The fundamental differences between photoelectric and Auger lines, means that a spectrum cannot properly be understood without both the energy range over which the intensities were recorded and the characteristic energy of the excitation source. Unfortunately it is not uncommon for acquisition systems to omit information about the source under the (limiting) assumption that binding energy will be sufficient for XPS spectra.

Satellite Peaks

The spectra shown in Figure 19•1 on page 321 and Figure 19•2 on page 322 are recorded using monochromatic - i.e. essentially monoenergetic - X-ray sources: this is becoming more and more the "norm" for routine analysis. In an X-ray source, accelerating electrons onto the target (anode) material produces a range of X-ray energies characteristic of that material. In XPS analysis anode materials, usually magnesium or aluminium, are chosen because of a dominant, strong resonance in its X-ray spectrum, but for these (and all) elements other X-ray resonances are also excited which contribute to the complexity of a recorded spectrum by "repeating" features at low intensity.

> The X-radiation used to generate a photoelectron spectrum is usually produced in a Henke type multiple target source. The X-ray emission spectrum of a target coating in such a source exhibits a series of characteristic lines including, in addition to the main transition, a series of minor components at higher energies. In a magnesium X-ray source for example, the main $MgK\alpha_{1,2}$ emission at 1254eV is a close doublet resulting from the $2p_{3/2,\,1/2} \rightarrow 1s$ transition. The magnesium $K\alpha_3$ and $K\alpha_4$ "satellite" lines arise from less favoured two hole (KL) ionisations and the corresponding transitions. Similarly the $K\alpha_5$ and $K\alpha_6$ lines derive from triple hole (KLL) states. The $K\beta$ line comes from a valence band to K shell transition. All of these lines excite corresponding photopeaks in an irradiated sample, and in the quantitative analysis of materials containing low concentration components, an ability to distinguish satellite peaks, and to resolve them, for example by a peak fitting procedure, can be crucial.

In an unusually helpful quirk of nature, X-rays with wavelengths related by integral multiples of the aluminium $K\alpha$ X-ray line can be selected via Bragg diffraction using a quartz crystal to produce so called monochromatic X-ray sources. One of the advantages in using diffracted (monochromatic) X-rays is that the distribution of the photon energies selected is narrow compared to the "natural" X-ray line, which improves the resolution of the photoelectric peaks in the XPS spectrum. A further consequence of monochromating X-rays prior to irradiating the sample is that the minor resonance lines (X-ray satellites) in the X-ray spectrum are removed from the irradiating beam. X-ray satellite lines in an XPS spectrum, which appear at kinetic energies characteristic of the energy separation between the primary X-ray line and the minor transition, (Figure 19•4, and in more detail, Figure 19•5) actually have the same binding energy as the primary line, but without numerically altering the spectrum, this fact is difficult to accommodate using a binding energy scale for their display.

It is important to distinguish X-ray satellites - usually unwanted - from "intrinsic satellites" caused by "shake-up", "shake-off" and similar processes, which can provide

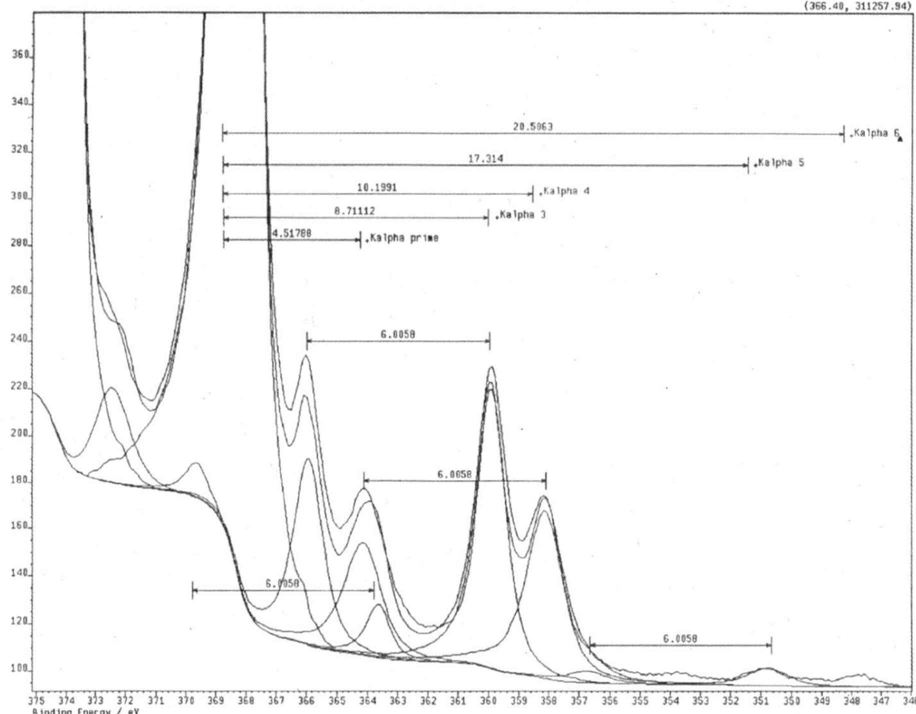

FIGURE 19•5 The complexity of magnesium satellite structure for a silver spectrum (see also "X-Ray Energies" on page 344)

useful structural (and analytical) information (see following topic). Note also that although the overall intensity of a monochromatic spectrum is in general lower than that obtainable from a normal non monochromatic multi-anode source, (because of the diffraction process) the background is also lower, which can improve detection limit. Unmonochromated X-ray sources also produce photons at energies up to the incident electron beam energy - a low level X-ray continuum called bremsstrahlung (Ger. "retarding radiation") and these X-rays excite electronic states out of the reach of the monochromatic source. Bremsstrahlung thus can have positive effects for the analyst by exciting relatively high energy Auger transitions characteristic of the sample. Electrons generated by highly energetic X-rays undergo inelastic scattering within the surface of the sample and appear as a background at energies within the measurement range. Background shape and intensity can be analytically useful also. "Extraneous" X-rays from multi-anode sources were useful in early instruments, where modern methods of charge control were not well developed, since they produce a flux of low energy electrons (from irradiation of X-ray windows and surrounding material) which helped to maintain charge balance in the sample under analysis. A final set of spectrum features sometimes referred to a "satellites" refers to the undesirable phenomena generated by worn targets, where "copper break-though" (exposure and irradiation of an underlying layer in a target) can give rise to significant "ghost" peaks and also "cross-over" (cross-talk) where a misaligned filament (for example) in a multi-anode source can produce X-rays from more than one target at once, but these artefacts should be rare in a well maintained instrument.

Backgrounds and Plasmon Peaks

The X-rays penetrate the surface to depths that exceed those surface layers responsible for photoelectric and Auger peaks in an XPS spectrum. Electrons emitted within the sample may undergo inelastic collisions thus altering the energy of the electron recorded by the detection system. These energy loss processes result in a background of counts that derive from electronic states other than the characteristic energies for the photoelectric lines, but moreover the shape of the background takes on a character determined by the probability distribution for electrons with a given kinetic energy undergoing some modification to their initial value. If this probability distribution also exhibits a resonance type structure, the convolution of a photoelectric peak with this bias in the energy loss distribution may give rise to peak-like structures that may be considered as part of the background. That is to say, peaks may appear in XPS spectra that are actually extrinsic in nature rather than intrinsic to the photo-excitation mechanism.

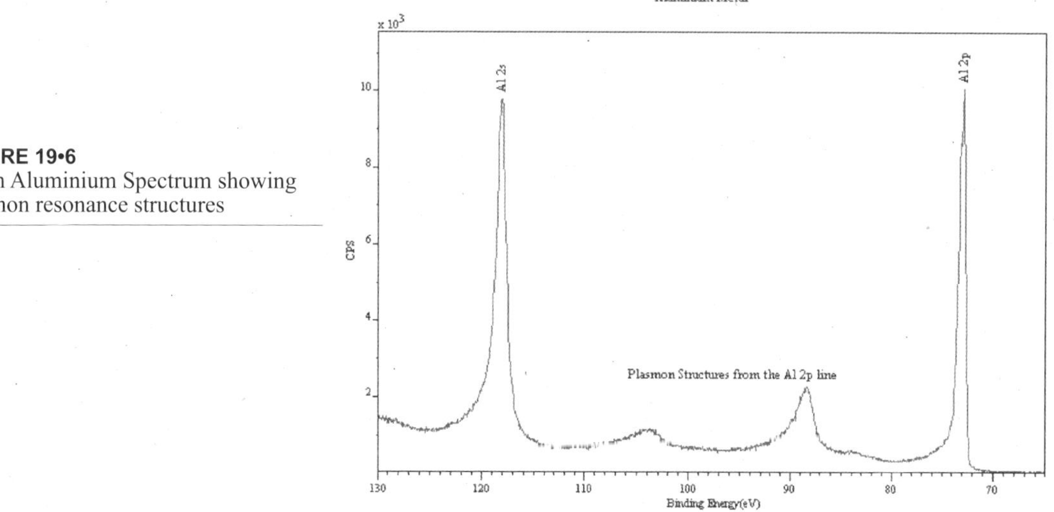

FIGURE 19•6
Clean Aluminium Spectrum showing plasmon resonance structures

The gold spectrum in Figure 19•2 on page 322 is an example of a material where the energy loss distribution is relatively broad and so the background rises to the lower kinetic energy side of the primary peaks positions, but without a pronounced energy loss shape. This is in contrast to the spectrum shown in Figure 19•6 where a spectrum from a clean aluminium sample includes peak shapes between the Al 2s and Al 2p photoelectric lines. These plasmon peaks are associated with energy loss events for electrons from the Al 2p line. In the case of aluminium the free electrons are constrained to move within energy bands that are characteristic of the material and these material properties influence the shape of the energy loss distribution, namely, scattering of the photoelectric electrons by free electrons with discrete energy bands produces energy loss distributions with relatively narrow structures. The convolution of an intense photoelectric peak with a relatively narrow energy loss distribution results in plasmon structures that could easily be mistaken for a primary line.

A further source for structure in an XPS spectrum is that of two intrinsic energy loss processes. So called "shake-up" peaks appear at energies characteristic of the excited states for an element with respect to the state measured by the zero loss intensity. Some of the photon energy is used to excite the ion out of the zero loss state whilst at the same instant ejecting the photoelectron with the remaining photon energy. Similarly, "shake-off" events, where more than one electron is ejected at the time of photo-ionization, may result in broad structures to high binding energies in a spectrum.

Spikes

Any measurement process requires stability from the system used to measure the signal. Unfortunately, instruments that rely on high voltages will occasionally suffer an unwelcome event ("flashover", voltage breakdown) during an acquisition. The consequence to an XPS spectrum of such an event is either a loss of signal or spurious counts entering the data. The latter appears as a narrow peak on the spectrum and in the case of multiple detector systems a series of spikes appear, arranged in accordance with the geometry of the analyzer and detectors.

XPS Instrumentation - an overview

Modern XPS instruments combine technology from a range of fields including electron optics, X-ray optics, digital/analog electronics, mechanical engineering and software engineering. The complex mixture of these disciplines is challenging to manage at the best of times but the ever-increasing demands of a maturing technique coupled with the need to satisfy both research and application-orientated customers (from a comparatively small market) means that state-of-the-art instruments must come at a relatively high price. It also requires users to be relatively sophisticated in order to exploit the technique to its full potential since they too must appreciate many of the details wrapped up in the systems. With this in mind, exploring the features of a typical system helps to gain some insight into how measurements are performed and what the results mean.

FIGURE 19•7
A commercial XPS research instrument (Specs GmbH) incorporating extensive sample handling, a monochromator, and imaging system

The logical components of an XPS instrument are shown in Figure 19•8 on page 330. X-rays illuminate an area of a sample causing electrons to be ejected with a range of energies and directions. The electron optics, which may be a set of electrostatic and/or magnetic lens assemblies, collects that proportion of the emitted electrons defined by the system geometry and its apertures and focuses them onto the analyzer entrance slit. Electrostatic fields are established within the energy analyzer (the instrument in Figure 19•7 employs a classical hemispherical analyser (HSA) - other arrangements are possible) to allow electrons only of a given energy (the so called Pass Energy - PE) to arrive at the detector slits and travel on into the detectors themselves.

Electrons of a specific initial kinetic energy are measured by setting voltages for the lens system that both focuses electrons of the required initial energy onto the analyzer entrance slit and retards their velocity so that their kinetic energy after passing through the transfer lenses matches the pass energy of the analyzer. To record a spectrum over a range of initial excitation energies it is necessary to scan the voltages applied to these transfer lenses and the prescription for these lens voltages is known as the set of lens functions. These lens functions are typically stored in some configuration file used by the acquisition system.

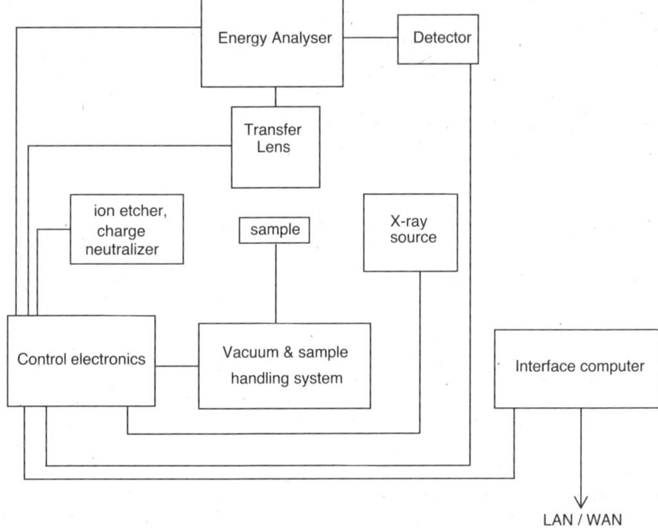

FIGURE 19•8 The logical components of an XPS Instrument.

The efficiency with which electrons are sampled by a spectrometer is very dependent on these lens functions and without properly tuned data set for the functions the performance of an instrument can be severely impaired. Even with a well-tuned system the collection efficiency varies across the many operating modes and it is necessary to characterize an instrument using a corresponding transmission function for each of the lens modes and energy resolutions.

A hemispherical analyzer and transfer lenses can be operated in two modes, namely, Fixed Analyzer Transmission (FAT), also known as Constant Analyzer Energy (CAE), or Fix Retard Ratio (FRR) also known as Constant Retard Ratio (CRR). In FAT mode, the pass energy of the analyzer is held at a constant value and it is entirely the job of the transfer lens system to retard the given kinetic energy channel to the range accepted by the analyzer. Most XPS spectra are acquired using FAT. The alternative mode, FRR, scans the lens system but also adjusts the analyzer pass energy to maintain a constant value for the quantity "initial electron energy" / "analyzer PE". This mode is typically used for Auger spectra since the energy interval accepted by the detection system (i.e. peak width) increases with kinetic energy and recovers weak peaks at high kinetic energies while restricting the intense low energy

background that could do damage to the detection system.
Overall performance of an XPS spectrometer is thus a combination of design, engineering quality - the care and precision with which an instrument is assembled, and electronic skills which provide stable power and accurate field modelling. Needless to say, bringing all of this together in a competitive, commercial context is not an insignificant task.

Energy Resolution

A number of factors influence the energy resolution achieved within a spectrum. The diameter of the analyzer, its pass energy, slit widths, and (not least) the spread of energies in the X-ray source, all play a major role in determining the full width half maximum (FWHM) for a given photoelectric line (which in turn has its own "natural" linewidth). Sample dependent considerations are also important where localized charging may broaden lines regardless of the precision built into the instrument and thus effective charge neutralization is an important part of any system.

FIGURE 19•9 Monochromator / HSA geometry.

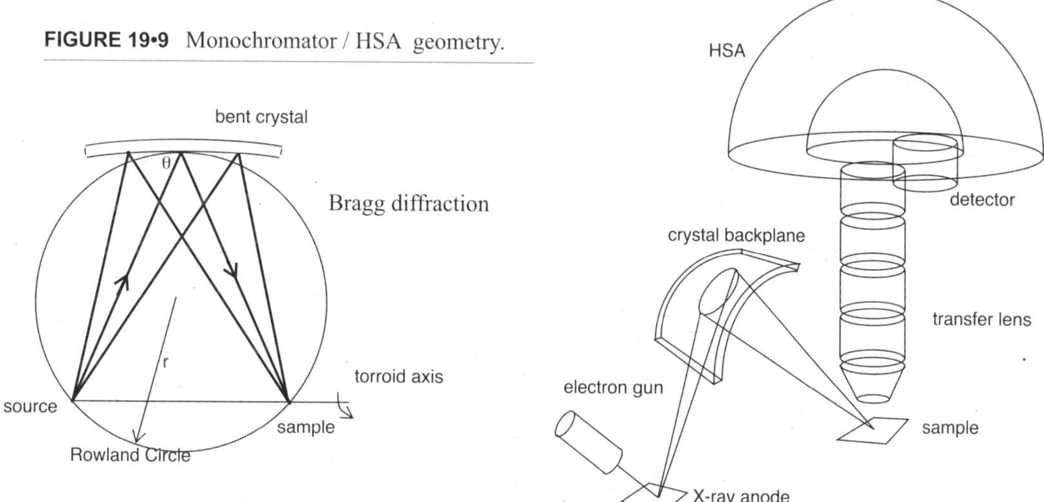

As mentioned above, a "normal" non monochromated X-ray source relies on narrow resonance transitions in the X-ray spectrum for the anode material and this limits the energy resolution possible for a photoelectric line excited by such a source. Monochromatic X-ray sources provide improved energy resolution by selecting a narrower band of X-rays using the diffraction properties of a quartz crystal (Figure 19•9) enabling certain wavelengths to be reinforced. Thin crystals can be shaped to "reflect" these wavelengths into a spot which can be directed at a sample from distance and also, with careful design of the electron gun generating the X-rays and the crystal shape (focusing properties), can be rastered over the sample (as in the PHI Quantum and Quantera instruments). Natural quartz has the right atomic lattice dimensions to select a band of energy (set of wavelengths) some 0.26eV wide from the "normal" X-ray emission of aluminium at 1486.6eV.

The fact that multiples of this diffracted wavelength are also reinforced by a quartz crystal means that other anode materials are possible sources for monochromatic X-rays, such as silver and chromium. Commercial X-ray monochomators have been offered using both two (Ag Lα, Kratos, VG) and four (Cr Kβ, Scienta) times the wavelength (energy) of aluminium X-rays.

There has been a recent tendency (regrettable, in this author's view) to designate classical, non monochromatic, dual anode XPS as "achromatic". Whilst undoubtedly there is a naming difficulty for this radiation, and this is a less cumbersome term, it does not seem appropriate to describe a strongly "coloured" source thus. Perhaps it is just a current fashion, and will, given time, disappear. [2]

Small Area Analysis

It has been said that the (surface) analyst has only three questions: "what is it?" "where is it?" and, "how much of it is there?" Given a suitable sample, XPS can make a pretty good stab at the first and third of these, but what of the second? The "holy grail" - production of surface chemical maps on the (sub-) micron scale for any material, conducting or otherwise - has been sought since the realization that fine focused electron beam (Auger) techniques could not really offer all the information chemists required. In most XPS instruments, electrons are dispersed through a hemispherical analyzer so that different energy electrons arrive at different positions in the radial direction, but they are also *spatially* dispersed (imaged) around the circumference of the sphere. This relationship was first exploited by the Ebel group in Vienna and later, commercially, by Scienta in their ESCA300, where images can be recorded that show both energy dispersion and spatial information in the form of sets of line-scans.

Before people started delving into the internals of XPS analysers, though, "selected area" analyses had been attempted (and achieved) by much simpler methods, including isolating physically small specimens in space (so that collected photoelectrons could only come from a small area: appropriate for wires and fibres) and using a physical collimator (a tube) in front of the first lens. Count rates (sensitivity) for these methods were, needless to say, limited. Efforts to improve both sensitivity and

2. Like all fashions, something must replace it: perhaps an acronym? NMX for Non-Monochromatic XPS? After all there is nmr...

instrument capability in general led to development of the so-called 'multimode' transfer lens, origninally designed to maintain high Auger count rates from the small illuminated area. It soon became clear that the system could be used as a "virtual probe" accepting signal from a small part of a widely illuminated area, and with the practical development of scanning techniques for these lenses (as foreseen in early patents) the possibilities of detailed surface mapping began to flower and move towards "real" imaging. Some of these early developments are set out below

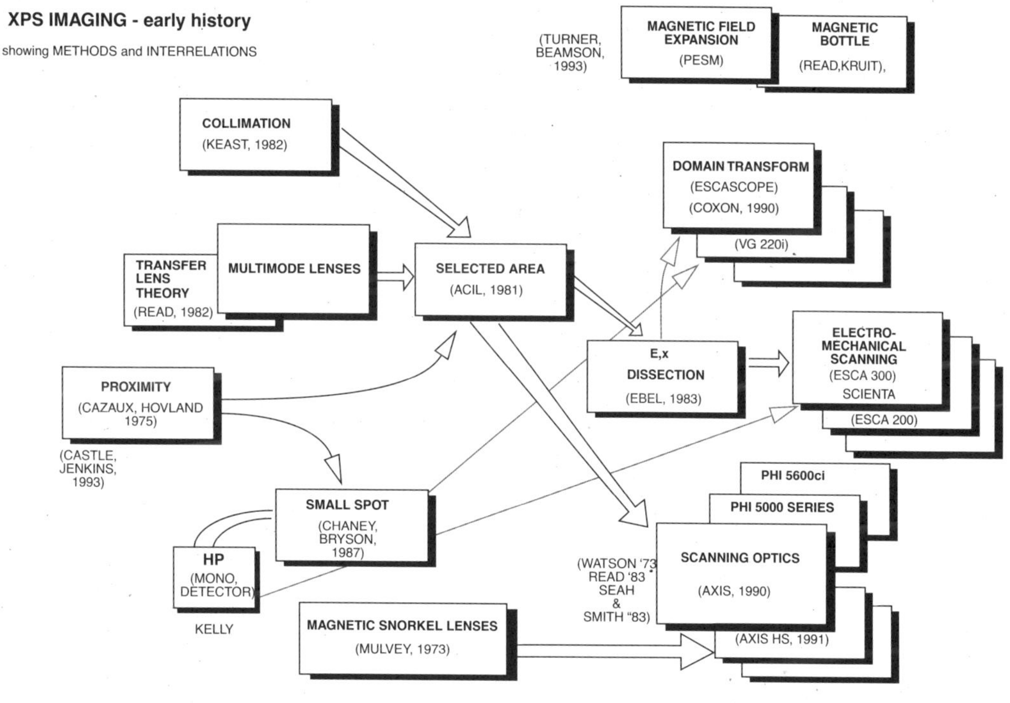

Moving from "spatially keyed spectroscopy"(small area analysis) to fully-fledged (parallel acquisition) chemical imaging in a practical context requires application of sophisticated optics and electronics, none of which comes cheap.

Imaging systems

The earliest instrument to provide parallel XPS images was the "ESCASCOPE" from VG. This (and later variants) used additional lenses at the entrance slit and be-

fore an imaging detector, which perform what was essentially a Fourier transform and the inverse operation to allow an energy resolved stigmatic image to be recorded through the hemispherical analyzer whilst operating in deflection mode. Although count rates in practice were somewhat limited because of the complexity of the system, it had the advantage of being a truly parallel imaging arrangement.
Development of the "serial" (probe) type instrument continued. Spatially resolved images can be recorded by restricting the field of view for the transfer lens system using a combination of magnetic and electrostatic lenses plus apertures, and then scanning the field of view using deflection plates. This technique has been used by Kratos Analytical in their Axis HS/Axis165 range and has since been enhanced by use of an additional hemispherical analyzer operating as a mirror analyzer (SMA) rather than in deflection mode. Stigmatic energy-resolved images are acquired through the mirror hemispherical analyzer which offers a fast XPS image acquisition mode.
An alternative means of acquiring XPS images is to restrict the X-ray source and raster the small spot of X-rays over the sample. This probe-oriented method is used by Physical Electronics (PHI) in their Quantum range of instruments, where the monochromator is used to produce a small spot of X-rays. The X-rays are rastered over the sample by scanning a focused electron gun across the X-ray anode material whilst compensating for any changes in photon energy through the control system.
XPS images provide the means for assessing a sample, but quantitative analysis is performed using spectra, even if these spectra derive from stacks of images. Small area analysis using direct spectroscopy acquisition modes requires the use of imaging techniques applied to a single pixel.

Matching the type of imaging for a sample can be an important factor in the choice of system. Many samples degrade under exposure to X-rays and there are costs and benefits associated with both field-of-view verses probe oriented methods. The probe method allows X-rays to be pointed at a specific location and therefore damage to the surrounding sample surface is reduced. On the other hand, stigmatic imaging involves irradiating a larger area but parallel sets of spectra are possible and charge compensation has to operate in a wider, global context.
The "spectra-from-images" approach enabled by the sorts of tool provided within CasaXPS promises to provide a suitable compromise applicable to many situations.

XPS images can contain reliably interpretable, quantifiable chemical information, and the drive to provide this information for larger and larger samples in finer and finer detail seems set to continue, alongside the improvement of data processing techniques.

Trends

XPS is a mature analytical technique with a great stock of practical and theoretical expertise available both in the academic literature and commercially. The technique is useful to those who have little or no interest in anything other than its ability as a practical problem solver and thus we see a trend (in commercial offerings) towards separation of the design of research and routine machines: on the one hand highly automated, throughput orientated systems for industrial contexts, as against more flexible, often "multitechnique", instruments for more detailed work. The PHI Quantum / Quantera series and Kratos Axis NOVA typify the modern industrial tool.

FIGURE 19•10 Kratos Axis NOVA

As a natural extension of the automation of routine instruments, some considerable effort has been expended on work towards "expert systems" with capabilities for automated analytical interpretation as well as automated instrument set up and optimisation. None of this has yet bourne any real (commercial) fruit, but it seems only a matter of time before it does. For the fundamentals of the technique, new sources and new analyzers are continually sought after, with the greater probability of success (given the difficulties in producing an X-ray laser) resting with the latter. Time of flight (ToF) energy analysers, as in other spectroscopies, appear to offer some promise, but for the moment, it seems that our XPS machines will remain as rather large, fairly expensive, quite complicated beasts, requiring all the care and devotion that a skilled analyst can provide.

For further details on XPS instrumentation see the chapter by I.W. Drummond in *"Surface Analysis by Auger and X-ray Photoelectron Spectroscopy"(D. Briggs and J.T Grant (Eds), (ISBN: 1 901019 04 7) IM Publications & Surface Spectra Limited (2003)* page 117.
This volume also contains chapters on XPS imaging (by Kateryna Artyushkova and Julia Fulghum: page 677) and on AES instrumentation (Masato Kudo, page 145), amongst many other topics.

Section 5 - Appendices

Appendices

Appendix 1: ISO 14976 format files annotated

1. Simple single block XPS (region) spectrum

Item Description	Line	ISO 14976 Format Item (in PET-C1s.vms file)
format identifier	(1)	VAMAS Surface Chemical Analysis Standard Data Transfer Format 1988 May 4
institution identifier	(2)	Acolyte
instrument model i.d.	(3)	Kratos AXIS-HS
operator i.d.	(4)	AC
experiment i.d.	(5)	PET C1s Test Spectrum
# comment lines	(6)	1
(comment lines)	(7)	Acolyte copyright 1999
experiment mode	(8)	NORM
scan mode	(9)	REGULAR
# regions	(10)	1
# experiment variables	(11)	0
# parameter exclusion entries	(12)	0
# manual items	(13)	0
# future experiment items	(14)	0
# future block entries	(15)	0
# blocks	(16)	1

block i.d.	(17)	C1s 1
sample i.d.	(18)	PET
year (4 digits)	(19)	1999
month	(20)	06
day	(21)	24
hour	(22)	14
minute	(23)	45
second	(24)	00
# hours + GMT	(25)	1
# lines in block comment	(26)	0
technique	(27)	XPS
source	(28)	Al
source energy	(29)	1486.6
source strength	(30)	300
source width x, (mu)	(31)	500
source width y, (mu)	(32)	500
source polar angle	(33)	54
source azimuth	(34)	180
analyser mode	(35)	FAT
analyser resolution characteristic	(36)	20
magnification of analyser	(37)	1
analyser work function	(38)	4.5
target (sample) bias	(39)	0
analysis width x, (mu)	(40)	200
analysis width y, (mu)	(41)	200
analyser axis polar angle	(42)	0
analyser azimuth	(43)	180
species	(44)	C
transition state	(45)	1s
charge of analysed particle	(46)	-1
abscissa label	(47)	kinetic energy
abscissa units	(48)	eV
abscissa start	(49)	1191.6
abscissa increment	(50)	0.05
# corresponding variables	(51)	1
corresponding variable label	(52)	counts per channel
corresponding variable units (d= none)	(53)	d
signal mode	(54)	pulse counting
signal collection time/channel	(55)	0.5
# scans for this block	(56)	1
signal time correction	(57)	400E-9
sample normal tilt	(58)	0
sample normal azimuth	(59)	0
sample rotation angle	(60)	0

# additional params	(61)	0
# ordinate values	(62)	401
min. y	(63)	42
max. y	(64)	2250
(first data point)	(65)	179
(399 data points)	( )	( )
(last data point)	(465)	59
terminator	(466)	end of experiment

2. Multiple block depth profile composed of XPS (wide scan) spectra

Note: the line numbers for the "experiment header" and "block 1" have been deliberately left the same as in example 1, so that differences stand out clearly. Extra lines (variables and values) in this example are denoted with a + sign.

Item Description	**Line**	**ISO 14976 Format Item** (in DATA2.vms file)
format identifier	(1)	VAMAS Surface Chemical Analysis Standard Data Transfer Format 1988 May 4
institution identifier	(2)	Not Specified
instrument model i.d.	(3)	AXIS-165 (SAC)
operator i.d.	(4)	kratos (VISION User)
experiment i.d.	(5)	/export/home/kratos/data/2043it1.dset
# comment lines	(6)	0
experiment mode	(8)	NORM
scan mode	(9)	REGULAR
# regions	(10)	1
# experiment variables	(11)	1
experiment variable label	+	Etch Time
experiment variable units	+	s
# parameter exclusion entries	(12)	0
# manual items	(13)	0
# future experiment items	(14)	0
# future block entries	(15)	0
# blocks	(16)	117
block i.d.	(17)	OSnIn a a/2
sample i.d.	(18)	OSnIn/2
year (4 digits)	(19)	1999
month	(20)	12
day	(21)	22
hour	(22)	17
minute	(23)	40
second	(24)	2

# hours + GMT	(25)	0
# lines in block comment	(26)	3
"comment" lines used internally	+	Lens Mode: Magnetic Resolution: Pass energy 80 Anode: Al(180 W)
by Kratos Vision system	+	Step(meV): 300.0 Dwell(ms): 136 Sweeps: 1
CasaXPS adds "history" here	+	Acquired On: Tue Dec 22 17:40:02 1998
technique	(27)	XPS
experiment variable value	+	0
source	(28)	Al
source energy	(29)	1486.6
source strength	(30)	180
source width x, (mu)	(31)	1E+37
source width y, (mu)	(32)	1E+37
source polar angle	(33)	1E+37
source azimuth	(34)	1E+37
analyser mode	(35)	FAT
analyser resolution characteristic	(36)	80
magnification of analyser	(37)	1E+37
analyser work function	(38)	1E+37
target (sample) bias	(39)	0
analysis width x, (mu)	(40)	1E+37
analysis width y, (mu)	(41)	1E+37
analyser axis polar angle	(42)	1E+37
analyser azimuth	(43)	1E+37
species	(44)	Wide
transition state	(45)	none
charge of analysed particle	(46)	-1
abscissa label	(47)	kinetic energy
abscissa units	(48)	eV
abscissa start	(49)	926.7
abscissa increment	(50)	0.3
# corresponding variables	(51)	2
corresponding variable label	(52)	Intensity
corresponding variable units (d= none)	(53)	d
	+	Transmission
	+	d
signal mode	(54)	pulse counting
signal collection time/channel	(55)	0.136
# scans for this block	(56)	1
signal time correction	(57)	0
sample normal tilt	(58)	1E+37
sample normal azimuth	(59)	1E+37
sample rotation angle	(60)	1E+37
# additional params	(61)	0
# ordinate values	(62)	878
min. y	(63)	427

max y	(64)	15744
(first data point)	(65)	2967
(first transmission value)	+	1
(second data point)	(66)	2954
(second transmission value)	+	1
(875 total data & transmission points)	(+)	( )
(last data point)	(940)	458
(last transmission value)	(941)	1
block i.d.	(942)	OSnIn a a/6
sample i.d.	(943)	OSnIn/6
year (4 digits)	(944)	1999
month	(945)	12
day	(946)	22
hour	(947)	17
minute	(948)	43
second	(949)	27
# hours + GMT	(950)	0
# lines in block comment	(951)	3
	+	Lens Mode: Magnetic Resolution: Pass energy 80 Anode: Al(180 W)
	+	Step(meV): 300.0 Dwell(ms): 136 Sweeps: 1
	+	Acquired On:Tue Dec 22 17:43:27 1998
technique	(955)	XPS
experiment variable value	+	30
source	(957)	Al
source energy	(958)	1486.6
source strength	(959)	180

.

and so on, as above for this block (block 2)

.

(last data point)	(1884)	504
(last transmission value)	(1885)	1
block i.d.	(1886)	OSnIn a a/11
sample i.d.	(1887)	OSnIn/11

.

and so on, as above for this block (block 3)

.

and also for another 114 blocks,

.

.

until

.

(last data point)	(15894)	504
(last transmission value)	(15895)	1
terminator	(15896)	end of experiment

Appendix 2: ISO 14976 and the World Wide Web

The ISO 14976 file format is designed for simple, unambiguous communication of information derived from surface chemical analysis experiments. The World Wide Web is one of the most important communications links for experimentalists, and so it is appropriate that means are provided for transport and display of ISO data over the web. Standard methods of text file compression and transfer work well with the format because it is text based. Display of spectra in a web environment - on a browser page for example, is a (non trivial) matter which is of interest in many contexts, and there exists (at least one) simple Java applet which enables display of XPS spectra "live" on a web page. The applet is also designed to provide a basic practical mechanism for "platform independent" (that is "universal") transfer and display of XPS spectra across the world wide web.

See, for example, "http://www.npl.co.uk/nanoanalysis/eviewerindex.html"
and "http://www.acolyte.co.uk/JISO"

"MIME" type. MIME types (Multimedia Internet Mail Extensions) are the designators which enable applications such as web browsers and their "helpers" to decide how to handle (interpret) an otherwise "unknown" file type. The ".vms" file extension is used for the ISO 14976 spectra and is in process of submission through the IETF (Internet Engineering Task Force - part of the World Wide Web Consortium - W3C) as a "Chemical MIME" type. The present status is "experimental" (i.e. the type/sub-type) "chemical/x-vamas-iso14976" should be set. Since the format is solely text based, however, http systems (servers and browsers communicating with them) will in general react correctly if configured to serve (or read) unknown MIME types as "text/plain". In most cases this will be the default action

The .vms file extension. Within CasaXPS, experiment files are stored with the ".vms" extension (file type) and this should be maintained for data exported from the system.

Appendix 3: Definitions and Formulae

Common peak shape models

Gaussian/Lorentzian Product Form

$$GL(x, F, E, m) = \frac{\exp\left((-4\ln 2)(1-m)\frac{(x-E)^2}{F^2}\right)}{1 + 4m\frac{(x-E)^2}{F^2}}$$

Gaussian/Lorentzian Sum Form

$$SGL(x, F, E, m) = \exp\left(-4\ln 2\frac{(x-E)^2}{F^2}\right)(1-m) + \frac{m}{1 + 4\frac{(x-E)^2}{F^2}}$$

Doniach Sunjic:

$$DS(x, \alpha, F, E) = \frac{\cos\left(\frac{\pi\alpha}{2} + (1-\alpha)\operatorname{atan}\left(\frac{x-E}{F}\right)\right)}{(F^2 + (x-E)^2)^{(1-\alpha)/2}}$$

X-Ray Energies

TABLE 20•1. Anodes used in XPS

X-rays	Energy,eV	Width,eV
Y Mζ	132.3	0.47
Zr Mζ	151.4	0.77
C Kα	277	
Cu Lα	929.7	
Mg $K\alpha_{1,2}$	1253.6	0.7
Al $K\alpha_{1,2}$	1486.6	0.85
Si $K\alpha_{1,2}$	1739.5	1.2
Zr Lα	2042.4	
Ag Lα	2984	
Ti $K\alpha_1$	4511	1.4
Cr $K\alpha_1$	5414.7	1.8
Cu $K\alpha_1$	8048	2.5

TABLE 20•2. High Energy Satellite Lines from Mg and Al Anodes

X-Ray Line	Separation from $K\alpha_{1,2}$ (eV) and relative intensity (%)			
	Mg		Al	
$K\alpha_{1,2}$	0.0	100%	0.0	100%
Kα'	4.5	1.0%	5.6	1.0%
$K\alpha_3$	8.4	9.2%	9.6	7.8%
$K\alpha_4$	10.0	5.1%	11.5	3. 3%
$K\alpha_5$	17.3	0.8%	19.8	0.4%
$K\alpha_6$	20.5	0. 5%	23 .4	0. 3%
Kβ	48.0	2.0%	70.0	2.0%

The above values taken from M.O. Kause and J.G. Ferreira, *J. Phys., B*, **8** (1975) 2007.
Note that peak widths vary and intensity values change with x-ray anode voltage
Determinations by A. Carrick & K.S. Robinson, Proc. ECASIA, 1993 (Catania, Italy) gave, for magnesium, the values shown in Table 20•3 on page 345

TABLE 20•3. Magnesium radiation satellites (14kV x-rays)

Line	Position*	Height†	FWHM§
$K\alpha_{1,2}$	0	1 00%	1.00
$K\alpha'$	4.56eV	0.95%	1.29
$K\alpha_3$	8.39eV	7.30%	1.08
$K\alpha_4$	10.17eV	4.40%	1.43
$K\alpha_5$	17.26eV	0.50%	1.43
$K\alpha_6$	20.47eV	0.30%	1.40

*Position from $K\alpha_{1,2}$ fitted maximum. Uncertainty ±0.03eV, apart from Ka',where uncertainty is - 0.06, +0.03eV.
† Relative to $Ag3d_{5/2}$ at 100%, ±0.25%
§ Relative to $Ag3d_{5/2}$ at 0.84ev= 1

Glossary of terms

Block (ISO). A sub unit of an Experiment (file) (q.v.) which consists of a header - a set of parameters that only apply to that block - followed by a series of ordinate values which may represent a curve (e.g. a depth profile), a spectrum, or a map. Note that a block is **not** an invariant entity (it is *not* simply a spectrum or "region") and can contain a complete set of determinations (see Chapter 3, and also annotation, page 339)

Experiment (ISO). A complete ISO 14976 file which defines the context of and data acquired during a surface chemical analysis determination. Consists of a header containing of a series of parameters which apply to the measurement procedure as a whole, followed by one or more blocks (q.v) of data

VAMAS. the **V**ersailles project on **A**dvanced **M**aterials **A**nd **S**tandards

(see also ISO 18115:2001 (Vocabulary), and Seah, M.P. *Surf. Interface Anal.* 2001; 31: 1048–1049)

Appendix 4: References and other Resources

World wide web links:

CasaXPS	http://www.casaxps.com
UK Surface Analysis Forum	http://www.uksaf.org
Kratos Analytical	http://www.kratos.com
Physical Electronics	http://www.phi.com
Thermo-VG	http://www.thermo.com
SPECS	http://www.specs.de
American Vacuum Society (AVS)	http://www.vauum.org
ECASIA	http://www.ecasia.org
Institute of Physics(UK)	http://www.iop.org
Royal Society of Chemistry(UK)	http://www.rsc.org
American Chemical Society (ACS)	http://www.acs.org
American Institute of Physics (AIP)	http://www.aip.org
Materials Research Society (USA)	http://www.mrs.org/
IUVSTA	http://www.iuvsta.org
VAMAS	http://www.vamas.org/
ISO	http://www.iso.org

Bibliography :

(books you should borrow from your library)

"ESCA: Atomic, Molecular and Solid State Sructure by means of Electron Spectroscopy"
K.Siegbahn, C. Nordling, A. Fahlman, R. Nordberg, K. Hamrin, J. Hedman, G. Johansson, T. Bergmark, S-E. Karlson, I. Lindgren and B. Lindberg,
Almqvist and Wiksells, Uppsala, Sweden (1967)
The ESCA book - difficult to find now, but worth it.

"Electron Spectroscopy: Theory, Techniques and Applications"
C.R. Brundle and A.D. Baker, (Editors)
Volume 1 (ISBN 0-12-137801-2) Academic Press (1977)
A fasinating insight on early techniques and detailed theory

"Surface Analysis by Auger and X-ray Photoelectron Spectroscopy"
D. Briggs and J.T Grant (Editors),
(ISBN: 1 901019 04 7) IM Publications & Surface Spectra Limited (2003)
Big, heavy, expensive: the modern bible of surface analysis with lots of current references and useful data.

Appendix 5: Some Questions and Answers

1) Why does my Custom Quantification report contain different percentage concentrations....?

Short Answer

The names used for quantification items, i.e. integration regions and synthetic components, are important in custom reports. If the same name has been used for, say, an oxygen quantification region and also for each peak in the set of synthetic peaks used to fit the same data, then a custom report using this name will sum all these intensities for the oxygen. Thus the intensity for the oxygen will be counted twice under these circumstances. The solution is to name the region with a different name from the synthetic components, e.g. the region name could be "O 1s envelope" and then each component could be named "O 1s" (Figure 20•1). Now, the intensity used in the custom quantification report can be chosen to be either of these names, which achieved by selecting the appropriate name in the formula list (Figure 20•2).

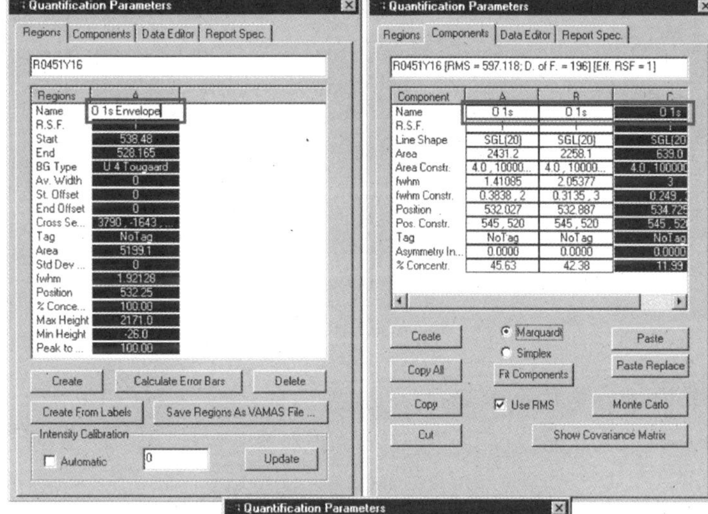

FIGURE 20•1. Custom reports reference intensities through the names given to the quantification items. Distinct names for regions and components stop the intensity for a fitted region being counted twice

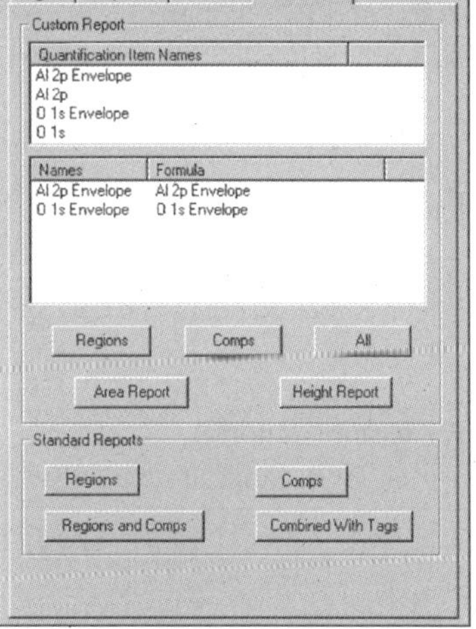

FIGURE 20•2. Custom Report where region and component items have been defined with different distinctive names.

Long Answer:

The standard reports treat each quantification item as a separate entity and therefore a percentage concentration is reported for each item used to quantify a sample. Custom reports, on the other-hand, sum the intensities for any quantification items defined with the same name. It is therefore important when using the Custom report option to label synthetic peaks with a different name from the region used to define the background for those peaks. Consider the following example:

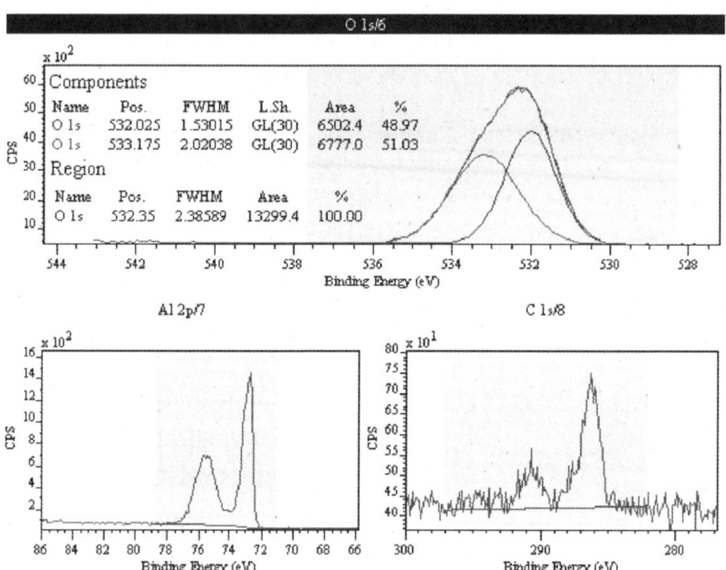

FIGURE 20•3. Three high-resolution spectra used to quantify a sample. The O 1s region is fitted using two synthetic components, whilst the C 1s and the Al 2p spectra are quantified using integration regions only.

Each spectrum has an integration region defined and the names for these regions are O 1s, Al 2p and C 1s. In addition to the integration regions, the O 1s spectrum is fitted using two synthetic components and both of these components are named using the same name as the integration region for the O 1s spectrum. A Standard report generated from these integration regions is as follows:

TABLE 20•4. Standard report generated from the quantification regions shown in Figure 20•3

Name	Energy	FWHM	R.S.F.	Area	% Conc.
O 1s	532.35	2.38589	0.78	13299.4	54.52
Al 2p	72.7	0.740902	0.193	2192.6	36.33
C 1s	286.3	1.39428	0.278	795.3	9.148

A custom report using the same names as those shown in Table 20•4 is as follows:

TABLE 20•5. Custom report using the same names as those used in Table 20•4.

O 1s CPSeV	Al 2p CPSeV	C 1s CPSeV	O 1s %	Al 2p %	C 1s %
34075.5	11360.8	2860.9	70.5538	23.5227	5.92353

Note that the O 1s percentage concentration is larger than the equivalent value reported by the Standard report. The problem lies in wrongly using identical names for more than one quantification item. The integration region for the O 1s spectrum has the same name as both the synthetic components and therefore the three quantities associated with the O 1s spectrum are summed to provide the intensity used in the custom report. This was not the intention. To prevent this type of error it is best to use a different name for the integration region from that used for the synthetic components. Figure 20•4 shows the same data where the synthetic peaks have been assigned a common name, which is different from the integration region. A Custom report generated from this newly named set of quantification items is shown in Table 20•6 and can be seen to be identical to the report in Table 20•4. .

TABLE 20•6. Custom report after the synthetic components have been renamed

O 1s CPSeV	Al 2p CPSeV	C 1s CPSeV	O 1s %	Al 2p %	C 1s %
17050.5	11360.8	2860.9	54.523	36.3287	9.14836

The value of the Custom report lies in the ability to combine intensities, such as those from the synthetic components used to model the O 1s region, with the intensities from the integration regions for the C 1s and the Al 2p regions. Table 20•7 has been calculated from a combination of the "Peak O 1s" intensities and the previously used Al 2p and C 1s regions. The real value comes when the O 1s region, say, has an interference peak which requires excluding from the final results by fitting multiple peaks to the data but only selecting a subset of these peaks for the purposes of the quantification report.

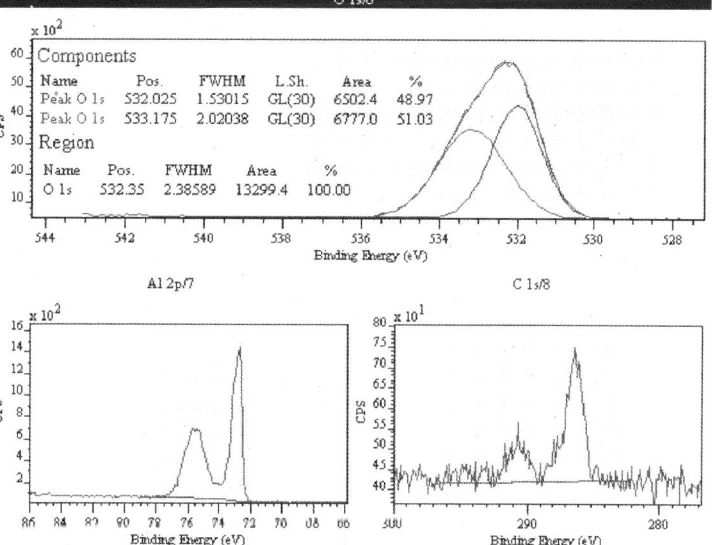

FIGURE 20•4. Synthetic peaks for the O 1s spectrum are now named differently from the integration region.(cf. Figure 20•3)

TABLE 20•7. **Custom report using the Synthetic Peaks for the O 1s intensity but the integration regions for the C 1s and Al 2p intensities.**

Peak O 1s CPSeV	Al 2p CPSeV	C 1s CPSeV	Peak O 1s %	Al 2p %	C 1s %
17024.9	11360.8	2860.9	54.4856	36.3585	9.15586

2) Why did the backgrounds to my quantification regions not propagate ...?

The background is calculated for a quantification region only when the start and end energies lie inside the range of energies recorded for a spectrum. The spectrum used to define the region(s) must be prepared so that the quantification region(s) is valid for all the spectra to which the region(s) is to be propagated. Spectra for which a background failed to be computed will require visual inspection and adjustment before backgrounds can be calculated. Alternatively the quantification region on the initial spectrum must be adjusted to accommodate all the spectra involved in the propagation procedure.

Note that even when data has been acquired over identical energy ranges, processing the spectra to compensate for charge correction may require adjustments to quantification region start and end points. Propagating a region can therefore fail even when the acquired data is all recorded using the same acquisition parameters.

3) Why does Print Preview lose font colours?

Print Preview changes the attributes of fonts to simulate the output on your printer. If your default printer is a black and white printer (or a FAX/Modem for example) then the Microsoft Print Preview mechanism may alter the font color to correspond to the output device. This unfortunately is not true for line colors, which unlike the fonts, are previewed using the original colors (usually) regardless of the printer type.

Solution:

Select a full color printer for your default printer

4) How do I overlay spectra from different files?

Spectra can only be displayed together when the data is in the same experiment-frame. If the data are found in different files then it will be necessary to copy the spectra for which an overlaid display is required into a new experiment frame as follows:

1. Load each file into CasaXPS.
2. Select the Window menu and choose the Tile menu option. This will arrange the maximized frames in an orderly array of windows within the CasaXPS program window.
3. Adjust the experiment frames so that the data browser window is displaying the data block names

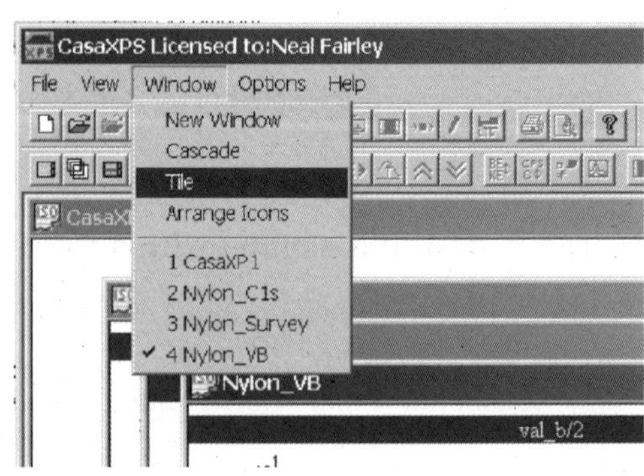

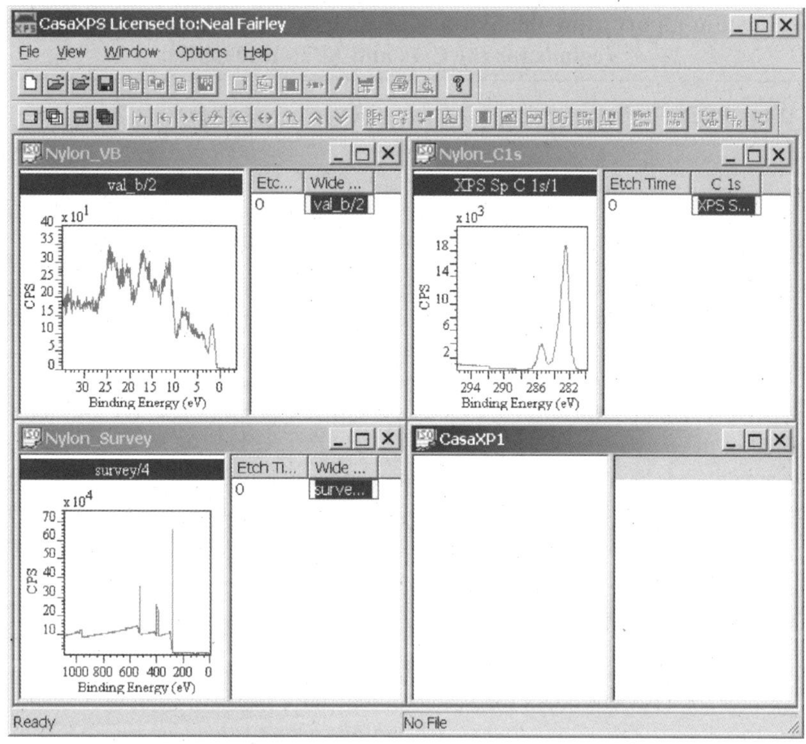

4. Select each data block name so that block name remains highlighted within each of the experiment frames. This is achieved by selecting the first block name using the left-hand mouse button, and then, whilst holding down the control key, select the additional block names. The Control key modifies the selection mechanism so that those blocks already selected remain selected when the left-hand mouse button is pressed subsequently.
5. Create an empty experiment frame by pressing the New toolbar button.
6. Move the cursor over the right-hand-side of the empty experiment frame and right-click the mouse. A dialog appears and the list of currently selected blocks confirms those block that will be copied into the empty frame when the OK button is pressed.

Once the spectra are located in the same experiment frame the standard overlay procedure may be used to view the data.

5) How do I fit a doublet with a fixed area ratio, e.g. Cr $2p_{3/2}$ and Cr $2p_{1/2}$?

Peaks with known relationships, either spin-orbit or stoichiometric, can be modeled using constraints applied to the peak parameters. Figure 20•5 shows a Cr 2p doublet, where the two peaks are governed by a 2:1 ratio. An area constraint of this nature is introduced by relating a parameter in one synthetic peak to the corresponding parameter in another peak using the alphabetical label on the Components property page shown in Figure 20•6 on page 352. The area in column B is constrained to be half that of the area in column A by entering the string A * 0.5 into the area constraint of column B. The FWHM and position for related peaks can be constrained is a similar fashion (see Figure 20•6).

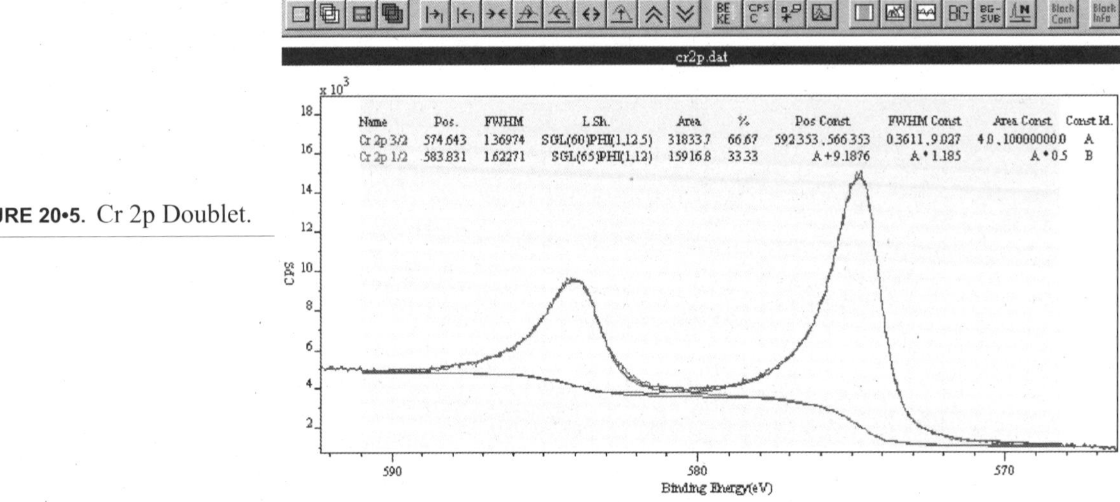

FIGURE 20•5. Cr 2p Doublet.

Notes:

- The line-shapes are constructed from a linear combination of Lorentzian and Gaussian profiles where the asymmetric adjustments to the SGL profile are defined using the PHI tail function PHI(scale,length).
- Constraining two peaks in this way is not just physically superior but helps to reduce computation time during optimization.
- The Component annotation table shown in Figure 20•5 includes the constraints as part of the quantification table. This is optional and can be removed from the table if desired.

Component	A	B
Name	Cr 2p 3/2	Cr 2p 1/2
R.S.F.	1	1
Line Shape	SGL(60)PHI(1,12.5)	SGL(65)PHI(1,12)
Area	31833.7	15916.8
Area Constr.	4.0 , 10000000.0	A * 0.5
fwhm	1.36974	1.62271
fwhm Constr.	0.3611 , 9.027	A * 1.185
Position	574.643	583.831
Pos. Constr.	592.353 , 566.353	A + 9.1876

FIGURE 20•6. A sections of Components Property Page showing the columns with constraints used to construct a rigid peak model for a Cr 2p doublet.

6) How do I automatically get the RSF values from the element library?

The Element Library contains a field for Relative Sensitivity Factors (RSF). There are a number of ways to access these values when an integration region or synthetic peak is created:

1. If the Element Library dialog window is active when a region or synthetic peak is created and an entry in the Element Library scrolled list has been previously selected, then any new region or synthetic peak will be created using the RSF from the selected entry in the Element Library.
2. If the Element Library is not active and the species/transition pair defined for the spectrum for which regions and/or synthetic peaks are required, match the Name field from an entry in the element library, then the RSF etc will be automatically loaded from the element library when a region or synthetic peak is created.
3. If both the Region (Component) property page is active at the same time as the Element Library Element Table property page, and the Create When Line Selected tick-box on the Element Table property page is ticked, then every time a line from the Element Table is selected a new region (component) is created.
4. Regions may be created form peak labels defined using the Annotation dialog window (see the Quantification of a Survey Spectrum). Regions created in this way will automatically retrieve the RSF values from the Element Library.

7) How can I quantify spectra without using a background?

For CasaXPS v2.0.33 or higher a background type of Zero will allow spectra to be quantified without a background.

Intensity measurements are made using the Regions Property Page on the Quantification Parameters dialog window. A peak intensity is calculated from the background subtracted data so if the required intensity is that of the peak without a background, the background type must be set to None and the Start Offset field on the Regions Property Page set to 100. The Start Offset defines a percentage offset from the intensity at which the background would otherwise be tied to the spectrum. An offset set to 100% will ensure that the background is set equal to zero at that end point and the background type of None results in a constant background set equal to the lowest point determined from the data and the end points. The result is therefore a peak quantified without a background.

8) What is the difference between the Chi Square and RMS Figure-of-Merit?

Given a spectral envelope $\{s_i\}$ and a synthetic model for the data $\{m_i\}$,
then the chi-square is defined to be

$$\chi^2 = \sum \frac{(s_i - m_i)^2}{|s_i|}$$

The RMS is defined to be

$$RMS = \sqrt{\sum (s_i - m_i)^2}$$

The chi-square is computed assuming that noise within the data can be approximated by the square root of the signal recorded in each channel (so-called Poissonian statistics). The consequence of using a chi-square goodness-of-fit metric is to reduce the influence of the peak maximum when determining the parameters for the synthetic model. Notably, for high resolution data the use of chi-square to optimize the model may broaden the resulting FWHM compared to the same model optimized using the RMS.

Different goodness-of-fit metrics offer alternative views from which an optimum for a model can be established, however neither statistic is a measure for how well a model itself relates to the true chemical states present within the data.

9) How do I use the Transmission Correction options (PHI MultiPak Transmission Function Correction)?

PHI MultiPak binary files contain two coefficients labeled **IntensityCalCoeff: a b**.

These coefficients relate to the formula for the transmission function

$$T(E) = PE\left[\frac{a^2}{a^2 + \left(\frac{E}{PE}\right)^2}\right]^b$$

where E is the kinetic energy of the electrons collected by the detectors and PE is the Pass Energy of the analyzer.

The **Intensity Calib** option on the Spectrum Processing dialog window offers a means of both changing these coefficients and applying the transmission correction to the data. A transmission function appears in the data block as a second corresponding variable. That is to say, the first corresponding variable represents the spectrum intensities (ordinate index 0) at each of the analyzer energies used to measure the signal and for each of these spectral intensities a second entry appears corresponding to the transmission function value at the same analyzer energy. The VAMAS Ordinate field on the Intensity Calib property page allows the index for the transmission function to be altered from the default index of 1, however, as a rule CasaXPS converts transmission information into the default ordinate index and so this field should typically be left unaltered.

Apart from the NPL transmission correction procedure, it is believed that all other transmission correction mechanisms reference the recorded spectra to some reference mode of operation defined by the manufacturer of the instrument. After transmission correction, the intensities are calibrated so that the same sample measured with different pass energies results in the same intensity for a given line. The intensity, after transmission correction, is a relative value and therefore suitable for relative quantification rather than absolute measurements.

The counts per channel no longer relate to the true measurement process and so error estimates based on Monte Carlo simulation are not valid for corrected spectra.

Transmission correction may be applied in two ways:

1. The spectrum itself is adjusted on a point by point basis, or
2. Intensities measured using the un-corrected data are adjusted when calculating the quantification tables.

CasaXPS offers both methods for transmission correction. The Intensity Calib option on the Spectrum Processing dialog window offers a means of adjusting the spectrum directly. If the PHI coefficients are altered using the options shown in Figure 20•7 and the Change Phi TF button is pressed, the corresponding variable holding the transmission values will be updated with new values. To see the results of the new transmission function it will be necessary to re-apply the transmission function to the raw spectrum. This may require undoing any processing operations involving the Intensity Calib option before pressing the Apply button on the property page shown in Figure 20•7

Similarly, if a new transmission function is read from file (Figure 20•7 “Add Transmission Function from File”), the same considerations relating to the PHI transmission coefficients also apply.

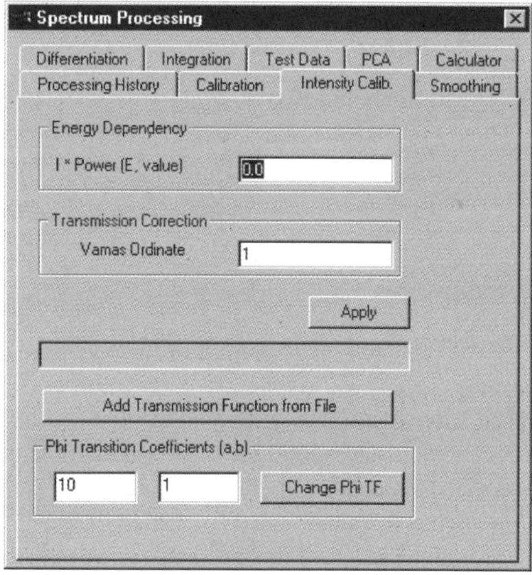

FIGURE 20•7. Intensity Calibration Property Page.

The second method for transmission correction, where the integrated areas are corrected for transmission, is available on the Regions property page of the Quantification Parameters dialog window (Figure 20•8). Note that only one method should be used to quantify data. Either the data should be adjusted via the Intensity Calib option or the integrated areas adjusted, but not both.

When the chosen method for transmission correction is to adjust the integrated areas (by ticking the Automatic check box shown in Figure 20•8 and pressing the Update pushbutton), each time a quantification report is created the transmission correction to the areas will be recalculated. This means that changes to the transmission function made via the PHI transmission coefficients or reading a new transmission file on the Intensity Calib property page will take effect when a new quantification report is generated. Note that if no mean-free-path correction is required, the text-field next to the Automatic check box should be set to "0". Otherwise the value for the mean-free-path correction should be entered in accordance with the expression for the intensity energy dependence seen in Figure 20•7

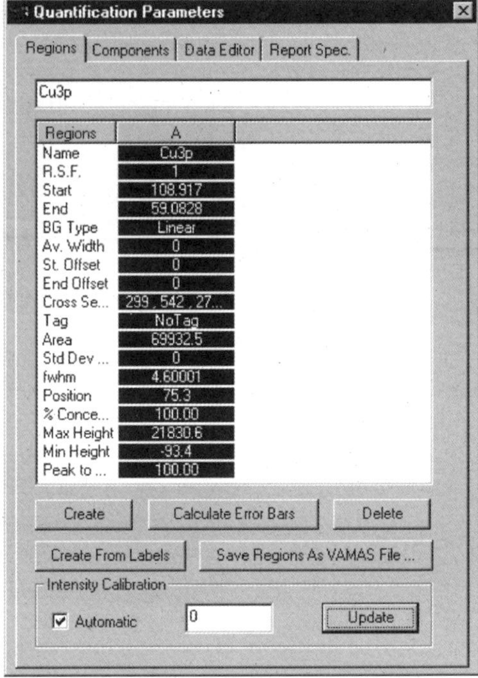

FIGURE 20•8. Regions property page showing the flag used to enable transmission correction without altering the underlying spectral data.

10) What about transmission correction and the Kratos Element Library?

Elemental quantification using XPS requires the use of relative sensitivity factors (RSF) coupled with a corresponding intensity correction. The intensity correction is intended to allow for variations in detected electrons as a function of initial kinetic energy, which is not dependent on the relative abundance of the element or the different transition probabilities for the emission of a electron under the influence of a given X-ray energy. Once the intensity of a photoelectric line is measured, these two adjustments, namely the RSF and the energy dependent effects, to the intensity must be made before the line can be compared in terms of elemental concentration.

The Kratos Axis range of instruments is delivered with an element library containing RSFs initially equivalent to the set of RSFs published in Briggs and Seah. The Kratos Applications Lab personnel, to accommodate the Axis range of instruments, have adjusted these empirically determined RSF table (originally established by Wagner *et al* for a Physical Electronics instrument) to offer elemental quantification based on spectra acquired using Pass Energy 80, Magnetic lens mode. Quantification for other acquisition modes is made self-consistent by calibrating an instrument with respect to a standard survey spectrum acquired from a gold sample at Pass Energy 80 in Magnetic lens mode. This is achieved by running a calibration set of spectra for each of the operating modes of the instrument on a clean gold sample and then generating a set of relative transmission curves. Once created, the appropriate relative transmission curves is stored in each spectra acquired on the calibrated instrument and is subsequently used together with the tabulated RSFs to determine the relative intensities used in the quantification step.

The use of empirically determined RSFs is not universal. SSI instruments, Scienta, VG and Kratos ESCA3400 systems use Scofield photoionisation cross-sections to quantify their spectra. These Scofield cross-sections are computed using a Hartree-Fock relativistic model for determining the initial and final states of the ionisation process. The Scofield tables provide relative sensitivity (with respect to the C 1s transition) for the transitions observed in XPS, however they do not allow for transmission or mean free path/escape depth considerations, which also vary with the kinetic energy of the electron as it first leaves the surface. Before these Scofield cross-sections can be used to quantify spectra, the spectra must be corrected to account for energy dependence of the measurement process. In the case of SSI systems the sensitivity factors are adjusted using an energy exponent of 0.7 (although the exponent is sometimes adjusted to calibrate the SSI system), while most VG users apply a similar adjustment but with an exponent of 0.5.

The relationship between a RSF and the transition to which it refers is a key issue for quantification. A major benefit of using the Scofield cross-sections is that the transitions are well documented in the Scofield paper and there is little doubt regarding where to place the background to a region. This is in contrast to the Kratos element library where it is not always clear, for example, whether both parts of a doublet should be included in the integration region when using a given RSF or not. Also the consistency of sensitivity factors for minor transitions is suspect for empirically determined RSF tables. Obviously there is a doubt regarding the applicability of the theory behind the Scofield cross-sections to the complex problems at the surface of a real sample, however in the case of clean gold Scofield cross-sections when corrected for energy dependence do result in consistent results across the energy range spanning Au 4s to Au $4f_{7/2}$.

A means of using the Scofield cross-sections within the framework of the Kratos Vision data system would be to replace the standard spectrum recorded on the original Axis Ultra at PE 80/Magnetic Lens mode by the equivalent spectrum calibrated to the NPL system for intensity calibration. This would have the benefit of a relatively consistent RSF table coupled with a published method for calibrating the instrument. The bottom line is that quantification from most XPS systems produces repeatable, but for the most part, incorrect atomic concentrations. In using Scofield cross-sections the advantage lies in belonging to the “theoretical” club rather than the “I did it my way” club. If accurate quantification is important, then RSFs from standard sample should be used rather than either of these tabulated RSF options.

Appendix 6: CasaXPS Installation Notes

CasaXPS is supplied as a compressed ".zip" file which contains the CasaXPS.exe program together with a set of library files and sub-directories holding configuration files and HTML help files. Installing CasaXPS requires simply that you decompress ("extract") the zip file to any suitable location on a hard drive of a PC. All the CasaXPS files stay within the extracted directory, and so removing (uninstalling) CasaXPS is equally straightforward and achieved entirely by deleting the CasaXPS directory.

Steps for Installing CasaXPS from CD ROM

1. Place the CD-ROM in CD-ROM drive. Figure 20•9 shows the disk contents at the top directory level.

FIGURE 20•9. Contents of CasaXPS CD ROM

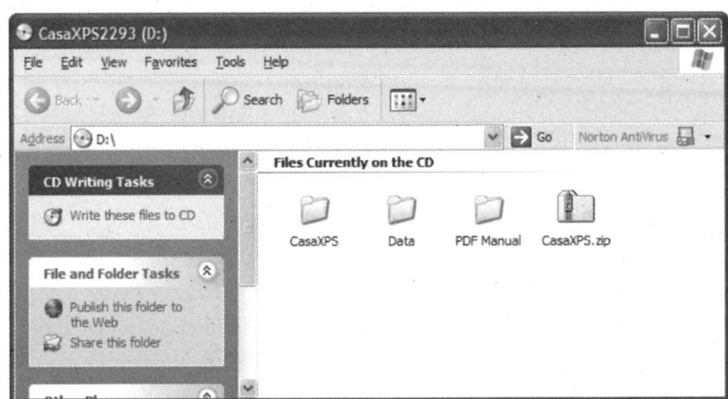

2. Double-click the CasaXPS.zip file. If your computer has a suitable decompression program installed, then the un-zipping (extraction) process will begin automatically, otherwise the CasaXPS directory is available for those without the means to extract the zip file. Using the CasaXPS.zip file is preferable to copying the CasaXPS directory because all the read/write privileges for the sub-directories and files are preserved by the un-zip operation, whereas the directory CasaXPS, while identical in content, will require additional steps to enable proper read/write access to the configuration files.
3. Follow the instructions for the decompression program to write the extracted version of CasaXPS to a location of your choice. It is important that the chosen location for the extracted CasaXPS directory has full read/write permission available for the CasaXPS directory itself and all the sub-directories and files it holds. To change the read/write attributes for a file, select the file/directory in Windows Explorer and right-click the mouse over the selected file/directory. Choose the Properties option from the menu and ensure the Read-only Attribute (bottom of the General tab) is ***not*** ticked before pressing the OK button. Changing the Read-only attribute is necessary only if the CasaXPS directory is directly copied from the CD-ROM to the hard drive (note 2, above).
4. Using Windows Explorer, move to the CasaXPS directory and create a shortcut for the CasaXPS program. To create a shortcut, right-click over the CasaXPS.exe file and select the option labelled Create Shortcut. A shortcut file will appear in the CasaXPS directory and this file can be dragged onto the Desktop or onto the Start button: dragging the shortcut icon over the Start button will add a menu item to the Start button menu.
5. Using the newly created shortcut, start CasaXPS and enter the license strings on the About Dialog window of CasaXPS then press the Update License pushbutton. Ensure the two strings are entered *exactly* as given including case of the characters. When accepted, the Username string will appear as part of the title for the CasaXPS main program window.

Index

Symbols

Numerics

A

B

Index

Index

Index

Index

Index

Index

Recipe

Oyster Tit-Bits

Ingredients

8 or 9 oysters, 8 or 9 round croûtes of fried bread 2 inches in diameter, bacon, anchovy paste, lemon-juice, butter, Krona pepper.

Method

Beard the oysters, place them between two plates with their own liquor and a small piece of butter, and warm in the oven or over a saucepan of boiling water. Spread each croûte with anchovy paste, cover with a stamped-out round of very thin, fried bacon, and place an oyster on the top of each. Sprinkle with lemon-juice and Krona pepper, and serve as hot as possible.

Time

About half an hour. Sufficient for 6 or 7 persons.

Note: Oysters are in season from September to April.

(*from* "Mrs. Beeton's Everyday Cookery",
Ward, Lock & Co., Limited, London and Melbourne, 1950)

Krona pepper is a condiment made from a mixture of paprika and dried, ground capsicum pods. It is similar to cayenne pepper but brighter red in colour and less pungent.

Colophon

The CasaXPS Cookbook was designed and produced within Framemaker 6.0 running under Mac OS 9.2 on an Apple Macintosh Cube, with occasional help from a G5 iMac, OS X (10.3.9), and Adobe Acrobat, which distilled the postscript to produce the PDF files from which the book was printed. The typeface throughout is (mostly) "Times New Roman" and the Figures and screen shots were created and reworked endlessly in Photoshop and Graphic Converter.

Alas, FrameMaker for the Mac is no more.

The characters who crop up at the section boundaries should be familiar to connoisseurs of Lewis Carroll's "Alice" books. All except the Cook are taken from "Through the Looking Glass" and are more or less as immortalized by (Sir John) Tenniel. Their aid was enlisted to provide some kind of clue to the nature of the section content. Haigha (Section 1 - try saying "hare" as two syllables), the Anglo Saxon messenger, brings a swift overview. The Cook (Section 2) does all the hard work (don't forget the Cheshire Cat). The White Knight (Section 3) brings "his own invention" (SVD sort), the Red Queen (Section 4) a little discipline, and the Carpenter (Section 5), urges us to be trotting home. You'll find a Borogove (only one "r") on the back and Hatta (the other messenger) elsewhere. The hope is that you too might become interested in learning more about these splendid people from Mr Dodgson himself (also a mathematician): the exercise of seeing how many more allusions and indications these and other Looking Glass characters bring to this work is left to the reader. Suffice it to say that it all began in Cheshire, and Daresbury is not far from here.... Try "The Annotated Alice" by Martin Gardiner (Allen Lane, London, 2000. ISBN 0-713-99417-7).

The front cover is another, slightly modified, Tenniel (Looking Glass, Chapter 2 - The Garden of Live Flowers). The help and kind permission of Macmillan in the use of all these illustrations is gratefully acknowledged.

About the Authors

Neal Fairley is a mathematician by training and the creator and driving force behind Casa Software: Alan Carrick is a chemist by trade and the hub of Acolyte Science. Between them, the authors have many years experience in practical surface analysis, computer based instrumentation, system design, programming and mathematical methods.
Both of them at one time worked for Kratos Analytical in Manchester.